15th Edition of the QENS Series and 10th of the WINS Workshops (QENS/WINS 2022)

Held online

EPJ Web of Conferences Volume 272 (2022)

San Sebastian, Spain
23 – 27 May 2022

Editors:

A. Arbe
J. Colmenero

ISBN: 978-1-7138-6620-6

Printed from e-media with permission by:

Curran Associates, Inc.
57 Morehouse Lane
Red Hook, NY 12571

Some format issues inherent in the e-media version may also appear in this print version.

For additional information, please contact EDP Sciences – Web of Conferences
at the address below.

EDP Sciences – Web of Conferences
17, Avenue du Hoggar
Parc d'Activité de Courtabœuf
BP 112
F-91944 Les Ulis Cedex A
France

Phone: +33 (0) 1 69 18 75 75
Fax: +33 (0) 1 69 28 84 91

contact-edps@webofconferences.org

Additional copies of this publication are available from:

Curran Associates, Inc.
57 Morehouse Lane
Red Hook, NY 12571 USA
Phone: 845-758-0400
Fax: 845-758-2633
Email: curran@proceedings.com
Web: www.proceedings.com

TABLE OF CONTENTS

QENS/WINS 2022
15th Edition of the QENS series and 10th of the WINS workshops

Organizers & Editors:
Arantxa Arbe and Juan Colmenero
Centro de Física de Materiales (CSIC-UPV/EHU) and Materials Physics Center MPC
San Sebastian, Spain

QENS/WINS 2022 took place during May 23-27, 2022, in hybrid format. Following the spirit of QENS Conferences since 1992, the main purpose of QENS ('Quasielastic Neutron Scattering') was to cover the broad spectrum of scientific activities related with the investigation of material dynamics using quasi-elastic neutron scattering techniques (accessing both, frequency and time domains). QENS included diverse topics like energy-related investigations, biology, polymers, soft matter in general, water and other liquids, magnetism and functional materials. It offered a platform for discussion and exchange of scientific ideas, and a general overview to newcomers about the capabilities of QENS in exploring atomic and molecular motions and relaxation processes of novel materials. WINS ('Workshop on Inelastic Neutron Scattering') focused on innovative aspects of neutron instrument design. As in previous editions, the theme of "New idea, New concept, New design, New instrumentation for New sciences" was followed. Recent developments in the application of polarization analysis, magnetic field, pressure, or improvements in sample environments for soft matter systems, as well as data analysis and software were covered.

The event gathered 120 in-person participants (Miramar Palace in San Sebastian) and 52 on-line attendants, from 26 different countries. Along QENS/WINS 2022, 45 invited talks were given, including 4 plenaries and 11 keynotes. In addition, 58 oral talks and 36 posters were presented. It was sponsored by the Basque Government, DIPC, SoftComp, ESS-Lund, ILL, J-PARC/MLF, ISIS, JCNS, MLZ, ESS-Bilbao, LLB, ANSTO, SoNS, Swiss Neutronics and PSI, and counted with the support of ENSA and SETN.

edp sciences

Web of Conferences

Statement of Peer review

In submitting conference proceedings to *Web of Conferences*, the editors of the proceedings certify to the Publisher that

1. They adhere to its **Policy on Publishing Integrity** in order to safeguard good scientific practice in publishing.
2. All articles have been subjected to peer review administered by the proceedings editors.
3. Reviews have been conducted by expert referees, who have been requested to provide unbiased and constructive comments aimed, whenever possible, at improving the work.
4. Proceedings editors have taken all reasonable steps to ensure the quality of the materials they publish and their decision to accept or reject a paper for publication has been based only on the merits of the work and the relevance to the journal.

Title, date and place of the conference

QENS/WINS 2022 : 15th EDITION OF THE QENS SERIES AND 10th OF THE WINS WORKSHOPS

Date : MAY 23-27, 2022

Place : SAN SEBASTIAN (SPAIN) & ONLINE

Proceedings editor(s):

ARANTXA ARBE and JUAN COLMENERO

CENTRO DE FISICA DE MATERIALES (CSIC-UPV/EHU) AND MATERIALS PHYSICS CENTER MPC, SAN SEBASTIAN (SPAIN)

Date and editor's signature

OCTOBER 28, 2022

The combination of neutron spin echo and dielectric spectroscopy to examine tube dilation

Paula Malo de Molina[1,2], *Angel* Alegría[1,3], *Jürgen* Allgaier[4], *Margarita* Kruteva[4], *Ingo* Hoffmann[5], *Sylvain* Prévost[5], *Michael* Monkenbusch[4], *Dieter* Richter[4], *Arantxa* Arbe[1], and *Juan* Colmenero[1,3,6*]

[1]Materials Physics Center (MPC), Centro de Física de Materiales (CFM) (CSIC-UPV/EHU), Paseo Manuel de Lardizabal 5, 20018 San Sebastián, Spain
[2]IKERBASQUE-Basque Foundation for Science, Euskadi Plaza 5, 48009 Bilbao, Spain
[3]Departamento de Polímeros y Materiales Avanzados: Física, Química y Tecnología, University of the Basque Country (UPV/EHU) PO Box 1072, 20018 Donostia, Spain
[4]Jülich Centre for Neutron Science, Forschungszentrum Jülich GmbH, 52425 Jülich, Germany
[5]Institut Laue-Langevin, 71 Avenue des Martyrs, CS 20156, Cedex 9 38042 Grenoble, France
[6]Donostia International Physics Center, Paseo Manuel de Lardizabal 4, 20018 Donostia, Spain

Abstract. The polymer dynamics in blends of long and short chains spans several decades in time and the understanding of the effect of the short chains on the relaxation mechanism of the long chains due to constraint release requires the combination of microscopic and macroscopic techniques. While the long-time dynamics can be accessed by mechanical or dielectric spectroscopy (DS), its relation to the microstructural details requires the application of theoretical models. In contrast, neutron spin echo (NSE) measures directly the dynamic structure factor reflecting the process of constraint removal at the molecular scale. Here the comparison of NSE and DS results in a model blend of short and long polyisoprene enables the exploration of the entire time regime showing that constraint release leads to a dilation of the confining tube. We show the description of the dynamic tube dilation using a simple model in which the time controlling the tube dilation for the long chain is the terminal time of the short chain.

1 Introduction

In polymer materials, additives are typically used to modify their mechanical properties or improve processability. In particular, the addition of small molecules to high molecular weight entangled polymers leads to lower viscoelasticity, which is very relevant in blends and polydisperse polymers. In general, the long-time dynamics of well-entangled linear monodisperse chains in the melt is well understood in terms of the tube model proposed by Edwards and De Gennes [1,2]. In this ideal tube model the surrounding chains act as topological constraints and confine the probe chain into a tube region, which restricts the lateral motion of the chain within its diameter, d. Then, the probe chain relaxes by reptation. However, in real systems other mechanisms compete with reptation, mainly contour length fluctuations (CLF), i.e., fluctuations of the length of the tube, and constraint release (CR), which refers to the probe chain motions induced by the motions of the surrounding chains [3,4].

Binary polymer blends composed of two monodisperse components with the same chemistry (isofrictional blends) and architecture, are simplified model systems to investigate the effect of polydispersity [5–13]. CR effects are more important in binary blends of short and long chains, where the rapid motions of the short chains release the constraints on the long chain, allowing the latter to relax faster than it would by its own reptation mechanism. Thus, the tube that confines the long chain reorganizes as a consequence of the release of constraints. This process can be, in principle, modelled by assuming Rouse-like motion of the tube, where the segmental motion is replaced by individual CR-events [14]. On the other hand, a simplified representation of the many-chain CR effects is the so-called dynamic tube dilation (DTD), where the diameter of the tube confining the long chain increases with time due to CR. The concept of the DTD was first proposed by Marrucci [15] for homopolymer monodisperse melts where the portion of the chain that has escaped from its original tube can act as a solvent for the remaining tube sections.

Most of the experimental studies of CR effects in polymer blends are carried out with macroscopic rheological or dielectric techniques, focusing on the terminal regime of the viscoelastic behavior. Yet, within a DTD mechanism, the tube should already have reached its final dilated value at that time-regime. Therefore, the time dependence of the CR or DTD mechanism at molecular level is little known. Some molecular dynamics (MD)-simulations studies have attempted to address this problem from a molecular

* Corresponding author: juan.colmenero@ehu.eus

point of view [16,17]. In a recent study [18] we have shown that the time-dependent tube dilation in isofrictional polymer blends can be directly observed by neutron spin echo (NSE) as an additional time dependence of the dynamic structure factor in the local reptation regime. By combining NSE with DS, we could identify the characteristic time driving tube dilation as the terminal time of the short component of the blend.

In general, the dynamic process of tube reptation occurs over a very wide range of time scales, which requires not only the time and length scale resolution provided by neutron spectroscopy but also the combination with other macroscopic techniques. In particular, dielectric spectroscopy in combination with neutron spectroscopy has been very helpful to shed light on the dynamic mechanisms of many systems such as water [19]. In this article, we show the dynamics of entangled chains blended with shorter chains along the entire time regime by combining NSE and DS.

2 Experimental

The single-chain dynamics of a polymer chain can be measured with neutron spin echo by selectively H-labelling a small part of the long chains immerse in a matrix of deuterated polymers. Nearly monodisperse long and short linear polyisoprene were synthesized by anionic polymerization. The molecular weight of the polymers were : M_W = 83.5 kDa and PDI=1.01 (long hPI); M_W = 90.2 kDa and PDI=1.03 (long dPI); M_W = 12.4 kDa and PDI=1.04 (short dPI). The entanglement mass of PI is $M_e \approx 5$ kg/mol. Long-linear PI had $Z = M/M_e \approx 16$ entanglements per chain while the short-linear PI had Z =2.4. The samples for neutron scattering experiments contained 10% of long protonated PI chains in the otherwise deuterated material. The NSE data were obtained in IN15 (ILL, Grenoble, France) [20]. A broadband dielectric spectrometer, Novocontrol Alpha analyzer, was used to measure the complex dielectric permittivity.

3 Results

Figure 1 shows the normalized chain dynamic structure factor $S(Q,t)/S(Q,0)$ corresponding to long chains in the melt and in the blend with the small linear additive (with a total long chain volume fraction $\phi_L = 0.5$) that were previously reported [18]. In both cases, the curves are characterized by a first decay at short times, related to motions of the chain within the tube, followed by a plateau characteristic of confinement. The Q-dependence of this plateau is related to the confining geometry, in this case the tube.

The dynamic structure factor of entangled polymers can be described in terms of the De Gennes model [21]:

$$\frac{S(Q,t)}{S(Q,0)} = \left[1 - \exp\left(-\frac{Q^2d^2}{36}\right)\right] S^{loc}(Q,t) + \exp\left(-\frac{Q^2d^2}{36}\right) S^{esc}(Q,t) \quad (1)$$

where $S^{loc}(Q,t)$ and $S^{esc}(Q,t)$ are related to the local reptation and escape process from the tube (creep motion), respectively. Generally, in the time window covered by NSE it can be assumed that $S^{esc}(Q,t) \approx 1$. We will show later that this is a good approach also in our case. On the other hand, $S^{loc}(Q,t)$ might be approximated by a phenomenological stretched exponential function. Then, the expression for the dynamic structure factor is reduced to:

$$\frac{S(Q,t)}{S(Q,0)} = \left[1 - \exp\left(-\frac{Q^2d^2}{36}\right)\right] \exp\left[-\left(\frac{t}{\tau}\right)^{\beta}\right] + \exp\left(-\frac{Q^2d^2}{36}\right) \quad (2)$$

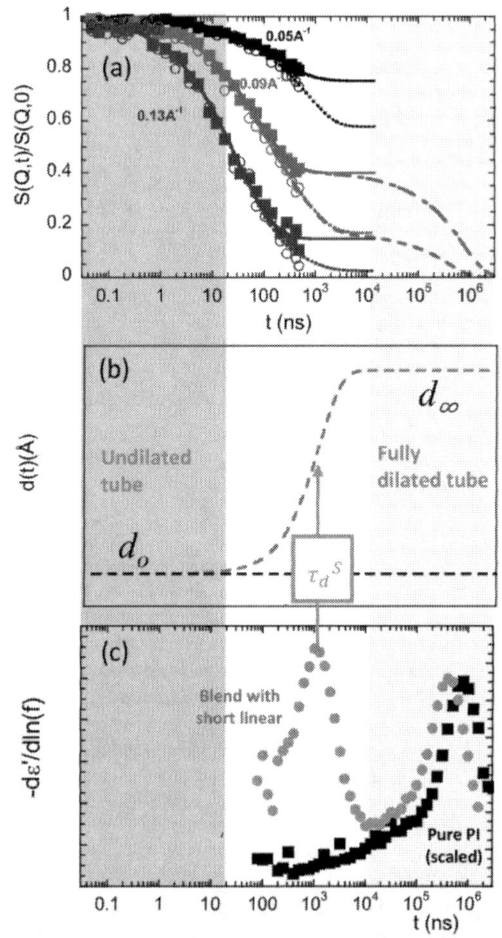

Fig. 1. a) Single-chain dynamic structure factor measured by NSE of long-PI chains in bulk (filled symbols) and in blends with short linear (empty symbols) at $\phi_L = 0.5$ and the three values of the scattering vector indicated. Lines are fits as explained in the text. **b)** Time dependence of the tube diameter. **c)** Normalized natural-log frequency (f) derivative of the real part of the dielectric permittivity as a function of the time variable t = 1/(2πf). All results correspond to 423 K.

Using eq. 2 to model the $S(Q,t)/S(Q,0)$ of the pure long chains (solid lines in Figure 1a), we obtain a value of the effective tube of $d_0 = 64$ Å — in good agreement with the values reported for polyisoprene [22]. We explicitly refer to this tube diameter as effective since it includes all effects of CR, CLF, etc. taking place in the homopolymer. This is our reference value.

The comparison between the data of the long/short blend and pure long melt reveals that the dynamic structure factor of the long chain is unaffected by the presence of the additives at times shorter than ~100 ns. At about 100 ns, a progressively more pronounced decay of $S(Q,t)/S(Q,0)$ is observed for long chains in the blends with respect to the pure long chains. Yet, the decay does not reveal a significant tube dilation. This suggests that up to 100 ns the confinement is the same in both systems and the tube starts to dilate at around 100 ns. Incidentally, this time is close to the entanglement time of polyisoprene, $\tau_e = 90$ ns at 423K.

If the additives were small (or fast) enough, the tube would be expected to be fully dilated in the NSE time window and the tube diameter obtained from eq. 2 and from macroscopic spectroscopy techniques should agree [23]. Here, the short chains are twice the entanglement molar mass, and NSE shows that the tube barely starts dilating within the accessible time window. Thus, to probe the tube dilation it is necessary to measure the long-time behavior with macroscopic relaxation techniques. In the case of polyisoprene that has both local dipole moments parallel and perpendicular to the chain backbone, the terminal relaxation range can be nicely assessed by dielectric spectroscopy (DS) through the so-called "normal mode" [24]. This relaxation reflects the fluctuations of the end-to-end vector of the chains and therefore it directly reveals the disentanglement time, which is related to the tube diameter with the tube theory.

Generally, DS results are measured as a function of frequency but, to compare with NSE, here it is more convenient to represent the data in terms of the time ($t = 1/(2\pi f)$). Figure 1c shows the time-dependence of the derivative of the real part of the complex dielectric permitivity, $d\varepsilon'(f)/dln(f)$, with $\ln f$ being the natural logarithm of the frequency f. The peaks observed in the experimental dynamic window at the high temperature explored correpond to the above introduced "normal mode". In the case of entangled polymer melts, the maximum of this peak corresponds to the disentanglement time, τ_d. In the pure melt, only one peak is observed. By contrast, the results corresponding to the blend show two peaks: one centered at shorter time associated with the terminal time of the short polymer (τ_d^S =1086 ns) and the long-time peak related to the normal mode of the long polymer in the blend, $\tau_d(\phi_L)$. The latter is shifted to faster times compared to the long chain in its pure matrix, owing to the acceleration of τ_d in the presence of the short-chain additives.

In the tube model, the tube diameter d is related to the reptation time as $d = [3R_E^6/(\pi^2 W\ell^4 \tau_d)]^{1/2}$, where R_E is the chain end-to-end distance and $W\ell^4$ the Rouse rate. Taking into account that the blend is isofrictional (with the same $W\ell^4$) and that the conformation of the long linear chains doesn't change upon blending (Gaussian, with R_E=24 nm) [18,25], the dilated tube diameter for the long PI component in the blend relative to the diameter of the pure long PI is given by $d_\infty/d_0 = [\tau_d/\tau_d(\phi_L)]^{1/2}$. Considering the value of the undilated tube as $d_0 = 64$ Å, then the fully dilated tube diameter in the blend is $d_\infty = 89$ Å.

Note that the dynamic structure factor of the chain in the blend cannot be described using eq. 2 with the dilated tube diameter obtained from DS. Thus, the dilation from d_0 to d_∞ happens at times between τ_e and τ_d. In dielectric spectroscopy, the effective dilation process is inaccessible because it is hidden by the normal mode relaxation of the short component. Therefore, it can only be accessed by NSE that probes the long single chain dynamic structure factor selectively. Then, in the spirit of the DTD, instead of a static tube, we use a time-dependent tube for the long PI chains in the blends growing from d_0 to d_∞ in the time range of the relaxation of the fast component of the blend. Accordingly, eq. 2 becomes:

$$\frac{S(Q,t)}{S(Q,0)} = \left[1 - \exp\left(-\frac{Q^2 d(t)^2}{36}\right)\right] \exp\left[-\left(\frac{t}{\tau}\right)^\beta\right] + \exp\left(-\frac{Q^2 d(t)^2}{36}\right) \quad (3)$$

As a first approximation, the time dependence of the tube diameter can be considered as an exponential function $d(t) = d_\infty + (d_0 - d_\infty)\exp[-t/\tau_d^S]$ with the time constant corresponding to the terminal time of the short additive in the blend, τ_d^S, obtained from DS (see Figure 1b).

Moreover, considering that the blend is isofrictional and the short-time behavior of the NSE results is insensitive to the presence of additives, the β and τ parameters of the first term in eq. 3 were fixed to the values obtained for pure long-linear PI. The fits using eq. 3 for the blend represented by dotted lines in Figure 1a show excellent agreement with the experimental data, in particular considering that there are no fitting parameters.

We note that a similar result was also found for star additives with 8 and 18 arms, where there the time governing the tube dilution was closer to the terminal time of the additive obtained from rheological measurements [25].

Finally, we can verify the validity of neglecting the escape dynamics using eqs. 2 and 3 to describe the NSE data since we know from DS the reptation time controlling the escape process from the tube $S^{esc}(Q,t)$. For that, in Figure 1a we calculated the full curves in a representative Q-value (Q=0.09 Å^{-1}) in an extended time window considering for the escape term the expression $S^{esc}(Q,t) = \frac{8}{\pi^2}\sum_{n-odd}\frac{1}{n^2}\exp\left(-n^2 t/\tau_d\right)$ [26]. This is obtained from the full general one given by Doi and Ewards [27] under the assumption $QR_E \gg 1$. The full model is represented by the dashed lines in Figure 1a showing that they overlap with eqs. 2 and 3 up to ~3·10^4 ns both in the case of the pure long polymer and the blend. In this way, we demonstrate that within the NSE time window ($t < 480$ ns), the expression $S^{esc}(Q,t) \equiv 1$ is valid.

4 Conclusions

The combination of neutron spin echo with dielectric spectroscopy in isofrictional polymer blends of short and long polyisoprene chains shows that at short times the tube is unaltered and at long times it is dilated. NSE can directly probe the process of time-dependent constraints removal at molecular scale. A model of dynamic tube dilation describes the NSE experimental data using the parameters obtained from dielectric spectroscopy showing that the time at which the tube dilates is the terminal time of the additive.

The authors acknowledge the financial support received from the IKUR Strategy under the collaboration agreement between Ikerbasque Foundation and Materials Physics Center on behalf of the Department of Education of the Basque Government. Financial support by MCIN/AEI/10.13039/501100011033 and "ERDF – A way of making Europe" (PID2021-123438NB-I00) and Eusko Jaurlaritza – Basque Government (IT1566-22).

References

1. S. F. Edwards, Proc. Phys. Soc. **91**, 513 (1967)
2. P. G. de Gennes, J. Chem. Phys. **55**, 572 (1971)
3. J. Klein, Macromolecules **11**, 852 (1978)
4. M. Daoud, P. G. De Gennes, J. Polym. Sci. Polym. Phys. Ed. **17**, 1971 (1979)
5. S. Wang, S.-Q. Wang, A. Halasa, W.-L. Hsu, Macromolecules **36**, 5355 (2003)
6. H. Watanabe, S. Ishida, Y. Matsumiya, T. Inoue, Macromolecules **37**, 6619 (2004)
7. H. Watanabe, Y. Matsumiya, E. van Ruymbeke, Macromolecules **46**, 9296 (2013)
8. M. E. Shivokhin, E. van Ruymbeke, C. Bailly, D. Kouloumasis, N. Hadjichristidis, A. E. Likhtman, Macromolecules **47**, 2451 (2014)
9. P. S. Desai, B.-G. Kang, M. Katzarova, R. Hall, Q. Huang, S. Lee, M. Shivokhin, T. Chang, D. C. Venerus, J. Mays, J. D. Schieber, R. G. Larson, Macromolecules **49**, 4964 (2016)
10. T. Ebrahimi, H. Taghipour, D. Grießl, P. Mehrkhodavandi, S. G. Hatzikiriakos, E. van Ruymbeke, Macromolecules **50**, 2535 (2017)
11. T. Shahid, Q. Huang, F. Oosterlinck, C. Clasen, E. van Ruymbeke, Soft Matter **13**, 269 (2017)
12. H. Lentzakis, S. Costanzo, D. Vlassopoulos, R. H. Colby, D. J. Read, H. Lee, T. Chang, E. van Ruymbeke, Macromolecules **52**, 3010 (2019)
13. R. Hall, B.-G. Kang, S. Lee, T. Chang, D. C. Venerus, N. Hadjichristidis, J. Mays, R. G. Larson, Macromolecules **52**, 1757 (2019)
14. D. J. Read, M. E. Shivokhin, A. E. Likhtman, J. Rheol. (N. Y. N. Y). **62**, 1017 (2018)
15. G. Marrucci, J. Polym. Sci. Polym. Phys. Ed. **23**, 159 (1985)
16. Z. Wang, R. G. Larson, Macromolecules **41**, 4945 (2008)
17. M. Langeloth, Y. Masubuchi, M. C. Böhm, F. Müller-Plathe, J. Chem. Phys. **141**, 194904 (2014)
18. P. Malo de Molina, A. Alegría, J. Allgaier, M. Kruteva, I. Hoffmann, S. Prévost, M. Monkenbusch, D. Richter, A. Arbe, J. Colmenero, Phys. Rev. Lett. **123**, 187802 (2019)
19. A. Arbe, P. Malo de Molina, F. Alvarez, B. Frick, J. Colmenero, Phys. Rev. Lett. **117**, 185501 (2016)
20. M. Kruteva, B. Farago, I. Hoffmann, P. Malo de Molina, *Data for "Influence of the Topology on the Polymer Dynamics: Blends of Long Linear Polymer Chains with Shortlinear and Star-like Polymers" Https://Doi.Ill.Fr/10.5291/ILL-DATA.CRG-2462* (Institut Laue-Langevin, Grenoble, 2018)
21. P. G. De Gennes, J. Phys. **42**, 735 (1981)
22. L. J. Fetters, D. J. Lohse, D. Richter, T. A. Witten, A. Zirkel, Macromolecules **27**, 4639 (1994)
23. B. J. Gold, W. Pyckhout-Hintzen, A. Wischnewski, A. Radulescu, M. Monkenbusch, J. Allgaier, I. Hoffmann, D. Parisi, D. Vlassopoulos, D. Richter, Phys. Rev. Lett. **122**, 088001 (2019)
24. C. Riedel, A. Alegría, P. Tordjeman, J. Colmenero, Rheol. Acta **49**, 507 (2010)
25. P. Malo de Molina, A. Alegría, J. Allgaier, M. Kruteva, I. Hoffmann, S. Prévost, M. Monkenbusch, D. Richter, A. Arbe, J. Colmenero, Macromolecules **53**, 5919 (2020)
26. D. Richter, M. Monkenbusch, A. Arbe, J. Colmenero, Neutron Spin Echo in Polymers Systems, Advances in Polymer Science **174** (Springer, Berlin, Heidelberg, New York, 2005)
27. M. Doi, S. F. Edwards, The Theory of Polymer Dynamics (Clarendon, Oxford, 1986)

Dynamics of bottlebrush polymers

Karin J. Bichler[1*], *Bruno* Jakobi[1], and *Gerald J.* Schneider[1,2†]

[1]Department of Chemistry, Louisiana State University, Baton Rouge 70803, Louisiana, United States
[2]Department of Physics and Astronomy, Louisiana State University, Baton Rouge 70803, Louisiana, United States

Abstract. Bottlebrushes are an interesting class of polymers which shows intriguing material properties often associated with dynamics. While dynamical phenomena in linear polymers are well understood and existing theories can describe them in a good way, bottlebrush dynamics have only rarely been investigated. Therefore, we performed dielectric spectroscopy and quasi-elastic neutron scattering to study the dynamics of polydimethylsiloxane-based bottlebrush polymers, PDMS-g-PDMS focusing mostly on the segmental dynamics of the side chains. Comparing the relaxation times of the α – relaxation, tracked with dielectric spectroscopy, of bottlebrush polymers with those of their respective linear side chains show a slowing down once the side chains are attached to the backbone. This effect diminishes and finally vanishes with increasing side chain length. The time and length scale, offered by quasi-elastic neutron scattering, fits for the segmental dynamics together with faster processes. The Q-dependence of the segmental relaxation times allows to classify bottlebrush polymers as heterogenous including a non-Gaussian character. For such a dynamical system, the mean square displacement needs to be separated into single processes before an overall mean square displacement can be generated by applying the time temperature superposition principle.

1 Introduction

Graft polymers consist of linear chains chemically attached to a linear backbone. Depending on the grafting density, i.e., the number of side chains per backbone repeating unit, the resulting polymer is either a comb-, or a bottlebrush polymer. At the higher grafting densities of bottlebrushes extraordinary properties like supersoftness, or hyperelasticity emerge.[1] Additionally the entanglement molecular weight shifts to higher values[2] and due to permanent grafting micro or full separation of side chains and backbone is avoided.

Most dynamical studies concentrate on rheology and dielectric spectroscopy.[3-11] A recent dielectric study by Iwawaki *et al.* focuses on bottlebrush polymers having polyisoprene (PI) as backbone and polystyrene (PS) as side chains.[4] A 4-fold increased mean square end-to-end distance, $\langle \vec{R}^2 \rangle$, compared to linear PI with the same number of repeating unites at the bottlebrush backbone was deducted from the dielectric relaxation strength. Viscoelastic spectra on the same PI-g-PS samples are similar to those of linear PI, but with the PI backbone in the bottlebrush being stretched and stiffened.[4] Grigoriadis *et al.* compared bottlebrush polymers with PBiBEM backbone and PBA side chains with the linear counter parts. Hereby, the segmental dynamics of the side chains are slowed-down, while the backbone dynamics are plasticized by the attached side chains.[3] Rheological studies based on grafting density and overall molecular weight were investigated by Haugan *et al.* proposing relationships of the ratio of the diameter of gyration of the side chains, d_{sc}, and the average backbone contour length between grafted chains, L_g, for thin flexible chain behaviour, $\frac{d_{sc}}{L_g} < 1$, and for a behaviour similar to thick semiflexible cylinders, $\frac{d_{sc}}{L_g} > 1$.[9]

Generally, bottlebrush polymers can be achieved via three different synthetic routes, I. Grafting-to, II. Grafting-from, and III. Grafting-through:[10, 12, 13]

While grafting-to (**Fig.1**) is the only method which allows to characterize side chain and backbone separately and independently based on their molecular weight, it is also challenging to achieve high grafting densities due to steric hindrances caused by already grafted side chains.

Fig. 1. Illustration of the grafting-to process.

Grafting-from (**Fig. 2**) leads to synthesising the backbone first, with the possibility to determine the molecular weight, and subsequently, the side chains are polymerized directly from the functionalized backbone. This could lead to lower than expected grafting densities and uncertainties in the determination of the molecular

* Corresponding author: kbichler@lsu.edu,
† Corresponding author: gjschneider@lsu.edu

weight and molecular weight distribution of the side chains.

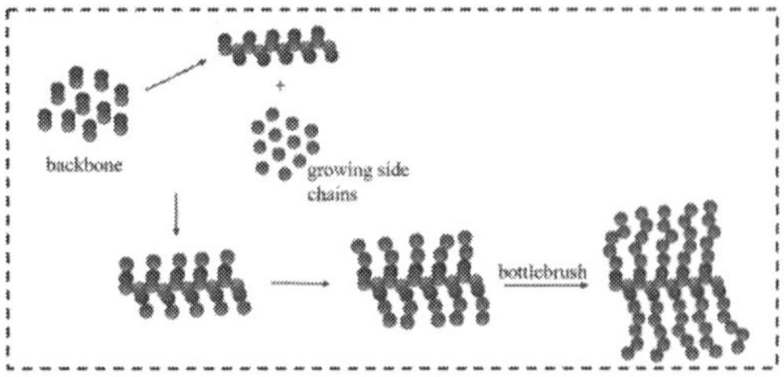

Fig. 2. Illustration of the grafting-from process.

The most promising way is the grafting-through (**Fig. 3**) method. It is based on synthesising macromonomers and enables 100% grafting densities, whereby the molecular weight of the side chains can be determined in advance.

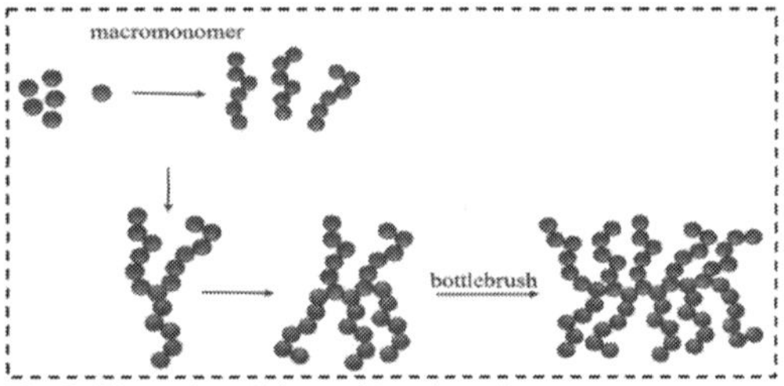

Fig. 3. Illustration of the grafting-through process.

As seen, each method has advantages and disadvantages. Additionally, to that, the choice of synthesis route is restricted by the targeted polymer. In our case, we have used polydimethylsiloxane (PDMS) for side chains and backbone which limits us to the grafting-to method. Using anionic ring opening polymerization in combination with chlorination of the backbone and a mild grafting reaction of silanol terminated side chain precursors results in the PDMS-g-PDMS bottlebrush polymers as shown in **Fig. 4**.[14, 15]

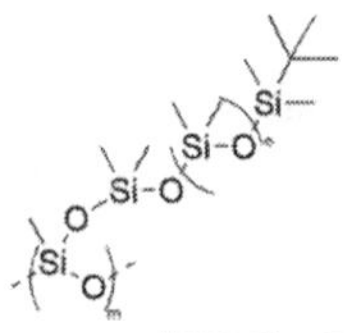

Fig. 4. Chemical structure of PDMS-g-PDMS bottlebrush polymer.[15]

For the structure, by changing the ratio of the degree of polymerization, DP, from backbone and side chains results in a shape transition from elongated objects for $\frac{DP^{side\ chain}}{DP^{backbone}} < 1$, to more spherical ones in case of $\frac{DP^{side\ chain}}{DP^{backbone}} > 1$.[16, 17] This means, by increasing the side chains length, while decreasing the backbone length different shapes are induced.

For our experiments, we have chosen three different combinations of side chain to backbone lengths, while the backbone was roughly constant for all samples. As shown in table 1, Sample 1 has the shortest side chains, resulting in an elongated object.[15] Going to Sample 3 the side chain length increases to be comparable with the length of the backbone. This leads to more spherical objects, confirmed by small-angle neutron scattering (SANS).[15]

Table 1. Number average molecular weight, M_n, of side chains, backbones, and the entire bottlebrush polymer.

Name	$M_n^{side\ chain}$ (g/mol)	$M_n^{backbone}$ (g/mol)	$M_n^{bottlebrush}$ (kg/mol)
Sample 1	298	16500	95
Sample 2	1800	13500	157
Sample 3	11500	13500	1106

2 Dynamics

2.1 Broadband Dielectric Spectroscopy

Dynamical studies have been performed using broadband dielectric spectroscopy (BDS). Hereby, the sample was placed between two gold electrodes with 0.1 mm thick glass spacers for ensuring a constant sample thickness. Depending on the frequency, the fluctuations of the permanent dipole moments are detected, giving relaxation peaks in the imaginary part of the complex permittivity, $\epsilon''(\omega)$ vs. ω. The dynamical information obtained from BDS depends on the orientation of the permanent dipole moments. In case of the parallel component, information about the fluctuations of the chain end-to-end vector, also known as normal mode, are obtained. The component perpendicular to the polymer chain, gives information about the dynamics of the segments, i.e., α – relaxation.[18]

The used polymer, PDMS, has only a perpendicular component, thus the tracked dynamical process is the segmental dynamics. Most of the dipole moments are located within the side chain of the bottlebrush polymers, therefore, the measured signal is dominated by the dynamics of the side chains.

As seen in **Fig. 5**, with increasing temperature, the relaxation peak shifts to higher frequencies.

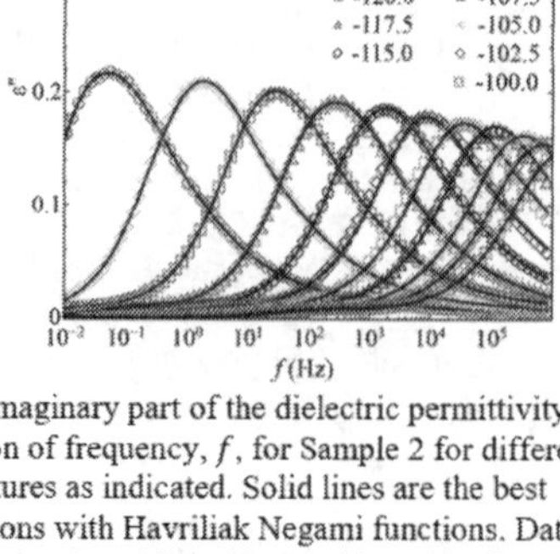

Fig. 5. Imaginary part of the dielectric permittivity, ϵ'', as a function of frequency, f, for Sample 2 for different temperatures as indicated. Solid lines are the best descriptions with Havriliak Negami functions. Data has been previously published in Jakobi *et al.*[15]

Due to the connection of relaxation time and frequency, $\tau \propto 1/f$, it is clearly seen that the relaxation time decreases, thus the relaxation itself speeds up. This experiment has been done for each bottlebrush polymer and for each respective single linear side chain. For a better comparison of the relaxation behaviour, within each set, we compare the temperature dependence of the relaxation times as shown in **Fig. 6**.

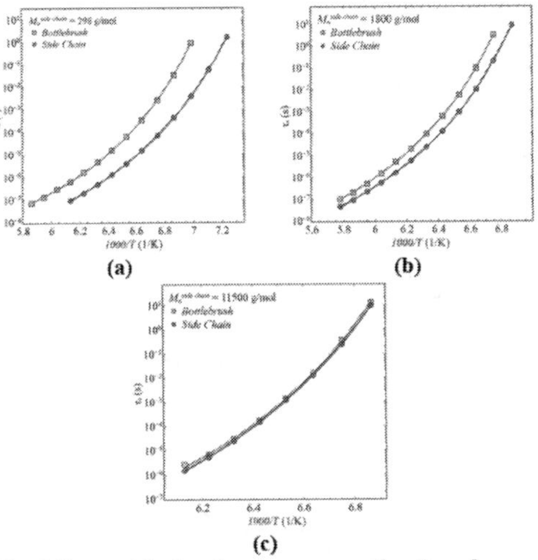

Fig. 6. Segmental relaxation time, τ_S, as a function of $1000/T$ for each bottlebrush polymer in comparison with their respective single linear side chain. **(a)** Sample 1, **(b)** Sample 2, and **(c)** Sample 3. Solid lines are the best description with the VFT equation. Data has been previously published in Jakobi *et al.*[15]

Comparing the shortest single side chains with the bottlebrush (Sample 1), a reduction of the segmental relaxation time, τ_S, of more than one order of magnitude is visible in the investigated temperature range. Increasing the side chain length across the three samples, leads to approaching segmental relaxation times with roughly half a decade difference (Sample 2) and finally to an almost identical relaxation time behaviour of both, single side chain and bottlebrush polymer (Sample 3) over the entire temperature range. This indicates that the inner segments of the side chains are influenced the most by the connection to the backbone leading to the observed slowed down process of the segmental relaxation time. With increasing side chain length, this effect levels off, since with increasing distance a decreasing correlation is expected. Hence, outer segments have the lowest correlation. As Figure 6 c) illustrates this leads to a relaxation time identical to freely moving chains, which implies that the number of segments that have lost correlation to the grafting point dominates.

Additionally, dielectric spectroscopy allows to extract the sample specific glass transition temperature, T_g, using the relationship of $\tau_s(T_g) = 100$ s.[19] Accounting for the different T_g of our samples by scaling the temperature axis to $T - T_g$ reveals differences in relaxation behaviour beyond those associated with different glass transition temperatures (**Fig. 7**).

As seen in **Fig. 7**, sample 2 and sample 3 show almost identical relaxation behaviour over the entire temperature range, while relaxation times of sample 1 indicate some minor differences but only at higher temperatures. Even after eliminating effects associated with T_g, the shortest side chains appear to be faster than the associated bottlebrush polymers. This observation seems to be similar to the strong influence of branching and functional end groups on the glass transition temperature as reported by Khalyavina *et al.*[20]

2.2 Quasi-Elastic Neutron Scattering

Stimulated by the results of dielectric spectroscopy (**Fig. 6**), quasi-elastic neutron scattering (QENS) experiments have been performed to elaborate on the segmental dynamics of bottlebrush polymer more in detail with the advantage of time and length scale information simultaneously. To maximize time coverage, a combination of three different spectrometers – EMU and Pelican at ANSTO, Sydney, Australia, and IRIS at ISIS in Didcot, UK – in the time domain was used, whereby the sample was placed in annular aluminium sample cans with sample thickness of 0.1 mm. QENS is sensitive to the incoherent scattering cross section which is most pronounced for protons. Most of the protons in our samples are located within the side chains, thus the dominating contribution to the scattering signal originates from dynamical processes within the side chains of the bottlebrush polymer. These processes include methyl group rotation at very low temperatures and segmental dynamics, both confirmed by the elastic fixed window scan prior to the quasi-elastic scans.

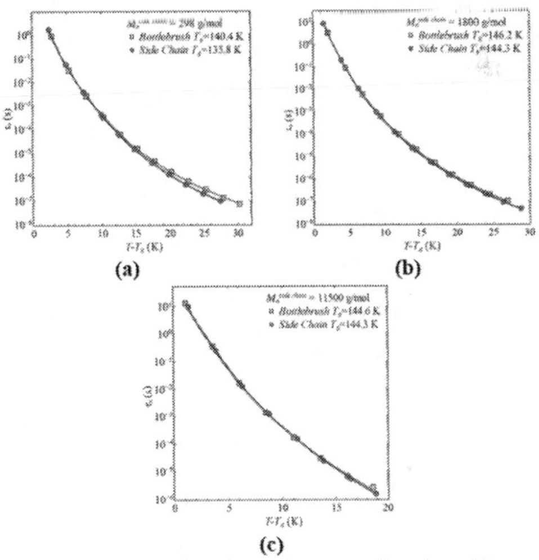

Fig. 7. Segmental relaxation time, τ_S, as a function of $T - T_g$ for each bottlebrush polymer in comparison with their respective single linear side chain. **(a)** Sample 1, **(b)** Sample 2, and **(c)** Sample 3. Solid lines are the best description with the VFT equation.

As typical for localized motion like the methyl group dynamics, a very weak decay with Q-independent relaxation time is observed for the intermediate scattering function, $S(Q,t)$, as displayed in **Fig. 8**.

With increasing the temperature from $T = 75$ K to $T = 175$ K, a plateau establishes for $t > 0.1$ ns which indicates confined motion for the methyl group dynamics. Hereby, the experimental data were described by

$$S_l(Q,t) = A \cdot DWF \times \left(EISF + (1 - EISF) \cdot \exp\left(-\left(\frac{t}{\tau_l}\right)^{\beta_l} \right) \right) \quad (1)$$

including the Debye-Waller factor, DWF, accounting for fast vibrations and the Elastic Incoherent Structure Factor, $EISF$, describing the confined nature of the localized motion. This allowed to quantify the confinement of the methyl group rotation as the three-fold jump, typical for that relaxation.[14, 21]

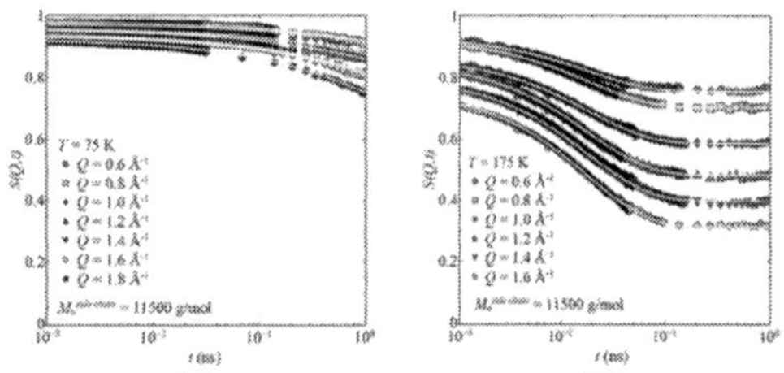

(a) (b)

Fig. 8. Intermediate scattering function, $S(Q,t)$, as a function of time, t, for Sample 3 for two selected temperatures, **(a)** $T = 75$ K, and **(b)** $T = 175$ K. Solid lines are the best description with the model function, equation (1). Data has been previously published at Bichler *et al.*[14]

Considering the relaxation times of this localized motion, τ_l, shows an independence on the side chain length additionally to the independence on the momentum transfer as seen in intermediate scattering function, $S(Q,t)$. The temperature dependence of the localized motion shows a well pronounced Arrhenius behaviour, indicative for thermally activated processes. An activation energy, E_A, of ~6 kJ/mol needs to be applied per three-fold jump, a value comparable in order of magnitude with those reported for other methyl group rotations.[14]

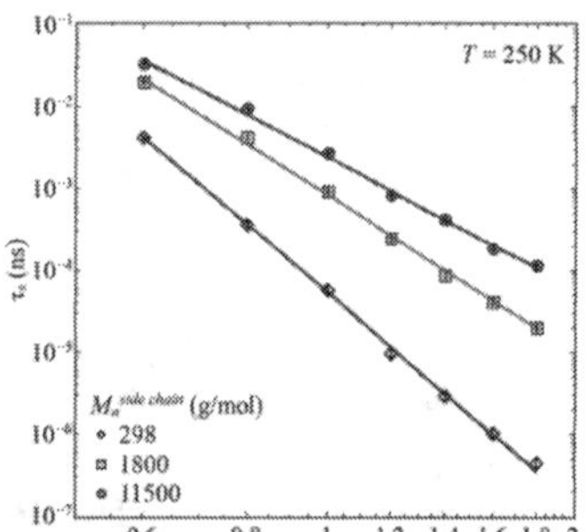

Fig. 9. Segmental relaxation time, τ_s, as a function of momentum transfer, Q, for all three samples at T = 250 K. Solid lines are the descriptions with the power law with exponents, −s = 7.2 (blue diamonds, Sample 1), −s = 5.0 (green squares, Sample 2), and −s = 4.2 (red circles, Sample 3). Data has been previously published at Bichler *et al.*[14]

Increasing the temperature to $T = 250$ K, shifts the segmental relaxation inside the available time range. Hereby, the intermediate scattering function is fully decaying to zero with an Q-dependent relaxation time. After describing the data with

$$S_s(Q,t) = A \cdot DWF \cdot \exp\left(-\left(\frac{t}{\tau_s}\right)^{\beta} \right) \quad (2)$$

the obtained relaxation time, τ_s shows a well pronounced power law dependence on the momentum transfer, changing with side chain length (**Fig. 9**).

With increasing side chain lengths, the relaxation time slows down together with increasing power law dependence on Q, reaching from −7.2 to −4.2, directing to faster dynamics at shorter length scales. This substantial decrease can be better understood by a comparison with the momentum transfer dependence observed for linear chains. Typically, $\tau_s \propto Q^{-4}$, corresponding to a stretching parameter of $\beta_s = 0.5$ is observed, which implies that sufficiently long side chains may approach the values of linear chains. This observation is consistent with the relaxation times for the free and the side chains which become similar for longer side chains.

The power law dependence allows to classify our systems either as homogenous or heterogenous. In case of homogenous behaviour, $\tau_s \propto Q^{-s}$ with $s = 2/\beta_s$ needs to be fulfilled. This would require that we need to be able to describe our power laws depending on the stretching parameters, s vs. β_s, with $s = 2/\beta_s$, which would correspond to the green line in **Fig. 10**.

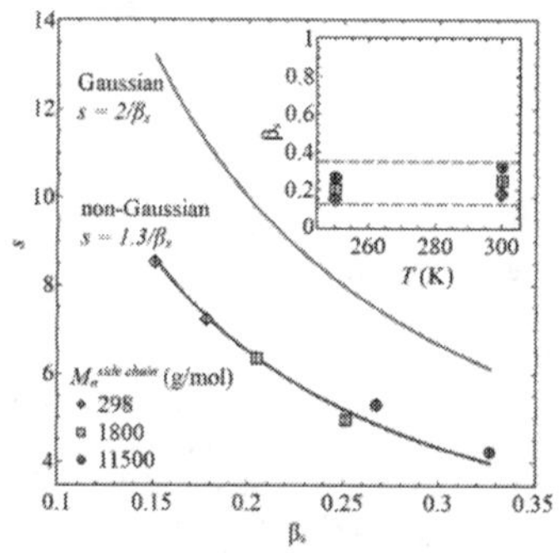

Fig. 10. Power law, s, as a function of stretching parameter, β_s, for all three sample and both temperatures, $T = 250$ K, and $T = 300$ K. Sold green lines corresponds to Gaussian behaviour for a homogenous system. Solid purple line corresponds to $s = 1.3/\beta_s$. Data has been previously published at Bichler *et al.*[14]

However, the obtained data are below the description for homogeneous system and therefore, pointing to a heterogenous system in all three cases. This heterogeneity is induced by two different origins. One is coming from the combination of different dynamical processes for momentum transfers $Q \geq 0.6$ Å^{-1}. The second heterogeneity is based on the intrinsic heterogeneity of the segmental relaxation which is enhanced by the branched nature of the bottlebrush polymer. Grafting side chains to a backbone broadens the distribution of segmental relaxation times, and thus increases the heterogeneity.

These heterogeneities are considered via the non-Gaussian parameter, $\alpha_2(t)$, which can be incorporated for the mean square displacement.

$$S(Q,t) = A \cdot \exp\left(-\frac{\langle r^2(t)\rangle}{6}Q^2 + \frac{\langle r^2(t)\rangle^2 \alpha_2(t)}{72}Q^4\right) \quad (3)$$

Hereby, $\alpha_2(t) \neq 0$ indicates deviations from a Gaussian distribution of relaxation times, and therefore deviations from homogeneous dynamics.

As seen in **Fig. 11** with increasing temperature, $\alpha_2(t)$ decreases which is attributed to a reduced heterogeneity at higher temperatures. In all cases, $\alpha_2(t)$ approximates zero for long times indicating the complete relaxation of all processes in the available time frame.

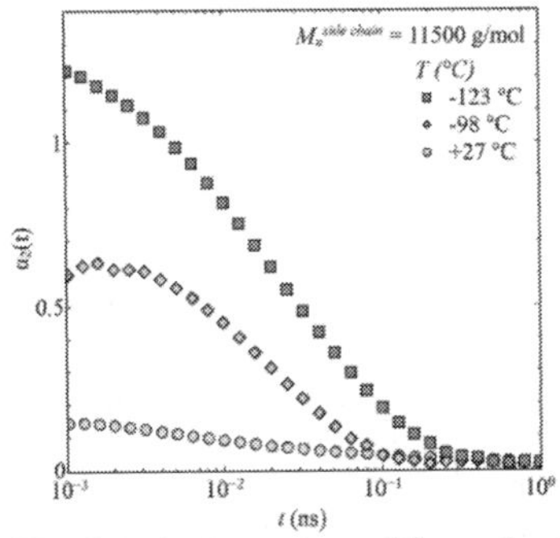

Fig. 11. Non-Gaussian parameter, $\alpha_2(t)$, as a function of time, t, for Sample 3 at selected temperatures. Data has been previously published at Bichler *et al.*[22]

Using equation (3) not only gives access to the non-Gaussian parameter, but we also obtain the mean square displacement, $\langle r^2(t)\rangle$, for each temperature and each sample, as shown in **Fig. 12** for Sample 3.

Clearly two distinct regions emerge, one at higher temperatures, describing the segmental dynamics and one at lower temperatures accounting for the methyl group dynamics. The plateau at long times indicates the confined nature of the methyl group rotations.

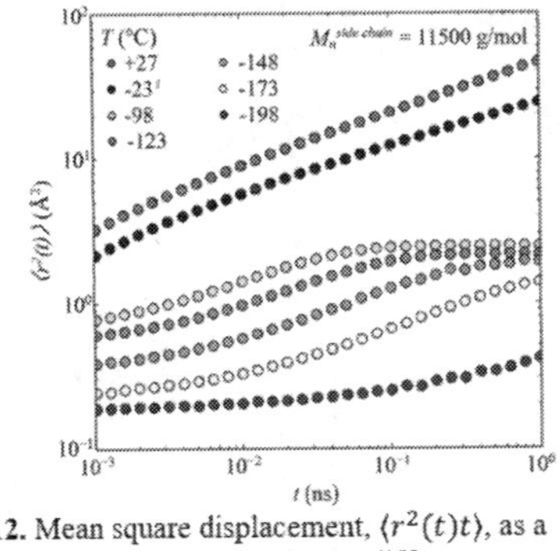

Fig. 12. Mean square displacement, $\langle r^2(t)t\rangle$, as a function of time, t, for Sample 3 at different temperatures as indicated. Data has been previously published at Bichler *et al.*[22]

Due to the similarity of each mean square displacement in each temperature region, time temperature superposition (TTS) principle has been applied to increase the available time range. However, simply shifting them together violates TTS due to discrepancies at short and long times for both regions.[22] This is caused by the mixture of different process of different length scales, causing the heterogeneity. Removing those faster contributions and reducing the mean square displacement to one single process, i.e., either pure methyl group or pure segmental relaxation, results in overlapping data over the whole time range (**Fig. 13**).

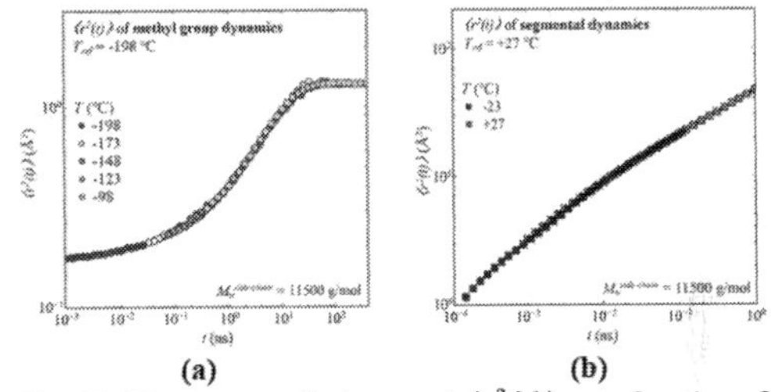

Fig. 13. Mean square displacement, $\langle r^2(t)\rangle$, as a function of time, t, for Sample 3. **(a)** pure methyl group dynamics. **(b)** pure segmental dynamics. Data has been previously published at Bichler *et al.*[22]

This leads to the conclusion, that TTS is fulfilled as long as only one process occurs in the observed time frame, without any contribution from others. Further, both single mean square displacements have been combined at one reference temperature by shifting them accordingly to the shift parameter, obtained either from dielectric spectroscopy or from the quasi-elastic scans. These two shifted $\langle r^2(t)\rangle$ have been added pointwise resulting in the combined mean square displacement containing methyl group and segmental dynamics.

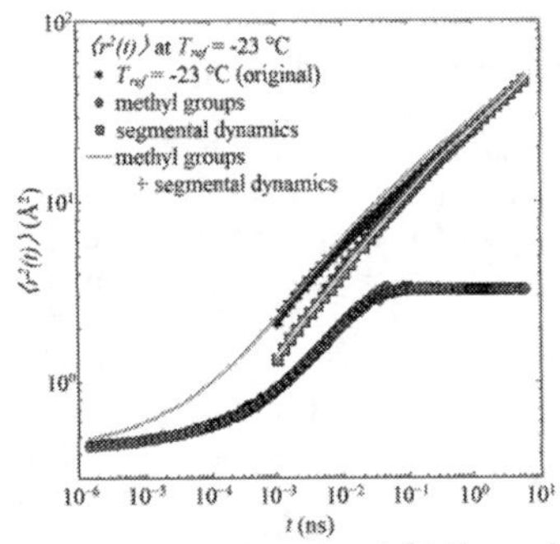

Fig. 14. Mean square displacement, $\langle r^2(t)\rangle$, as a function of time, t, for Sample 3. Data has been previously published at Bichler *et al.*[22]

As seen by the comparison of the generated mean square displacement, (**Fig. 14**, orange solid line), with the original $\langle r^2(t)\rangle$ obtained from $S(Q,t)$ at that specific reference temperature (**Fig. 14**, blue stars) the used approach can recover the original data well over an enlarged temperature range even across the glass transition temperature where usually the TTS is known to break down (**Fig. 14**).

Finally, the fast vibrations have been added for all three samples resulting in three different power law regions over the achieved time range as seen in **Fig. 15**. At very short times, accounting for vibrations and methyl group dynamics, the power law of $t^{0.1}$ does not change while changing the side chains length. Continuing to intermediate times, where a combination of methyl group and segmental dynamics occurs, the power law changes from $t^{0.35}$ to $t^{0.4}$. Since the relaxation of methyl groups is independent on the side chains length, this power law change could be due to the onset of segmental dynamics. At the long times the power law changes from $t^{0.2}$ to $t^{0.3}$. In this time range, the methyl group dynamics goes into a plateau and gives a constant contribution. Therefore, the change in power is entirely from the segmental dynamics. This change let's assume that with increasing side chain length, i.e., increasing distance from the backbone, the segmental dynamics has less confinement and eventually reaches the behaviour of a linear polymer chain ($t^{0.5}$).

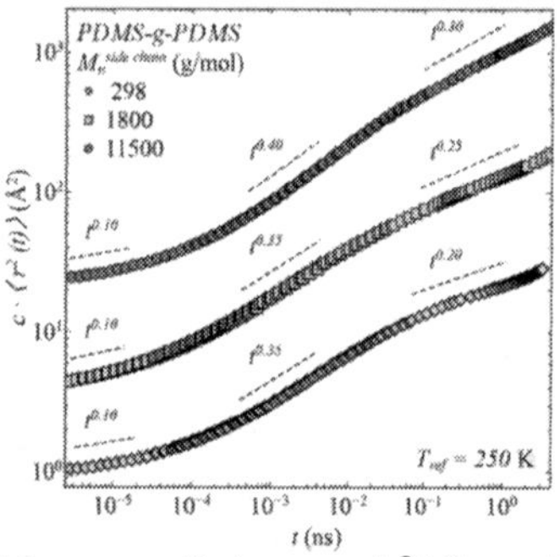

Fig. 15. Mean square displacement, $\langle r^2(t)\rangle$, as a function of time, t, for all three samples. Data has been previously published at Bichler *et al.*[22]

Summary and Conclusions

Dynamical studies on bottlebrush polymers using dielectric spectroscopy and quasi-elastic neutron scattering revealed more detailed insights to this polymer architecture. Since the majority of the permanent dipole moments, for dielectric spectroscopy, as well as most of the protons, for quasi-elastic neutron scattering, are located within the side chains, the information obtaining from the experiments is for the side chain dynamics. In our study, we focused on dynamical changes inherent with increasing side chain lengths of the bottlebrush polymers. This further induced the shape transition from elongated objects into more spherical ones. Based on dielectric spectroscopy, a slowed down process is observable by comparing side chains of the bottlebrush polymer with their respective single linear chains. This effect vanishes for sufficiently long side chains. Within the time and temperature range of quasi-elastic neutron scattering experiments, the methyl group and segmental dynamics of the side chains were tracked. Hereby, the methyl group rotations are independent on the length scale and the side chain length but confined on a characteristic three-fold jump. In contrast the segmental dynamics depends on the side chain length and shows a particular Q-dependence. This power law dependence allows to categorize the bottlebrush polymers as heterogenous systems with the non-Gaussian parameter incorporated into the mean square displacement analysis. Hereby, two distinct regions in $\langle r^2(t)\rangle$ representing the methyl group and segmental dynamics are obtained. Reducing the mean square displacement to one single dynamical process by separating faster processes opens the way to combine several processes by applying time temperature superposition principle. This is valid over an extended temperature range and even across the glass transition temperature where the TTS is usually known to break down.

We gratefully acknowledge funding by the U.S. Department of Energy (DoE) under grant no. DE-SC0019050.

References

1. L.-H. Cai, T. E. Kodger, R. E. Guerra, A. F. Pegoraro, M. Rubinstein, D. A. Weitz, Adv. Mater., **27**, 5132 (2015)

2. C. Gerstl, G. J. Schneider, A. Fuxman, M. Zamponi, B. Frick, T. Seydel, M. Koza, A. C. Genix, J. Allgaier, D. Richter, J. Colmenero, A. Arbe, Macromolecules, **45**, 4394 (2012)

3. C. Grigoriadis, A. Nese, K. Matyjaszewski, T. Pakula, H.-J. Butt, G. Floudas, Macromol. Chem. Phys., **213**, 1311 (2012)

4. H. Iwawaki, O. Urakawa, T. Inoue, Y. Nakamura, Y. Matsumiya, H. Watanabe, Macromolecules, **53**, 7096 (2020)

5. H. Liang, B. J. Morgan, G. Xie, M. R. Martinez, E. B. Zhulina, K. Matyjaszewski, S. S. Sheiko, A. V. Dobrynin, Macromolecules, **51**, 10028 (2018)

6. M. Hu, Y. Xia, G. B. McKenna, J. A. Kornfield, R. H. Grubbs, Macromolecules, **44**, 6935 (2011)

7. S. J. Dalsin, M. A. Hillmyer, F. S. Bates, ACS Macro Letters, **3**, 423 (2014)

8. S. J. Dalsin, M. A. Hillmyer, F. S. Bates, Macromolecules, **48**, 4680 (2015)

9. I. N. Haugan, M. J. Maher, A. B. Chang, T.-P. Lin, R. H. Grubbs, M. A. Hillmyer, F. S. Bates, ACS Macro Letters, **7**, 525 (2018)

10. M. Abbasi, L. Faust, M. Wilhelm, Adv. Mater., **31**, 1806484 (2019)

11. S. Alexandris, K. Peponaki, P. Petropoulou, G. Sakellariou, D. Vlassopoulos, Macromolecules, **53**, 3923 (2020)

12. R. Verduzco, X. Li, S. L. Pesek, G. E. Stein, Chem. Soc. Rev., **44**, 2405 (2015)

13. T. Pan, S. Dutta, Y. Kamble, B. B. Patel, M. A. Wade, S. A. Rogers, Y. Diao, D. Guironnet, C. E. Sing, Chem. Mater., **34**, 1990 (2022)

14. K. J. Bichler, B. Jakobi, V. G. Sakai, A. Klapproth, R. A. Mole, G. J. Schneider, Macromolecules, **53**, 9553 (2020)

15. B. Jakobi, K. J. Bichler, A. Sokolova, G. J. Schneider, Macromolecules, **53**, 8450 (2020)

16. K. J. Bichler, B. Jakobi, S. O. Huber, E. P. Gilbert, G. J. Schneider, Macromolecules, **53**, 78 (2020)

17. S. L. Pesek, X. Li, B. Hammouda, K. Hong, R. Verduzco, Macromolecules, **46**, 6998 (2013)

18. F. Kremer, A. Schönhals, *Broadband Dielectric Spectroscopy*. Springer Berlin Heidelberg: 2012.

19. J. Hintermeyer, A. Herrmann, R. Kahlau, C. Goiceanu, E. A. Rössler, Macromolecules, **41**, 9335 (2008)

20. A. Khalyavina, L. Häußler, A. Lederer, Polymer, **53**, 1049 (2012)

21. R. Zorn, B. Frick, L. Fetters, JCP, **116**, 845 (2002)

22. K. J. Bichler, B. Jakobi, V. G. Sakai, A. Klapproth, R. A. Mole, G. J. Schneider, Nano Lett., **21**, 4494 (2021)

Unexpected molecular dynamics of ethanol in hydrogen-bonded binary mixtures, ethanol-octanol and ethanol-water.

Ingo Hoffmann[1,*], *Firoz* Malayil Kalathil[1,2], *Tobias* Lopian[1], *Didier* Touraud[3], *Orsolya* Czakkel[1], *Marie* Plazanet[2,**], and *Christiane* Alba-Simionesco[4]

[1]Institut Laue-Langevin, 38000 Grenoble, France
[2]Laboratoire Interdisciplinaire de Physique, CNRS and Univ. Grenoble-Alpes, 38402 Saint Martin d'Hères, France
[3]Institute of Physical and Theoretical Chemistry, University Regensburg, Universitätstraße 31, 93053 Regensburg, Germany
[4]Laboratoire Leon Brillouin, Paris-Saclay University. CEA, CNRS, 91171 Gif-sur-Yvette, France

Abstract. In view of getting a quantitative picture of the dynamics in mixtures of hydrogen-bonded liquids, in particular the ternary water-ethanol-octanol, we examine in this paper the dynamics of two binary systems: ethanol-water and ethanol-octanol. Our multiscale investigation includes quasi-elastic neutron scattering for the characterization of the dynamics at the molecular scale, completed by NMR at the mesoscopic scale and eventually compared with macroscopic viscosity properties. We highlight the decrease of diffusivity in pure alcohols when increasing the lengthscale and conjecture its relation with the two processes measured in dielectric spectroscopy. While the behaviour of ethanol-water is well understood, unexpected inversion of the slowest component between the micro and the mesoscale are evidenced in the ethanol-octanol mixture. This effect could be at the origin of the negative viscosity excess in the mixture of alcohols.

1 Introduction

Hydrogen bonded liquids, such as water and alcohols, are literally vital components of all biological systems, and essential in pharmaceuticals, food products, cryo-protection agents, microwave chemistry, and many technologically relevant materials for liquid-liquid extraction. Many important features are linked to their strong polarity that stems from the -OH dipole. Therefore, the structure of these liquids is more complicated than other organic liquids, e.g. alkanes, where van-der-Waals interactions prevail and no supramolecular structures are identified. For these systems, the local organization depends on the competition between hydrogen-bonding interactions that tend to favour clustering, and steric hindrance between alkyl chains, that limits the extension of the H-bond network. Monoalcohols, with only one donor site, can develop a two-dimensional HB network of various sizes and shapes (chains, branched chains, rings), at variance to liquid water that shows a three-dimensional HB network since the H_2O molecule has two donor and two acceptor sites.

An increased complexity is observed by mixing these liquids and the corresponding processes are widely studied in the literature because they are found in many applications over the last decades. Beside the industrial relevance of water-ethanol mixture, for example the water-octanol mixture has been studied as model system to mimic interfaces and cell membranes [1] and aggregation in the form of a reverse micelle-like or worm-like structures was suggested. Ternary aqueous mixtures also received a lot of attention as promising green solvents. The aqueous mixture of ethanol and octanol has been extensively studied during the last decade since it presents a model system for the study of the nano and meso-organization, also presenting the so-called Ouzo-effect [2], observed when an emulsion is formed after a rapid aqueous dilution of a low water content mixture, entering deeply into the two-phase region close to the binodal line. It was actually shown that a *pre-Ouzo* zone is already present in the monophasic region, extending all around the critical point (see Figure 1a for graphical visualization) [3]. This work, on the three involved binaries of this system, is a first step toward the understanding of the molecular dynamics of the pre-Ouzo region in following a line crossing the phase diagram across the region of interest.

Monoalcohols containing up to three carbon atoms are known to be fully miscible with water in any proportion. Thermodynamic excess quantities, transport properties and structure of water-ethanol mixtures as a function of mole fraction of alcohol have evidenced thermal anomalies [8, 9]. In particular, the viscosity maximum is observed for 0.25 molar fraction of ethanol, i.e., three water molecules per alcohol molecule [6]. Around this concentration, the H-bond network of water is enhanced due to the hydrophobic interaction of the ethyl groups in ethanol. Water molecules are easily added until all hydrogen bond sites of the alcohol are occupied; by adding further water, at the viscosity maximum, mesoscopic nanophase separation with water droplets is observed but the system remains macroscopically miscible. For monoalcohols containing

*e-mail: hoffmann@ill.fr
**e-mail: marie.plazanet@univ-grenoble-alpes.fr

 https://doi.org/10.1051/epjconf/202227201003

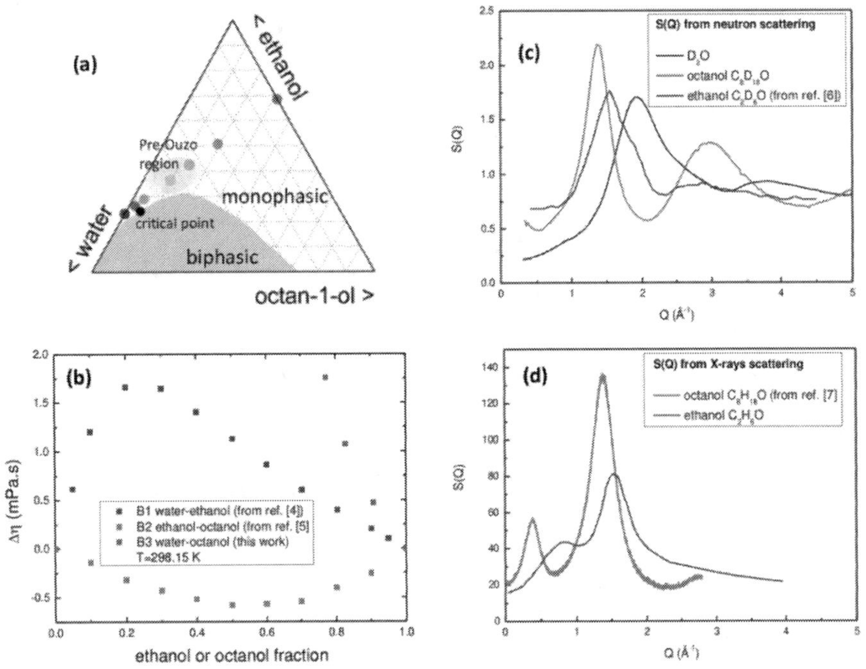

Figure 1. (a) phase diagram of the ternary mixture water-ethanol-octanol, expressed in mole fraction. The dots are distributed over the line of interest for the understanding of the ternary mixture. (b): excess viscosity for the three binary mixtures in ethanol or octanol mole fraction in their miscibility range. Data for the water-ethanol and ethanol-octanol are taken from references [4] and [5] respectively.(c) and (d) : structure factor of the pure compounds measured by neutron (our work, measurements on 7C2 at LLB, France and ref [6]) and X-ray diffraction (our work and ref. [7] respectively.

a longer alkyl chain, such as n-octanol, mixing is only possible in very limited proportions until aqueous saturation is reached: at room temperature, molar fraction solubility of water in octanol was found to be equal to 0.27 (48.91 $\mp$ 0.13 mg/g of water in octanol [10]). At variance to water-monoalcohol mixtures, monoalcohol mixtures, such as ethanol-octanol, exhibit a negative excess viscosity [5] (see Figure 1b), i.e., the system becomes less viscous than the ideal mixture by adding ethanol to octanol with a minimum close to a molar fraction of .5 (note that, for this latter case, one should also remember that the viscosity of monoalcohols increases with the number of carbons in the alkyl chain, ethanol being almost 8 times faster than n-octanol). Therefore, understanding the function of hydrogen-bonded liquids requires knowledge of the structure and dynamics of both the molecules and the dipoles.

The structure factor of the liquids studied in this work is reported in Figure 1 by neutron on panel (c) (our work and ref. [11]) and by X-Rays scattering on panel(d) (our work and ref. [7]). The mean peak (1-2 Å^{-1}) associated to intermolecular distances is shifted to higher Q's and broader for the smallest liquid, water. The occurrence of a prepeak (0-1 Å^{-1}), located beyond the usual short range liquid order at low wave vector is weakly seen by neutron scattering due to the low weight contribution of the O-O partial structure factor, but clearly noticeable by X-rays at room temperature for ethanol and n-octanol. Nothing equivalent is observed for water due to the smaller size of the molecule. Careful analysis of molecular dynamic simulations in terms of cluster sizes and shapes reported a fraction of ring aggregates of 27% for ethanol [11–14] and only 3.6% for octanol where chain like clusters are majority [1, 15]. The large dipole moment involved makes these associating liquids very good systems for studies by dielectric spectroscopy, a technique that has been extensively applied to this class of liquids [16–19]. It has been established for mono-alcohols, such as ethanol or octanol, and possibly water, that the dipoles relaxation occurs with a characteristic time that differ from those of the molecules. A prominent dielectric signal of Debye type appears at low frequency, i.e. with a long relaxation time slower than the relaxation of the remaining structure; it is attributed to the breaking of hydrogen bonds in molecular aggregates of 2-4-6 molecules associated in chain or ring like structures. As a result a slow mode due to the H bond network, the -OH dynamics, coexist with a faster one related to the structural relaxation and the mobility of the -OR part of the molecules.

In this framework, we investigated both binary mixtures at each end of the line crossing the ternary phase diagram, ethanol-water (B1) ethanol-octanol (B2) (0.21 and 0.65 mol fraction of ethanol in hydrogenated mixtures, respectively), corresponding moreover to a constant ethanol mass fraction. We present here measurements of the diffusion coefficient at different lengthscales: at the molecular scale using Neutron Spin Echo (NSE) and at the mesoscale with NMR. Although the diffusion coefficients of pure alcohols measured with both techniques are known from the literature, we emphasize here that the characteristic timescales of the dynamics vary with the lengthscale. Using specific deuteration, we were eventually able to disentangle the diffusion coefficients of the different molecular species, exhibiting different behaviors in a macroscopically homogeneous mixture. In light of the structural organization and dynamics characterized by dielectric spectroscopy, we conjecture that the origin of the lengthscale variation of the diffusion coefficient is directly related to the mesoscale cluster organization and propose that it is also at the origin of the different signs of the excess viscosities in the various mixtures.

2 Materials and methods

Anhydrous ethanol (purity > 99.9%) and octan-1-ol (purity > 99%) were provided by Carlo Erba Reagents. Water was prepared from a Millipore source with a resistivity of 18.2 MΩcm. D_2O (99.9% D) and d6-ethanol (C_2D_5-OD, 99.0% D, <0.3% water) were obtained from Eurisotop, d17-octan-1-ol (C_8D_{17}-OH, 98% D, where only the hydroxy labile hydrogen is not deuterated) from Cambridge Isotope Laboratories. All compounds were used without further purification. All measurements presented here, including the pure ethanol and octanol but not water, were performed during the same experimental runs.

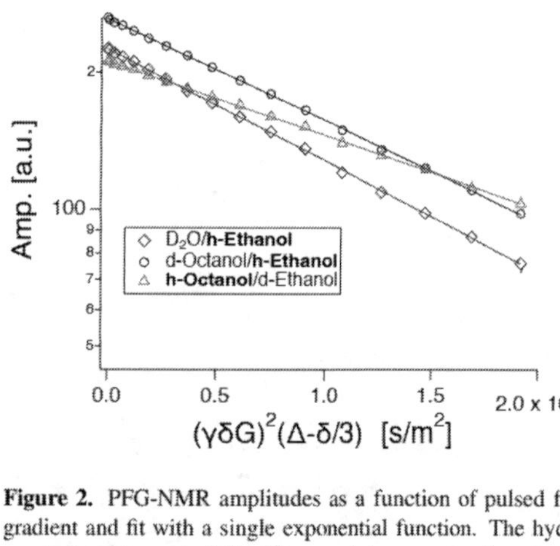

Figure 2. PFG-NMR amplitudes as a function of pulsed field gradient and fit with a single exponential function. The hydrogenated component of the binary mixture, accounting for the largest part of the signal, is in bold in the legend.

NSE measurements were performed on the instrument IN11C at a wavelength of 5.5 Å and the center of the detector positioned at an angle of 20 deg covering a Q range from 0.1 to 0.7 $Å^{-1}$ and allowing to access a maximum usable Fourier time of 1 ns. Additional NSE measurements were performed on the instrument IN15 using a wavelength of 8 Å and detector angles of 3.5, 6.5 and 9.5 deg covering a usable Q range from 0.05 to 0.145 $Å^{-1}$ and reaching maximum Fourier times of 70 ns.

Pulsed Gradient Field NMR (PGF-NMR) experiments were performed on a Bruker Minispec mq20 with a pulsed gradient unit for gradients up to 4 T/m using a standard Hahn Echo sequence with a delay between 90 and 180 deg pulse Δ=7.5 ms and a gradient pulse length δ=0.6 ms. Amplitudes vs field gradient were very well fitted using a single exponential decay, as shown in the figure 2. The lengthscale probed with NMR, equivalent of 'Q^2', is $(\Gamma\delta G)^2$. With the proton gyromagnetic ratio Γ = 26.7E7 rad/T/s, δ=0.6 ms and G from 0 to 3.2 T/m, gives Q=0.5 μm^{-1} at 3.2 T/m gradient, therefore 3 to 4 orders of magnitude larger than NSE typical lengthscale. Deuterated compounds were used in order to disentangle the different coefficients, based on the measurement of proton diffusion i.e. of the diffusion of the hydrogenated component. The extracted diffusion coefficients were therefore calculated, for each species, assuming a complete exchange of the labile hydrogen/deuterium between the hydroxyl groups. To reach this goal, two equations were coupled such as :

$$D_{meas1} = x_{11}D_1 + x_{12}D_2 \quad (1)$$

$$D_{meas2} = x_{21}D_1 + x_{22}D_2 \quad (2)$$

with x_{11} and x_{12} are the molar fraction of protons in component 1 and 2 respectively, in the measurement 1, while x_{21} and x_{22} are the molar fractions protons in component 1 and 2 respectively in the measurement 2. Equations are then inverted to extract the diffusion coefficient D_1 of the component 1 and D_2 of the component 2. This assumes that exchange of labile protons is slow compared to the time scale of the experiment and that all protons contribute equally to the NMR signal. We also neglect differences in viscosity between deuterated and hydrogenated components and we further assume 100% deuteration for the nominally deuterated components. This correction is however quite small. Results are directly presented in the section 4.

All measurements were performed at room temperature 298 K.

3 Results from NSE

Neutron Spin Echo enables the measurement of the translational diffusion coefficient on a lengthscale of a fraction of nm and a timescale smaller than a few ns. Mixtures of protonated and deuterated components were used in order to extract the dynamics of each component. Since NSE measures the difference between coherent and 1/3 of the incoherent neutron scattering signal, the data were fitted according to the equation

$$S(Q,t) = A_{coh}(Q)S_{coh}(Q,t) - A_{inc}(Q)S_{inc}(Q,t) \quad (3)$$

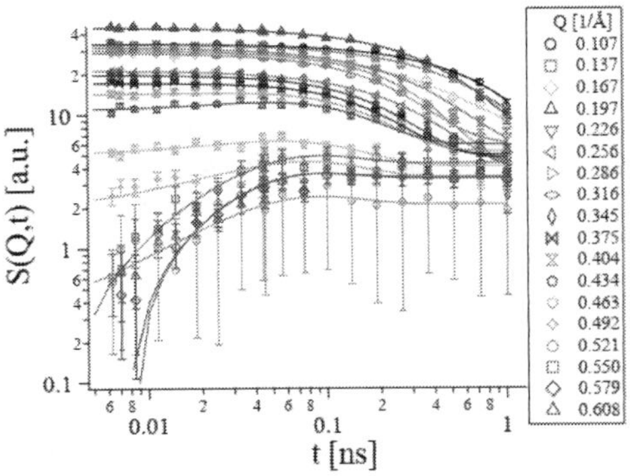

Figure 3. NSE data and fit for octanol in the B2 mixture d-ethanol/h-octanol (0.61 mol of ethanol), according to eq. 3 for the different values of Q. Data are measured on the spectrometer IN11 at the ILL (France).

where A_{coh} and A_{inc} are the (positives) coherent and incoherent amplitudes which are calculated from the polarised intensity in up and down direction assuming 100% flipper efficiency. S_{inc} is the normalised incoherent intermediate scattering function given by $S_{inc}(Q,t) = \exp(-D_{inc}Q^2t)$, where the incoherent diffusion coefficient D_{inc} is constant over the whole Q range. This last approximation therefore assumes a Fickian diffusion over the lengthscales covered by NSE, a posteriori justified by the quality of the fit. The normalised coherent intermediate scattering function S_{coh} is given by

$$S_{coh}(Q,t) = x_{inel}(Q)\exp(-\Gamma_{inel}t) + (1 - x_{inel}(Q))\left[\left(1 - x_{bkg}(Q)\right)\exp\left(-D_{coh}(Q)Q^2t\right) + x_{bkg}(Q)\right] \quad (4)$$

where D_{coh} is the Q dependent coherent diffusion coefficient, x_{bkg} is the fraction of elastic background and x_{inel} is the fraction of fast coherent dynamics [20]. Γ_{inel} has been set to a value of 300 1/ns based on the relaxation times observed in an all-deuterated sample. At this value, $\exp(-\Gamma_{inel}t)$ has essentially decayed after 8 ps. For a given Q in a given data set, we have D_{coh}, x_{bkg}, x_{inel}, in addition there is a single Q independent D_{inc} for the whole data set. The relatively large number of free parameters is eased by the fact that both x_{bkg} and x_{inel} are small corrections accounting for at most 2% of the coherent signal. The coherent (collective) dynamics is expected to be sensitive to the structural organization and therefore is allowed to vary over the investigated Q range. The IN15 data were fitted with the same model and constrained to the D_{inc} extracted from the IN11 data which covers higher Q, where the coherent signal has mostly decayed and the signal incoherent scattering is more visible. Because of the very large scattering cross section of hydrogen, we expect the incoherent signal to be dominated by the hydrogenated component. The coherent contribution is proportional to the square of the contrast, here between the hydrogenated and deuterated molecules. We assume in this analysis that the measured coherent dynamics is also the one of the hydrogenated component, justified by the good agreement between D_{coh} and D_{inc} at high Q.

Both data and fits according to eq. 3 are shown in the figure 3 for the mixture B2 where the octanol is hydrogenated. At small Q, the signal is dominated by the coherent diffusion. The incoherent contribution becomes dominant for Q larger than c.a. 0.4 $Å^{-1}$ when the small angle intensity I(Q) drops to zero and before the structure factor increases again above 1 $Å^{-1}$ because of to intramolecular correlations.

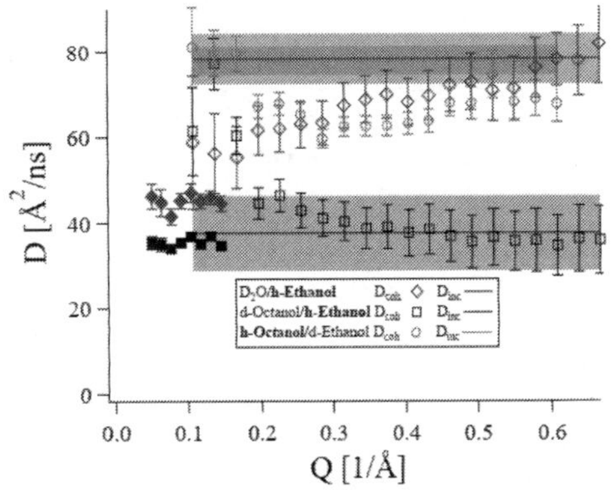

Figure 4. Coherent (empty squares) and incoherent (constant line) diffusion coefficients extracted from IN11C measurements over the whole Q range three mixtures: B1 d-water/h-ethanol, B2 d-ethanol/h-octanol, B2 h-ethanol/d-octanol, in the respective fractions of 0.14, 0.61, 0.66 mol of ethanol. Full squares are extracted from IN15 measurements. The colored zones represent the error bar on D_{inc}.

The extracted diffusion coefficients for the B1 and B2 mixtures are shown in the figure 4. D_{coh} and D_{inc}, for each sample, are quite similar and tend to agree at large Q (small distances). Deviations are observed at small Q, when probing longer distances. The collective diffusion of ethanol in the water/ethanol mixture tends to slow down over larger distances, as confirmed by the smallest Q data measured on IN15. This effect can be assigned to the presence of clusters of a few molecules as previously discussed, close to the maximum of the thermodynamic anomalies. If self and collective diffusion are similar very short distances, measuring the collective translational dynamics over larger distances (small Q) implies probing the dynamics of these aggregates. D_{coh} therefore deviates from D_{inc} and becomes slower, the diffusion of ethanol is hindered by the aggregation.

Due to the limited range of Fourier times covered by IN11C, during which barely any decay of S(Q,t) is visible at low Q, the increase of D_{coh} observed at the low Q end of the IN11C measurements for h-ethanol/d-octanol is merely an artefact as the measurements from IN15, which cover significantly longer times are clearly indicating constant values of D_{coh} at low Q. The ethanol-octanol behavior will be further discussed in the next section in regards to NMR measurements.

4 Discussion

The diffusion coefficients measured by NSE and NMR of the B1 and B2 mixtures on both sides of the pre-Ouzo region are presented in figure 5. NMR probes the individual molecular diffusion similarly to incoherent QENS but over a lengthscale larger by 5 orders of magnitude. This combination enables the study of the diffusion coefficients from the nm to the micron lengthscale: a slowing down of the dynamics when the lengthscale increases is well understood in complex systems [21].

Regarding the pure compounds, water presents a dynamics characteristic time, assigned to the translational diffusion coefficient, identical for the two techniques, while diffusion of ethanol and octanol slow down with increasing lengthscale. The values presented here are in very good agreement with values reported in the literature for neutron Time-Of-Flight spectroscopy in pure water and ethanol [22, 23] and PFG-NMR for water, ethanol and octanol[24–26]. At the molecular scale, water forms a 3-D hydrogen bond network that repeats over macroscopic distances. As discussed in the introduction, clustering due to hydrogen bonding of a few molecules occurs in monoalcohols because of the presence of hydrophobic chains. Dielectric spectroscopy measurements show two relaxation times, a fast one related to internal molecular motions and a slow one related to the cluster dynamics. We conjecture that the same causes apply here: internal motions within the cluster governing the dynamics over a short lengthscale measured by neutron scattering and cluster dynamics slowing down the diffusion over larger distances. Although the Debye relaxation is also observed, to a lower extent, in pure water, we do not observe any effect of large sale structure on the diffusivity in pure water.

We then measured the diffusivity of ethanol in the water-ethanol mixture. Overall, we measure a smaller diffusivity in agreement with the increase of viscosity. When probed over the full range of NSE lengthscales corresponding to the size of a few molecules, the ethanol mixed in water exhibits a small Q dependence. However, when going up to large distances as probed by NMR, no further slowing down of the dynamics is observed : for distances larger than the ethanol clusters, the diffusivity is governed by the 3-D organization of water and the diffusivity stays constant.

Eventually, the ethanol-octanol mixture presents the most unexpected behavior. While the diffusion coefficient of octanol decreases over large distances similarly to pure octanol, the one of ethanol, at the limit of the error bars, tends to increase.

Mixtures of alcohols are known to exhibit strong negative viscosity excess. Still, the diffusivity of ethanol is lower in the mixture than in the bulk, which is not the case of octanol. On one hand, the averaged diffusivity can be calculated on the basis of the diffusion coefficients of the components in the mixture, for both NMR and NSE : $D_{av}=x_1 D_{mix,1}+x_2 D_{mix,2}$ were $x_{1,2}$ are the molar fractions of each component. This gives numbers of ~ 57 and ~ 49 10^{-11} m^2s^{-1} for NSE and NMR respectively. On the other hand, one can compute an ideal diffusivity based on the diffusion coefficients of the pure compounds : $D_{ideal}=x_1 D_{bulk,1}+x_2 D_{bulk,2}$. This calculation instead leads to diffusivities of ~ 118 and 74 10^{-11} m^2s^{-1} respectively. Although these number indicate a negative excess of diffusivity that would lead to a positive excess of viscosity in the ethanol-octanol mixture, in opposition to its experimental value, the (absolute) diffusivity excess decreases from 60 to 25 with increasing lengthscale. A further evolution in this direction at macroscopic distances would explain the negative viscosity excess.

In this mixture, the hydrogen-bond network is not three dimensional because of the carbon chains and the larger importance of Van der Waals interactions. Ethanol is less mobile in the mixture than in the bulk at short distances but unlike the octanol, behaves more in a bulk-like way at long distances. The opposite is observed for octanol. The ethanol clusters may be hindered by the surrounding octanol molecules, but have a short enough lifetime so that the long range diffusion is less affected. Octanol molecules instead are fluidified by the presence of ethanol that locally disturbs the structural octanol organization. This last one may still persist at long distance, inducing the decrease of mobility.

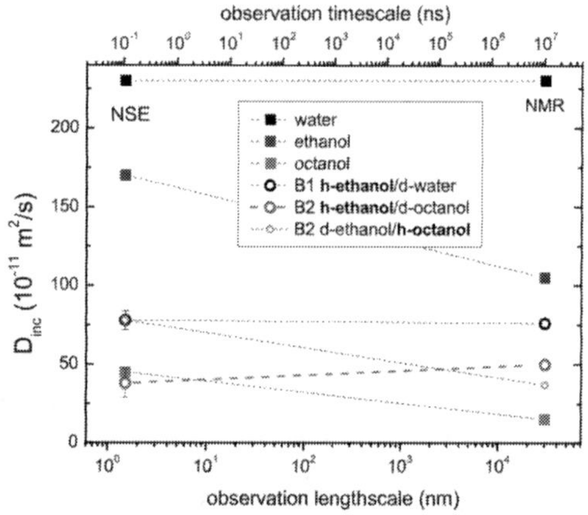

Figure 5. Self diffusion coefficients from NSE and NMR for the pure compounds and binary mixtures, measured at 298 K. In the binaries, the component characterized is in bold in the legend. Lines are only guides for the eye. Note the exceptional increase of the ethanol diffusion coefficient between NSE and NMR in the ethanol-octanol solution.

5 Conclusion and perspectives

In order to understand and rationalize the macroscopic behavior of hydrogen bonded mixtures of water and alcohols, we investigated the dynamics from the nano- to the micron scale. The use of Neutron Spin Echo in partially deuterated mixtures enables the distinction between coherent and incoherent diffusivity, giving further insight into the molecular dynamics over the scale of small aggregates. Combined with NMR, we could characterize the diffusion coefficients of individual species in molecular mixtures on the microscopic and mesoscopic lengthscale.

We emphasize that pure alcohols already present diffusivities that slow down over long distances with respect to molecular scale. This observation is in agreement with the different relaxations observed with dielectric spectroscopy on very different observables which are the dipoles.

In the binary systems, we observe complex behaviors of the molecular dynamics, changing with distance and time in a non intuitive manner. We propose that the positive excess viscosity measured in water-alcohols mixture is related to the long range extend of the hydrogen bond network due to the dominating water dynamics. In mixtures of alcohols, instead, the viscosity excess is negative. We indeed observe a singular behavior of ethanol that was not highlighted earlier and requires further investigations to be quantitatively understood, in relation to the long distance organization.
The two binaries investigated here are the most relevant of the ternary system water-ethanol-octanol, which exhibits nanoscale structuration. We expect the dynamics characterized here to be key processes in the organization of the ternary system.

Acknowledgments

We thank the ILL for beamtime and the use of PSM laboratories. We are grateful to Sylvain Prévost, Thomas Zemb and Olivier Diat for fruitful discussions.

References

[1] J.L. MacCallum, D.P. Tieleman, Journal of the American Chemical Society **124**, 15085 (2002), pMID: 12475354, https://doi.org/10.1021/ja027422o

[2] S.A. Vitale, J.L. Katz, Langmuir **19**, 4105 (2003)

[3] T.N. Zemb, M. Klossek, T. Lopian, J. Marcus, S. Schöettl, D. Horinek, S.F. Prevost, D. Touraud, O. Diat, S. Marčelja et al., Proceedings of the National Academy of Sciences of the United States of America **113**, 4260 (2016)

[4] B. González, N. Calvar, E. Gómez, Á. Domínguez, The Journal of Chemical Thermodynamics **39**, 1578 (2007)

[5] M.A.F. Faria, R.J. Martins, M.J.E.M. Cardoso, O.E. Barcia, Journal of Chemical & Engineering Data **58**, 3405 (2013), https://doi.org/10.1021/je400630f

[6] R. Böhmer, C. Gainaru, R. Richert, Physics Reports **545**, 125 (2014), structure and dynamics of monohydroxy alcohols—Milestones towards their microscopic understanding, 100 years after Debye

[7] K.S. Vahvaselkä, R. Serimaa, M. Torkkeli, Journal of Applied Crystallography **28**, 189 (1995)

[8] P.A. Artola, A. Raihane, C. Crauste-Thibierge, D. Merlet, M. Emo, C. Alba-Simionesco, B. Rousseau, Journal of Physical Chemistry B **117**, 9718 (2013)

[9] D. Rak, M. Sedlák, The Journal of Physical Chemistry B **123**, 1365 (2019), pMID: 30532969, https://doi.org/10.1021/acs.jpcb.8b10638

[10] B.E. Lang, Journal of Chemical and Engineering Data **57**, 2221 (2012)

[11] C.J. Benmore, Y.L. Loh, The Journal of Chemical Physics **112**, 5877 (2000), https://doi.org/10.1063/1.481160

[12] A. Vrhovšek, O. Gereben, A. Jamnik, L. Pusztai, The Journal of Physical Chemistry B **115**, 13473 (2011), pMID: 21916497, https://doi.org/10.1021/jp206665w

[13] S. Pothoczki, I. Pethes, L. Pusztai, L. Temleitner, K. Ohara, I. Bakó, Journal of Physical Chemistry B **125**, 6272 (2021), 2101.12142

[14] G. Matisz, A.M. Kelterer, W.M.F. Fabian, S. Kunsági-Máté, The Journal of Physical Chemistry B **115**, 3936 (2011), pMID: 21417299, https://doi.org/10.1021/jp109950h

[15] S. Stephenson, R. Offeman, G. Robertson, W. Orts, Chemical Engineering Science **62**, 3019 (2007)

[16] T. Sato, R. Buchner, The Journal of Physical Chemistry A **108**, 5007 (2004), https://doi.org/10.1021/jp035255o

[17] A. Arbe, P. Malo de Molina, F. Alvarez, B. Frick, J. Colmenero, Phys. Rev. Lett. **117**, 185501 (2016)

[18] J.S. Hansen, A. Kisliuk, A.P. Sokolov, C. Gainaru, Phys. Rev. Lett. **116**, 237601 (2016)

[19] S.K. Garg, C.P. Smyth, The Journal of Physical Chemistry **69**, 1294 (1965), https://doi.org/10.1021/j100888a033

[20] I. Hoffmann, Frontiers in Physics **8** (2021)

[21] Q. Berrod, F. Ferdeghini, J.M. Zanotti, P. Judeinstein, D. Lairez, V. García Sakai, O. Czakkel, P. Fouquet, D. Constantin, Scientific Reports **7**, 2241 (2017)

[22] J. Qvist, H. Schober, B. Halle, The Journal of Chemical Physics **134**, 144508 (2011), https://doi.org/10.1063/1.3578472

[23] T. Seydel, R.M. Edkins, K. Edkins, Physical Chemistry Chemical Physics **21**, 9547 (2019)

[24] K.C. Pratt, W.A. Wakeham, J. Chem. Soc., Faraday Trans. 2 **73**, 997 (1977)

[25] D. Bellaire, H. Kiepfer, K. Münnemann, H. Hasse, Journal of Chemical and Engineering Data **65**, 793 (2020)

[26] M. Iwahashi, Y. Ohbu, T. Kato, Y. Suzuki, K. Yamauchi, Y. Yamaguchi, M. Muramatsu, Bulletin of the Chemical Society of Japan **59**, 3771 (1986), https://doi.org/10.1246/bcsj.59.3771

Notes on Fitting and Analysis Frameworks for QENS Spectra of (Soft) Colloid Suspensions

Christian Beck[1,2,*], *Kevin* Pounot[1,2,**], *Ilaria* Mosca[1,2], *Niina* H Jalarvo[3,4], *Felix* Roosen-Runge[5], *Frank* Schreiber[1], and *Tilo* Seydel[2,***]

[1]Institut für Angewandte Physik, Universität Tübingen, Auf der Morgenstelle 10, 72076 Tübingen, Germany.
[2]Institut Max von Laue - Paul Langevin (ILL), CS 20156, F-38042 Grenoble Cedex 9, France.
[3]Jülich Centre for Neutron Science (JCNS), Forschungszentrum Jülich GmbH, D-52425 Jülich, Germany
[4]Chemical and Engineering Materials Division, Neutron Sciences Directorate, and JCNS Outstation at the Spallation Neutron Source (SNS), Oak Ridge National Laboratory, P.O. Box 2008, Oak Ridge, Tennessee 37831, USA
[5]Department of Biomedical Sciences and Biofilms-Research Center for Biointerfaces (BRCB), Malmö University, 20506 Malmö, Sweden.

Abstract. With continuously improving signal-to-noise ratios, a statistically sound analysis of quasi-elastic neutron scattering (QENS) spectra requires to fit increasingly complex models which poses several challenges. Simultaneous fits of the spectra for all recorded values of the momentum transfer become a standard approach. Spectrometers at spallation sources can have a complicated non-Gaussian resolution function which has to be described most accurately. At the same time, to speed up the fitting, an analytical convolution with this resolution function is of interest. Here, we discuss basic concepts to efficient approaches for fits of QENS spectra based on standard MATLAB and Python fit algorithms. We illustrate the fits with example data from IN16B, BASIS, and BATS.

1 Introduction

The interpretation of QENS data is generally performed in terms of maximum likelihood methods [1], commonly known as "data fitting". By this curve fitting, the most probable from an infinite set of hypotheses expressed by a continuous variable, namely the fit parameter, is determined by finding the optimum of a suitable likelihood function [1–3]. For the generally assumed situation of Poisson-distributed data points, the maximum likelihood method is mathematically equivalent to the least-squares method [1], and thus, QENS data fitting is generally based on least-squares minimization. The data fitting procedure also involves the determination of the confidence bounds on the fit parameters as well as of the goodness of fit.

Along with the higher brightness and better signal-to-noise ratios of recent neutron spectrometers [4], the topic of associated data analysis software frameworks is of continuous interest [5–7]. Better scattering signals allow for more complex models to be tested. The switching between numerous models to be applied on a set of data has to be quick and easy. Here, script-based approaches seem to be unavoidable, and concepts for easily modifiable scripts gain interest. Models of QENS data routinely contain several Lorentzians that are coupled in intensity and momentum transfer [8]. An example consists in the model of coupled diffusive states that for instance describes the internal diffusion of proteins, accounting for superimposed protein backbone and side chain fluctuations [9].

The model is convoluted with the resolution function of the instrument - obtained on dedicated data - to describe a signal that can be fitted to the experimental data. In cases where the model consists only of Lorentzians, Gaussians, and elastic contributions or a sum thereof, an analytical convolution can be readily performed with the benefit of low computational cost if the spectrometer resolution function is described by a sum of an arbitrarily large number of Gaussian (and Lorentzian) functions. In this case, the scattering function simply consists of a sum of Voigt (and Lorentzian) functions. The Fourier transform of the Mittag-Leffler function [10, 11], namely the generalized Lorentzian function, constitutes another important function for fitting of QENS spectra, which contains the stretched exponential. This function cannot yet be readily included in a generalized Voigt function for an analytical convolution, although initial approaches to generalize the Voigt function seem to exist in the mathematical literature [12], and Padé-approximations can be applied near the elastic line [13]. Fourier transform methods can be used alternatively, but are subject to the well-known general limitations and maybe more difficult to apply for non-symmetric resolution functions [14, 15].

The present article addresses script concepts to test different models for the scattering function efficiently. It

*e-mail: christian.beck@uni-tuebingen.de
**current address: European Synchrotron Radiation Facility - 71, avenue des Martyrs, CS 40220, 38043 Grenoble Cedex 9, France.
***e-mail: seydel@ill.eu

does not, however, address Bayesian inference, for instance regarding the number of Lorentzians present in a data set. This topic has been dealt with previously by other work [16, 17]. Moreover, it is noted that several standard procedures for fitting are available inside standard software such as Mantid [18].

Subsequent to a brief outline of script concepts in MATLAB and Python, this article illustrates the analytical convolution with the spectrometer resolution function employing IN16B, BASIS, and BATS example data on protein solutions. The aim of this work is to show the general applicability of the analysis framework to different instruments but not the comparison of instruments. Therefore, different spectra of similar, but not identical samples are analyzed.

2 Analysis

The analysis of the collected QENS spectra often implies a separation of different contributions to the scattering signal from superimposed motions as well as from immobile fractions. The contribution of the sample holder can be removed based on a separate measurement. For the subtraction of its signal from other spectra, the angular dependence of absorption can be accounted for using methods such as the Paalman-Pings corrections [19] . Additionally, the contribution of some sample components can be reduced by (partial) deuteration (i.e., using D_2O as solvent). However, it should be mentioned that the deuteration can significantly influence the system investigated [20]. Similarly to the sample holder, the solvent can also be measured individually. However, depending on the solute, the solvent composition might change due to effects such as dissolving ions or sugars and H-D exchange. In addition, the solvent contribution has to be rescaled based on the excluded volume by the solute [21]. The reduced solvent contribution can either be subtracted from the measured spectra or can be included in the model as fixed contribution. For the later option, a corresponding model for the solvent has to be applied, which might also depend on the energy and momentum transfer range probed [21].

The signal from dissolved soft ("floppy") colloids such as proteins consists of several different components. While a translational and rotational center-of-mass diffusion [21] can be observed due to the nature of the dissolved objects, also internal fluctuations, such as protein backbone and side chain fluctuations, and mesoscopic domain motions such as lobe motions in an antibody protein might be observed [8, 22, 23]. In order to separate these different contributions, an approach with relatively few assumptions based on fitting each spectrum for each momentum transfer q individually, may not be sufficient. A simultaneous fit approach, taking into account both energy and momentum transfer, is then necessary. A detailed description of the procedure, along with an interpretation of the models, can be found in the corresponding literature [8].

Given the recent technical improvements of the instruments allowing to access high energy transfers with an improved energy resolution, it becomes more important to have the possibility to fit more complex models, since the energy or time scales investigated are increased and diffusive contributions, which were masked previously either by the broad resolution function or by the small energy range investigated, become visible. Such a global fit offers on the one hand the possibility to extract more reliable parameters; on the other hand, it requires a higher level of modelling. Since several parameters, such as the parametrization of the resolution or the solvent contribution depend on q, a global fit needs to have q-dependent fixed parameters, as well as q-independent and q-dependent fit parameters. In the event that not only one single spectrum, but several spectra are analyzed simultaneously, e.g., to extract temperature independent entropy and enthalpy values instead of free energies [24], the complexity of the model increases even further, since the background signal dominated by the solvent is also temperature dependent.

The complexity of the system can also be increased by crowding through the addition of macromolecules in the aqueous suspension. Not only the solvent, but also parts of the crowders can be deuterated to mimic more complex environments, such as living cells, and to obtain a scattering signal which is dominated by the contribution of the tracer particle of interest [25–27].

Similarly to approaches already employed in small angle scattering data analysis [28, 29], the separation can be performed based on the knowledge of the sample composition. Since the signal is typically dominated by incoherent scattering, the q-dependence mainly influences the different dynamic features. While this approach includes the separation of the (apparent) global diffusion from internal diffusive processes [8], the same procedure can be applied to separate the scattering signal of the crowders from the one of the protein of interest [23], the contributions of monomers and crystals [30], the contributions of two different proteins [31], or to distinguish the scattering signal of proteins from the one coming from the detergent used for purification [32].

3 "Wrapping" in Python for flexible models

To keep the readability of the analysis code as good as possible as well as generally applicable, it is important to be able to easily change the model for the resolution function to apply the code to different spectrometers. This flexibility implies an automated convolution of the model function for the samples with the resolution function. Additionally, a change between different models can easily be performed in this way. Without going into any further detail of coding, such a flexible fit function call can for instance be written as follows using *python* and *scipy.curve_fit*:

```
popt, pcov = curve_fit( lambda x, *p: \
wrapper_function( x, q, n, r, len( f0 ), \
p, **keyword_parameters ), \
x, y, p0=f0, bounds=(l,u), sigma=dy )
```

Therein, the concept of the *lambda* function allows to point to external functions, and the *wrapper_function* contains the fit function that can be modified by a choice from a library of functions or a building-block syntax for functions. This choice from a library or selection of components is passed via suitable keyword arguments ***keyword_parameters*. These keyword arguments can take the form of an arbitrarily long list contained for instance in a *python* dictionary (*dict*). Besides the energy axis x, the q values and the used indices n on these q values can be passed as vectors. Moreover, the matrix r containing the parameters defining the analytical resolution function can be provided. $f0$ denotes the initial guess, i.e., the start values of the fit, and l and u the lower and upper bounds, respectively, and dy the standard errors on the spectral signal y. Importantly, the vector of fit parameters p can have a variable length, which is identical to the length of the initial guess *len(f0)*. The fit returns the optimum parameters *popt* and the covariance matrix *pcov*. The *wrapper_function* with the interface

```
def wrapper_function( x, q, n, r, N, \
*args, **kwargs ):
    ...
```

extracts the fit parameters into a vector *fp* usable subsequently via

```
a  = list( args[0][:N] )
fp = np.asarray( a )
```

Similar or technically simpler coding solutions can be achieved in other software such as MATLAB, as briefly outlined in the following section.

4 Fits in MATLAB

In MATLAB, simultaneous fits for all q at once to the data y at the points in energy x can be performed using the *lsqnonlin* command from the optimization toolbox:

```
lsqnonlin(model,f0,l,u,options,x,y);
```

Therein, as in the *python* version, $f0$ is the initial guess, and l and u are the lower and upper bounds, respectively. Importantly, x and y can be matrices. *model* is a function handle pointing to the function *EvaluatedModel* which returns a function of the model evaluated for the parameters fp. This "function of the model" has to take care of the weighting by the errors dy on the data, since *lsqnonlin* does not provide an option for explicit error weighting:

```
model=@(fp, xdata, ydata)EvaluatedModel(...
dy, q, fp, xdata, ydata, N, fixedp);
```

This syntax establishes that *EvaluatedModel* can take more parameters – namely all parameters in the second pair of round brackets – than the function handle *model* is allowed to accept – namely only the parameters in the first pair of round brackets – and therefore allows to hand over the number of q-independent fit parameters N as well as the fixed parameters and resolution parameters *fixedp*, and the q values, to calculate the scattering function depending on q and $\hbar\omega$. Hence, *model* in this MATLAB implementation takes the equivalent role of the *wrapper_function* outlined in the preceding section describing the *python* implementation, and the function handle is defined by the "@" symbol in MATLAB as opposed to the *lambda* function in *python*. The fit algorithm performs $\min_{fp} \|f(fp)\|_2^2$ with the fit parameters *fp* and, thus, represents a least-squares minimization.

Table 1: Specifications of the spectra shown in Fig. 1.

Subfigure	a	b	c
Instrument	IN16b	BASIS	BATS
number of q values/groups	18	18	19
number of $\hbar\omega$ values	1024	500	4126
mean spacing q [Å^{-1}]	0.096	0.010	0.094
mean spacing $\hbar\omega$ [µeV]	0.058	0.4	0.1
sample can outer ⌀[mm]	23	23	14

5 Analytical convolution

In Figure 1, we show the resolution functions for the different NBS Spectrometers IN16b (ILL) in its standard energy resolution high flux mode with Si(111) analyzers, of BASIS (SNS), and of IN16b in its BATS option (using the resolution setting lres4). All the resolution functions $\mathscr{R}$ were approximated with a sum of several Gaussian functions $\mathscr{G}(\sigma, x)$:

$$\mathscr{R} = \sum_n a_n \mathscr{G}(\sigma_n, \omega - \omega_n) \tag{1}$$

using $n = 2$ for the different options of IN16b and $n = 5$ for the BASIS resolution and a flat background with a_n, σ_n and ω_n being q-dependent scaling parameters, variances and center positions of the Gaussian function n, respectively. These parameters might depend on the binning applied to the q and $\hbar\omega$ space.

Any model of a scattering function containing only Lorentzians, including an elastic contribution, Gaussians, and Voigt functions, can be easily represented by an analytical convolution with these resolution functions.

Conveniently, the Voigt function $\mathscr{V}$, as the convolution of Gaussian with Lorentzian function, is available in *python* via the Faddeeva function *scipy.special.wofz*[33], $f(z) = \exp(-z**2)*\mathrm{erfc}(-i*z)$:

```
import numpy as np
from scipy.special import wofz

def V( x, sigma, gamma ):
return np.real( wofz( (x + 1j*gamma) \
      / sigma / np.sqrt(2.) ) )       \
      / sigma / np.sqrt( 2.*np.pi )
```

The voigt function $\mathscr{V}$ also exists as *scipy.special.voigt_profile* since *scipy* 1.9.1.

In MATLAB, the Voigt function can be implemented as C++ code – for instance via the Faddeeva package available from *mathworks.com* – and linked as compiled *mex* file.

In Figure 2, the spectra of different aqueous (D_2O) solutions containing the dissolved protein *bovine serum*

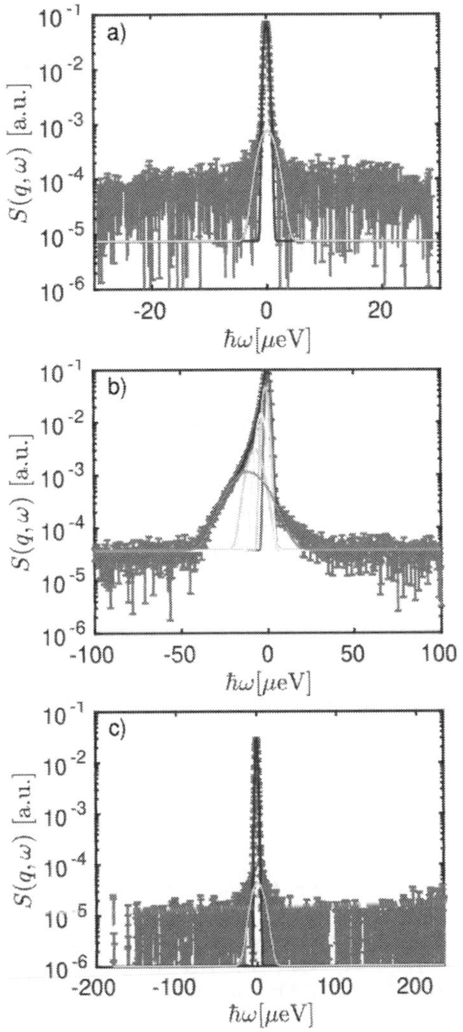

Figure 1: Resolution functions of different instruments (red symbols) and corresponding fits with the different Gaussian functions (colored lines). Different vanadium samples were measured with different statistics as well as different geometries, and different q and $\hbar\omega$ binnings are applied. A direct comparison between the instruments is therefore not possible. In contrast, the present and subsequent figure illustrate the generality of the analysis approach. For the different instruments, a different number n of Gaussian functions was used in addition to a flat background. a) IN16b: $n = 2$; $q = 1.03\,\text{Å}^{-1}$ b) BASIS: $n = 5$; $q = 1.05\,\text{Å}^{-1}$ c) BATS: $n = 2$; $q = 1.03\,\text{Å}^{-1}$. Details of the binning applied are given in Table 1.

albumin (BSA) at concentrations between $c_p = 100 - 200$ mg/ml are shown which are measured at the temperatures T=280 K and T=295 K. A direct comparison of the spectra is not possible due to the temperature dependence [22, 34] and the volume fraction scaling [34]. A model using two Lorentzian functions $\mathscr{L}$ to describe the apparent global center-of-mass diffusion coefficient and to summarize the internal diffusive processes of the proteins, which is established for different proteins in the small energy transfer range [8, 23, 25, 30, 35], and one Lorentzian function for the solvent contribution is convoluted with the resolution function to obtain the final fit model

$$\begin{aligned} S(q,\omega) &= \mathscr{R} \otimes \Big(A_0 \cdot \mathscr{L}_{\gamma}(\omega) + \\ &\quad (1-A_0)\cdot \mathscr{L}_{\gamma+\Gamma}(\omega) + \beta_{D_2O}\cdot \mathscr{L}_{\gamma_{D_2O}}(\omega)\Big) \qquad (2) \\ &= A_0 \sum_n a_n \mathscr{V}_{\sigma_n}(\gamma, \omega-\omega_n) + \\ &\quad (1-A_0) \sum_n a_n \mathscr{V}_{\sigma_n}(\gamma+\Gamma, \omega-\omega_n) + \\ &\quad \beta_{D_2O} \sum_n a_n \mathscr{V}_{\sigma_n}(\gamma_{D_2O}, \omega-\omega_n) \qquad (3) \end{aligned}$$

with a_n, σ_n, ω_n being the q dependent parameters fixed by the resolution function from Equation 1, $\gamma_{D_2O}, \beta_{D_2O}$ the parameters fixed based on the measurements of the solvent. For the apparent global diffusion, the individual contributions due to the different Gaussian functions used for the resolution function are displayed, while the internal diffusive processes and the solvent contribution are shown as a sum of the corresponding Voigt functions. By investigating the different contributions, it becomes visible that especially in the case of non-symmetrical resolution functions, it is important to consider the additional contributions. It should be mentioned that the internal diffusive processes are better accounted for by two coupled Lorentzian functions describing the motion of the protein backbone and side chains individually [22, 24]. For reasons of comparability, the same model is applied for the different energy ranges. The spectra shown illustrate the applicability of the approach for spectra from spectrometers located at spallation sources such as BASIS and also for spectrometers at continuous sources (IN16b, BATS) which are often characterized by more symmetrical resolution functions.

6 Global optimization

The complexity of the models used to fit the experimental data implies that the fit function is not necessarily strictly convex, but multiple local minima can be found. Hence, the least-square algorithm is sensitive to the initial values for the parameters given by the user. To circumvent this problem, common sense and experience can help discarding solutions for which the fitted parameters are aberrant. However, multiple solutions can make sense in regard to the system studied and more robust methods are needed to find the different minima and identify the global minimum. To this end, global optimization methods can be used, such as the genetic algorithm [36] or basin-hopping [37]. The basin-hopping algorithm consists in running

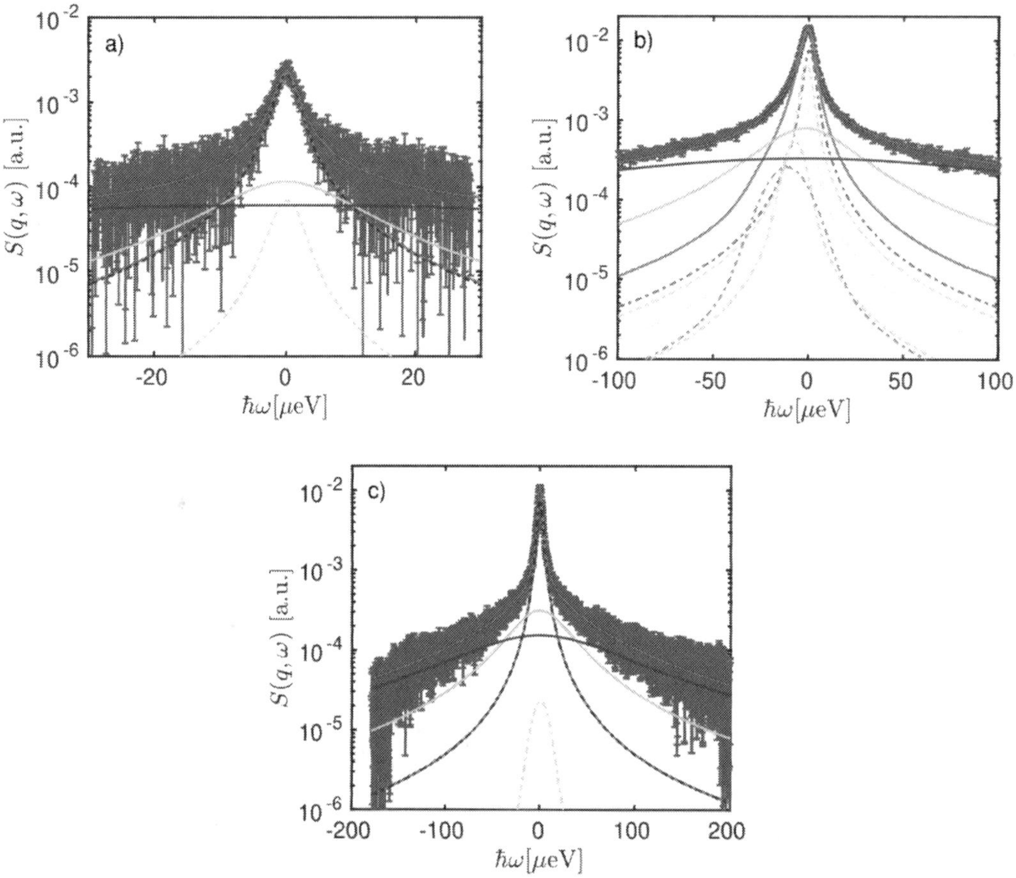

Figure 2: Spectra from aqueous (D_2O) BSA protein solutions recorded on different spectrometers (symbols) and associated fits (solid red lines superimposed on the symbols). The individual Voigt components associated with the convoluted apparent center-of-mass diffusion (magenta line) corresponding to each Gaussian in the resolution model are displayed as dash-dotted lines, with the colors corresponding to the ones used in Figure 1. In addition, the rescaled water contribution and the internal diffusion contribution are depicted as solid brown and green lines, respectively, without displaying their individual Voigt components. The protein concentrations in the samples as well as the sample temperatures differ for the spectra displayed: a) IN16b: c_p = 100 mg/ml; T = 280 K; q = 1.0 Å^{-1}; b) BASIS: c_p = 200 mg/ml; T = 295 K; q = 1.05 Å^{-1}; c) BATS: c_p = 150 mg/ml; T = 280 K; q = 1.05 Å^{-1}. Details of the different binning applied are given in Table 1.

several iterations of least-square minimization where the initial parameters are randomly updated. The new solution is typically accepted or discarded using the so-called Metropolis criterion from the Monte-Carlo method [38]. Numerous global optimization algorithms are available in *Python* via the *scipy.optimize* module. It is noteworthy that the signature and return types of these functions differ from the *curve_fit* function described above. For instance, the *basin-hopping* function call will read:

```
from scipy.optimize import basinhopping

result = basinhopping(
    fit_function,
    x0,
    minimizer_kwargs={
        args: additional_arguments
    }
)

optimal_params = result.x
```

where *x0* is a list of initial parameters for the first iteration and the *fit_function* has the signature (using Python type hints):

```
def fit_function(
        p: list[float],
        *additional_arguments) -> float:
    # compute the cost function
```

```
    # (can be chi-square function or other)
    cost = ...
    return cost
```

where *additional_arguments* can contain p, x, q and r as described above.

7 Future developments

In case the relevant diffusive processes of the sample investigated cannot be covered by one single instrument, different instruments or instrumental setups can be used to access different time and length scales. As a first approach, the corresponding spectra can be analyzed individually and the corresponding results can be further analyzed with models covering the different time scales [32]. In case it is desired to perform fits combining different spectrometers, it is necessary to employ frameworks which not only take into account the specific quantities for different instruments, but which also use a weighting between the different instruments based on data quality, instrument flux and amount of collected data points. A correct treatment of the error propagation is even more crucial in this case.

8 Conclusions

We have presented conceptual notes on MATLAB and Python implementations allowing for efficient scripting and fast fitting of neutron backscattering data, due to an analytical convolution. We have illustrated this fit procedure on example protein solution spectra. Spectrometer-design specific resolution functions can be routinely handled even in the case of strongly non-Gaussian resolution functions. Due to the computing time-saving analytical convolution, rather complex fits, including simultaneous fits of the spectra at all momentum transfers at once, can be carried out rapidly.

9 Code and Data Availability

Matlab analysis code performing multidimensional fits for individual and several spectra are available from the corresponding author on reasonable request. python code is available from *https://github.com/seydel/QENS_utilities* as well as from *https://github.com/kpounot/nPDyn*. The experimental data collected at IN16b are curated by the ILL and can be accessed via the experiments 9-13-952 [39] and 1-20-69 [40]. The BASIS data were collected during the experiment IPTS 18578.1.

C.B., K.P, I.M., F.R.-R. and T.S. thank the organizers of the QENS/WINS 2022 conference for an excellent meeting in San Sebastian. We thank M. Hennig and M. Grimaldo for earlier versions of the employed MATLAB code. We thank the DFG, ANR (ANR-21-CE06-0047), BMBF (ErUM-pro 05K19VTB and 05K22VTA), and InnovaXN (EU MSCA COFUND 847439) for funding. A portion of this research used resources at the Spallation Neutron Source, a DOE Office of Science User Facility operated by the Oak Ridge National Laboratory. The authors gratefully acknowledge the financial support provided by JCNS to perform the neutron scattering measurements at the SNS.

References

[1] J. Orear, *Notes on statistics for physicists* (University of California, Lawrence Radiation Laboratory; printed for the US Atomic Energy Commission, 1958)

[2] E.L. Frome, M.H. Kutner, J.J. Beauchamp, Journal of the American Statistical Association **68**, 935 (1973)

[3] S. Baker, R.D. Cousins, Nuclear Instruments and Methods in Physics Research **221**, 437 (1984)

[4] M. Appel, B. Frick, Review of Scientific Instruments **88**, 036105 (2017)

[5] N. Martinez, F. Natali, J. Peters, EPJ Web of Conferences **83**, 03010 (2015)

[6] S. Mukhopadhyay, B. Hewer, S. Howells, A. Markvardsen, Physica B: Condensed Matter **563**, 41 (2019)

[7] E. Mamontov, R.W. Smith, J.J. Billings, A.J. Ramirez-Cuesta, Physica B: Condensed Matter **566**, 50 (2019)

[8] M. Grimaldo, F. Roosen-Runge, F. Zhang, F. Schreiber, T. Seydel, Quarterly Reviews of Biophysics **52**, e7 (2019)

[9] F. Roosen-Runge, D.J. Bicout, J.L. Barrat, Journal of Chemical Physics **144**, 204109 (2016)

[10] G. Kneller, K. Hinsen, The Journal of Chemical Physics **121**, 10278 (2004)

[11] G.R. Kneller, Physical Chemistry Chemical Physics **7**, 2641 (2005)

[12] M. Pathan, M. Garg, S. Mittal, East West Math **8**, 49 (2006)

[13] A.N. Hassani, A.M. Stadler, G.R. Kneller, The Journal of Chemical Physics **157**, 134103 (2022)

[14] D. Gottlieb, C.W. Shu, SIAM review **39**, 644 (1997)

[15] G. Wertheim, Journal of Electron Spectroscopy and Related Phenomena **6**, 239 (1975)

[16] D. Sivia, C. Carlile, W. Howells, S. König, Physica B: Condensed Matter **182**, 341 (1992)

[17] D. Monserrat, A. Vispa, L. Pardo, R. Tolchenov, S. Mukhopadhyay, F. Fernandez-Alonso, Journal of Physics: Conference Series **663**, 012009 (2015)

[18] O. Arnold, J. Bilheux, J. Borreguero, A. Buts, S. Campbell, L. Chapon, M. Doucet, N. Draper, R.F. Leal, M. Gigg et al., Nuclear Instruments and Methods in Physics Research Section A: Accelerators, Spectrometers, Detectors and Associated Equipment **764**, 156 (2014)

[19] H. Paalman, C. Pings, Journal of Applied Physics **33**, 2635 (1962)

[20] M.K. Braun, M. Wolf, O. Matsarskaia, S. Da Vela, F. Roosen-Runge, M. Sztucki, R. Roth, F. Zhang, F. Schreiber, The Journal of Physical Chemistry B **121**, 1731 (2017)

[21] M. Grimaldo, F. Roosen-Runge, N. Jalarvo, M. Zamponi, F. Zanini, M. Hennig, F. Zhang, F. Schreiber, T. Seydel, EPJ Web of Conferences **83**, 02005 (2015)

[22] M. Grimaldo, F. Roosen-Runge, M. Hennig, F. Zanini, F. Zhang, N. Jalarvo, M. Zamponi, F. Schreiber, T. Seydel, Physical Chemistry Chem-

ical Physics **17**, 4645 (2015)

[23] A. Girelli, C. Beck, F. Bäuerle, O. Matsarskaia, R. Maier, F. Zhang, B. Wu, C. Lang, O. Czakkel, T. Seydel et al., Molecular Pharmaceutics **18**, 4162 (2021)

[24] C. Beck, M. Grimaldo, M.K. Braun, L. Bühl, O. Matsarskaia, N.H. Jalarvo, F. Zhang, F. Roosen-Runge, F. Schreiber, T. Seydel, Soft Matter **17**, 8506 (2021)

[25] M. Grimaldo, H. Lopez, C. Beck, F. Roosen-Runge, M. Moulin, J.M. Devos, V. Laux, M. Härtlein, S. Da Vela, R. Schweins et al., The Journal of Physical Chemistry Letters **10**, 1709 (2019)

[26] D.B. Anunciado, V.P. Nyugen, G.B. Hurst, M.J. Doktycz, V. Urban, P. Langan, E. Mamontov, H. O'Neill, The Journal of Physical Chemistry Letters **8**, 1899 (2017)

[27] J. Nickels, H. O'Neill, L. Hong, M. Tyagi, G. Ehlers, K.L. Weiss, Q. Zhang, Z. Yi, E. Mamontov, J. Smith et al., Biophysical Journal **103**, 1566 (2012)

[28] C. Neylon, European Biophysics Journal **37**, 531 (2008)

[29] D.I. Svergun, M.H.J. Koch, Reports on Progress in Physics **66**, 1735 (2003)

[30] C. Beck, M. Grimaldo, F. Roosen-Runge, R. Maier, O. Matsarskaia, M. Braun, B. Sohmen, O. Czakkel, R. Schweins, F. Zhang et al., Crystal Growth & Design **19**, 7036 (2019)

[31] C. Beck, M. Grimaldo, H. Lopez, S. da Vela, B. Sohmen, F. Zhang, M. Oettel, J.L. Barrat, F. Roosen-Runge, F. Schreiber et al., The Journal of Physical Chemistry B **126**, 7400 (2022)

[32] A. Cisse, A.L. Schachner-Nedherer, M. Appel, C. Beck, J. Ollivier, G. Leitinger, R. Prassl, K. Kornmueller, J. Peters, The Journal of Physical Chemistry Letters **12**, 12402 (2021)

[33] P. Virtanen, R. Gommers, T.E. Oliphant, M. Haberland, T. Reddy, D. Cournapeau, E. Burovski, P. Peterson, W. Weckesser, J. Bright et al., Nature Methods **17**, 261 (2020)

[34] F. Roosen-Runge, M. Hennig, F. Zhang, R.M.J. Jacobs, M. Sztucki, H. Schober, T. Seydel, F. Schreiber, Proceedings of the National Academy of Sciences (USA) **108**, 11815 (2011)

[35] C. Beck, M. Grimaldo, F. Roosen-Runge, M.K. Braun, F. Zhang, F. Schreiber, T. Seydel, The Journal of Physical Chemistry B **122**, 8343 (2018)

[36] B. Hartke, WIREs Computational Molecular Science **1**, 879 (2011)

[37] D.J. Wales, J.P.K. Doye, The Journal of Physical Chemistry A **101**, 5111 (1997)

[38] Z. Li, H.A. Scheraga, Proceedings of the National Academy of Sciences (USA) **84**, 6611 (1987)

[39] L. Colin, M. Appel, C. Beck, C. Grapentin, O. Matsarskaia, K. Pounot, F. Roosen-Runge, F. Schreiber, T. Seydel, *Diffusive dynamics in crowded solutions containing two types of proteins* (2021), `https://doi.ill.fr/10.5291/ILL-DATA.9-13-952`

[40] F. Roosen-Runge, O. Matsarskaia, T. Seydel, F. Schreiber, C. Beck, *Developing new analysis methods for fixed window scans for soft colloidal suspensions* (2021), `https://doi.ill.fr/10.5291/ILL-DATA.1-20-69`

Impact of composition on the crystal texture and on the dynamics of P(THF-*co*-ECH) copolymers

Jon Maiz[1,2*], *Ester* Verde-Sesto[1], *Isabel* Asenjo-Sanz[1], *Fanni* Juranyi[3], *José A.* Pomposo[1,2,4], *Arantxa* Arbe[1], and *Juan* Colmenero[1,4,5]

[1]Centro de Física de Materiales (CFM) (CSIC-UPV/EHU)-Materials Physics Center (MPC), Paseo Manuel de Lardizábal 5, 20018 Donostia-San Sebastián, Spain
[2]IKERBASQUE-Basque Foundation for Science, Plaza Euskadi 5, 48009 Bilbao, Spain
[3]Laboratory for Neutron Scattering, Paul Scherrer Institut, CH-5232 Villigen, Switzerland
[4]Departamento de Polímeros y Materiales Avanzados: Física, Química y Tecnología, Universidad del País Vasco-Euskal Herriko Unibertsitatea (UPV/EHU), 20018 Donostia-San Sebastián, Spain
[5]Donostia International Physics Center, Paseo Manuel de Lardizábal 4, 20018, Donostia-San Sebastián, Spain

Abstract. We present a combined study by quasielastic neutron scattering (QENS), differential scanning calorimetry (DSC) and wide angle X-ray scattering (WAXS) on poly(tetrahydrofuran-co-epichlorohydrin) copolymers, to see how their composition can be used to tune their crystallizability and to elucidate the impact of this factor on the dynamical properties. QENS reveals a strong effect on the local dynamics upon cooling down, where the local motions of a sample that remains in the supercooled state at lower temperatures are less Gaussian and slower than those in a sample that crystallizes a few degrees below. This can be attributed to the enhancement of local heterogeneities in the former, which could be a determining factor preventing crystallization.

1 INTRODUCTION

The structure of polymer materials is dictated by the crystallization and chain organization, and, in general, these two factors are important to understand the final properties of the system. In industry, during the last years multiphasic polymer materials have been developed in order to reach suitable final properties with the aim of extend them in different fields. To achieve this goal, researchers are trying to implement and/or develop new multiphasic polymers or complex systems, among them, polymer blends, polymer composites (or nanocomposites), polymer gels and different types of copolymers [1].

One of the most common type of copolymers are random copolymers. This type of copolymers are characterized by a statistical order of the repeating units along the backbone of the chain. The main objective of the synthesis of this kind of polymers is to achieve a homogeneous system with tunable properties depending on their composition. In the case of random copolymers, an average in the physical and chemical properties is observed e. g. in the value of the glass transition temperature (T_g). Moreover, random copolymers display a single-phase morphology; the sequences are normally too short to induce a phase separation morphology.

Over last years our group has developed different works related with poly(tetrahydrofuran-*co*-epichlorohydrin) P(THF-*co*-ECH) copolymers [2-5]. The main motivation of this interest was to produce functionalized linear chains to be used as precursors for the synthesis of Single Chain Nano-Particles (SCNPs); therefore, along this work, we will use the term 'precursor' to refer to these polymers. Polymer melts in general are complex systems where the relevant dynamic processes depend on the length scale of observation [6,7]. The dynamics is naturally intrinsically connected to the structural properties of the system. In the case of the copolymers here investigated, their composition can be used to tune their crystallizability. While PTHF has a strong tendency to cristallize, PECH is easily obtained in the glassy state upon cooling. Precursors rich in THF comonomer could form crystalline phases. Semicrystalline polymers have a partial regular chain structure, where the regular part is able to crystallize and the irregular part remains in the amorphous state. The crystalline part can adapt its conformation to the lowest energy state forming the denominated crystal whereas the chain defects are normally confined in the amorphous region [8,9]. It is well known that crystallinity of polymeric materials largely affects the properties of the remaining amorphous phase, in terms of structure and dynamics. Many questions remain unsolved in this type of semicrystalline polymers or complex systems. In this context, scattering techniques and particularly neutron scattering techniques are especially appropriate. Neutron scattering encompasses a set of non-destructive experimental techniques that enable the molecular

* Corresponding author: jon.maizs@ehu.eus

analysis of complex systems over a broad range of length and time scales.

In the present work, we investigate the impact of the composition on the crystallization process and on the obtained structural arrangements of bulk systems of precursor chains containing different relative amounts of ECH and THF monomers in the chains. We have explored a sample with a low amount of ECH content, c_{ECH} = 13% (ePrec), where the presence of THF leads to crystalline portions in material, and a sample with higher amount of ECH content, c_{ECH} of 27% (EPrec), that remains amorphous upon cooling. Differential Scanning Calorimetry (DSC) and Wide Angle X-Ray Scattering (WAXS) techniques are used to investigate these features. In addition, we carried out QENS experiments to study the incoherent scattering function of the hydrogens in both samples. For this investigation, a time-of flight (ToF) spectrometer was used in order to determine the segmental relaxation in both samples.

2 EXPERIMENTAL SECTION

2.1 Materials

All reagents and solvents were purchased from Merck (Munich, Germany) and Scharlab (Barcelona, Spain). Tetrahydrofuran (THF, ≥ 99.9% (GC), from Scharlab) was dried using calcium hydride as drying agent (CaH_2, 95 % from Merck, Aldrich), degassed by three freeze-degas-thaw cycles and distilled under reduce pressure at 323 K. The purification of (±)-Epichlorohydrin (ECH, ≥ 99 % (GC) from Merck, Aldrich) was performed by distillation under reduced pressure at 323 K. Tris(pentafluorophenyl)borane ($B(C_6F_5)_3$, 95 %, from Merck, Aldrich) was purified by sublimation under reduce pressure at 333 K using a cold finger condenser. Anhydrous N,N-dimethylformamide (DMF, ≥ 99.8 %, Scharlab), sodium azide (NaN_3, ≥ 99.5 % from Merck, Aldrich), and methanol (MeOH, ≥ 99.9%, Scharlab) were used as received.

2.2 Synthesis Methods

Synthesis of THF and ECH copolymers (P(THF-co-ECH)): Two copolymers of THF and ECH (amorphous and semicrystalline) were synthesized following a similar synthetic procedure previously described by our group [10]. Both polymerization reactions were carried out in bulk and under argon atmosphere. Amorphous P(THF-*co*-ECH): $B(C_6F_5)_3$ (200 mg, 0.4 mmol), THF (25.0 mL, 308.0 mmol) and ECH (6.3 mL, 81.0 mmol) were added into a Schlenk flask of 100 mL. Then, the mixture of reaction was stirred at room temperature (r.t.) for 48 h. The crude product of the reaction was precipitated three times in cold MeOH (3 x 600 mL) in order to remove unreacted monomers. The obtained copolymer was dried at 313 K under vacuum for 24 h, producing a sticky transparent material (P(THF-*co*-ECH, 22.1 g, 74 % of yield). The ECH content was determined by 1H Nuclear Magnetic Resonance (1H NMR) and estimated to be 27 mol %. By Size-Exclusion Chromatography (SEC), M_w = 38 kg/mol with polydispersity Đ = M_w/M_n = 1.27 were determined. The preparation of P(THF-*co*-ECH) semicrystalline was carried out following a procedure similar to that described above but with a different ratio of reagents: $B(C_6F_5)_3$ (44 mg, 0.09 mmol), THF (12.5 mL, 154.00 mmol) and ECH (1.33 mL, 17.1 mmol). We obtained 9.5g of copolymer, 75 % of yield. 1H NMR: 13 % ECH. SEC: 48 kg/mol, Đ = 1.32.

Synthesis of the azide-containing precursors (EPrec and ePrec) by azidation of the obtained P(THF-*co*-ECH)s: a solution of 2.0 g of P(THF-*co*-ECH) in 80 mL of anhydrous DMF was prepared into a round-bottom flask of 500 mL. Then, NaN_3 (925 mg) was added into the solution, and the obtained suspension was stirred for 24 h at 333 K. The mixture of the reaction was filtrated to remove NaCl generated and the unreacted NaN_3 and evaporated under reduced pressure until to get ca 15 mL. Then, the resulting crude product was precipitated in a cold mixture of 200 mL of 1:4 H_2O/MeOH and dried at 313 K under vacuum for 24 h to obtain the precursor (EPrec: 1.5 g, 72 % yield). The molecular weight M_w was 38.5 kg/mol with polydispersity Đ = M_w/M_n of 1.23 (determined by Size-Exclusion Chromatography (SEC)). Composition by elemental analysis (EA) calculated: C (%) cal. 58.94, H (%) cal. 9.61, N (%) cal. 3.36 and composition by EA found: C (%) exp. 58.56, H (%) exp. 9.88, N (%) exp. 3.81. The synthesis of ePrec was carried out following the same synthetic procedure described for EPrec: 2.5g of P(THF-*co*-ECH) crystalline, 100 mL of DMF and 0.65 g of NaN_3. We obtained 2.1 g of ePrec, 84% of yield. SEC: 47.5 kg/mol, Đ = 1.32. Composition by EA calculated: C (%) cal. 61.89, H (%) cal. 10.14, N (%) cal. 2.41 and composition by EA found: C (%) exp. 61.43, H (%) exp. 10.18, N (%) exp. 2.53.

2.3 Sample Preparation

Two different precursor bulk samples were studied in this work, EPrec and ePrec. For neutron scattering experiments, first solutions containing the sample and 2 mL of CH_2Cl_2 were stirred until completed dissolution, and then solutions were drop-casted onto the aluminum flat holders. Finally the solvent was completely evaporated, first slowly in the fume hood and subsequently in an oven under vacuum conditions at 343 K for 24 h.

2.4 Thermal Analysis

Calorimetric values were studied by differential scanning calorimetry (DSC). A TA instruments Q2000 under ultrapure nitrogen flow was used to carried out the DSC experiments. The protocol followed for each sample was: first, the samples were heated up until 350 K and kept at this temperature for 3 min in order to erase any previous thermal history. Thereafter, they were cooled down until 170 K and subsequently they were again heated up until 350 K. The cooling and heating rate used in the whole procedure was 20 K/min. Around 5 mg of the polymer mass was used in each sample.

2.5 Structural Analysis

Wide angle X-ray scattering (WAXS) technique on a Bruker D8 Advance diffractometer equipment was used to study the short-range order of the systems. The equipment works in parallel beam geometry with CuK_α transition photons of wavelength λ = 1.54 Å, in reflection mode (θ - 2θ configuration). The scattering angle 2θ was varied from 10° to 30° with steps of 0.05°. Two different temperatures were explored, at 233 and 300 K. The scattered intensities were recorded as a function of the scattering vector Q, $Q = 4\pi\lambda^{-1} \sin\theta$.

2.6 Quasielastic Neutron Scattering (QENS) Analysis

The QENS experiments were carried out on the SINQ time-of-flight FOCUS spectrometer. An incident wavelength of 6 Å was used yielding an energy resolution of around 40 μeV. Cuts at constant Q-value were performed in the energy range ±1meV. The effective Q-range covered was between 0.2 and 1.8 Å^{-1}. Five different temperatures were studied: 250, 290, 320, 360 and 400 K. A low temperature measurement (~ 40 K) of each sample was used for the accurate normalization of the Fourier transformed and deconvoluted results in each sample. Samples were filling flat aluminum sample holders, and with thicknesses calculated to yield a transmission of 90%.

3 RESULTS

First, in Fig. 1 the calorimetric results are shown.

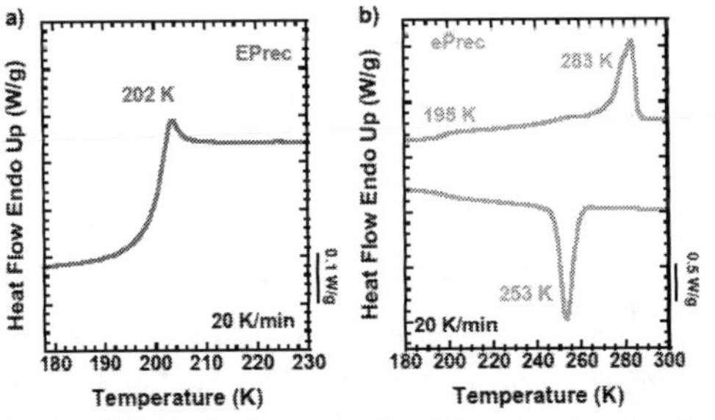

Fig. 1. DSC (a) heating scan for EPrec sample after the previous cooling process and (b) cooling and heating scans for ePrec sample at 20 K/min.

In Fig. 1a the DSC heating scan that corresponds to EPrec is presented. The T_g value calculated from the onset value for this sample is 202 K. Meanwhile, in Fig. 1b for ePrec both cooling and heating scans at 20 K/min are shown. This sample shows a crystallization process occurring at a crystallization temperature T_c of 253 K and a melting process taking place at the melting temperature T_m observed at 283 K. In addition, a T_g value of 195 K can be also estimated for this sample, being this value lower than that obtained for EPrec sample.

The results of the WAXS experiments performed at 233 K and 300 K after a previous cooling scan at 20 K/min in both samples are displayed in Fig. 2. We note that 233 K is below the crystallization temperature for the ePrec sample and 300 K is above the crystallization temperature for the ePrec sample. Both temperatures are above the glass transition temperatures of the samples.

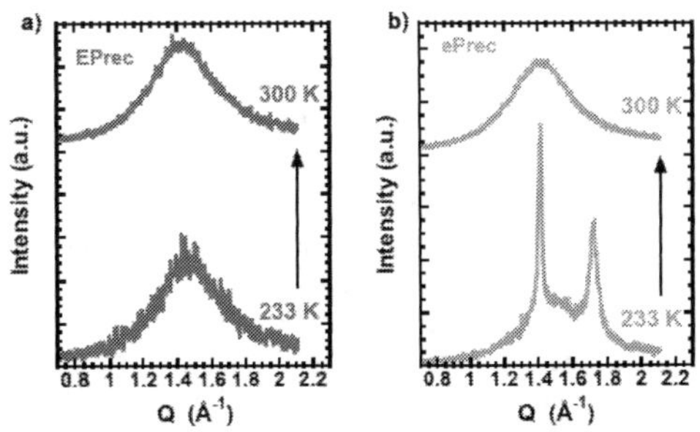

Fig. 2. WAXS measurements at low temperature (233 K) and at room temperature (300 K) for (a) EPrec and (b) ePrec samples after the previous cooling process at 20 K/min.

Fig. 2a shows that there are no appreciable differences on the short-range order for Eprec sample at these temperatures. One main amorphous halo is observed centred at Q ≈ 1.45 Å^{-1}, that slightly shifts with increasing temperature towards lower Q-values. This peak is associated with the inter-main chain correlations with average distances of about $d_{chain} = 2\pi/Q_{max} \approx 4.3$ Å. On the other hand, Fig. 2b shows WAXS data that correspond to the ePrec sample at 233 K (below its T_c value) and at 300 K (above its T_m value). The pattern at 233 K shows clear features of crystallinity for the ePrec sample, presenting two main diffraction peaks at 1.41 and 1.72 Å^{-1}. When the sample is heated up to 300 K, it melts and only an amorphous halo is observed centred at a similar Q-value as that of the EPrec sample.

If we move to the neutron scattering studies, Fig. 3 shows representative QENS spectra obtained from EPrec and ePrec samples, where the curves are normalized to their value at $\hbar\omega = 0$.

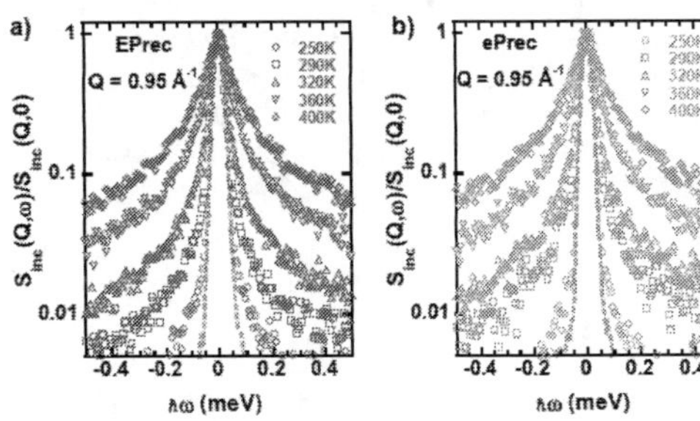

Fig. 3. Normalized FOCUS spectra obtained at Q = 0.95 Å^{-1}and the different temperature values indicated for (a) EPrec sample and (b) ePrec sample. The dotted line shows the instrumental resolution function.

The width of the quasielastic spectrum is related to the inverse of the characteristic time of motion probed by the instrument. For all the temperatures studied, at the representative Q-value chosen in Fig. 3 we observe a clear broadening with respect to the instrumental resolution function. For a given Q-value, this broadening becomes more pronounced with increasing temperature. In Fig. 3b, this increase is more pronounced when the temperature jump is from 250 to 290 K than in Fig. 3a, due to the transition in the ePrec sample from the crystalline material (at 250 K) to its molten state (already at 290 K). As we have confirmed before by DSC data, in Fig. 1b, at this temperature the ePrec sample is already melted (T_m = 283 K).

The spectra measured in the frequency domain are affected by instrumental resolution through convolution. Therefore, the quantitative analysis of the results was performed by Fourier transforming the data to the time domain and deconvoluting them from the instrumental resolution by division. In this way, it is possible to obtain the intermediate incoherent scattering function in the time domain $S_{inc}(Q,t)$. Fig. 4 shows the results obtained from this procedure for EPrec and ePrec samples at the same representative Q-value of 0.95 $Å^{-1}$ and at the five different temperatures studied.

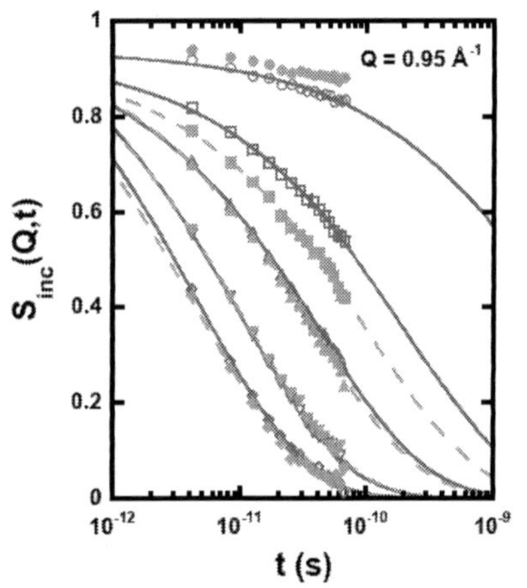

Fig. 4. Fourier-transformed and deconvoluted QENS spectra obtained from FOCUS on EPrec (unfilled) and ePrec (filled) samples at Q-value of 0.95 $Å^{-1}$ at the five temperatures investigated: 250 K (circle), 290 K (square), 320 K (up-triangle), 360 K (down-triangle) and 400 K (diamond). Lines are Kohlrausch-Williams-Watts (KWW) fits with β = 0.5.

The QENS results in glass-forming polymers are usually described by Kohlrausch-Williams-Watts (KWW) or stretched exponential functions

$$S_{inc}(Q,t) = A(Q)\exp\left[-\left(\frac{t}{\tau_s}\right)^{\beta}\right] \quad (1)$$

where β is the stretching parameter characterizing the deviations from a single exponential. The prefactor A (Q) accounts for the decay of the correlation function at shorter times due to vibrational and other fast contributions. In Fig. 4 we can observe that the description using Eq. 1 with β = 0.5 works well for both samples in the covered T-range for the Q-value shown. The exception is the sample ePrec at the lowest temperature investigated (250 K), where the tendency toward a plateau-like behaviour at long times prevents fitting with the same functional form as the other curves.

At this temperature, the ePrec sample –rich in THF comonomer– is in the semicrystalline form. This behaviour is attributable to the contribution of an elastic fraction reflecting the frozen dynamics of the crystalline portion of the material. Therefore, the KWW function (Eq 1.) was not applied for describing the result at 250 K. Fig. 5 shows the results obtained for EPrec and ePrec samples for four temperatures measured at four different representative Q-values.

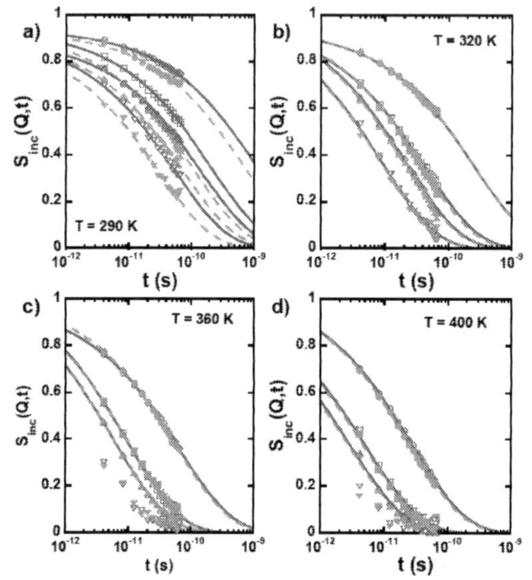

Fig. 5. Fourier-transformed and deconvoluted QENS spectra obtained by FOCUS on EPrec (unfilled) and ePrec (filled) samples at four Q-values investigated: 0.55 $Å^{-1}$ (circle), 0.95 $Å^{-1}$ (square), 1.15 $Å^{-1}$ (up-triangle) and 1.65 $Å^{-1}$ (down-triangle) and at the four temperatures studied: (a) 290 K, (b) 320 K, (c) 360 K and (d) 400 K. Lines are Kohlrausch-Williams-Watts (KWW) fits with β = 0.5.

The results, as can be appreciated in Figs. 4 and 5, can be in general reasonably well described by Eq. 1, using just a constant β-value of 0.5. We note, however, that for high Q-values and temperatures –as it is the case of 1.65 $Å^{-1}$ at 360 and 400 K shown in Fig. 5– the decay of the intermediate scattering function for both samples takes place with very short characteristic times. Actually, the function decays to values smaller than 1/e already at times shorter than 2—3ps. Therefore, we did not consider these results to fit them by KWW functions (we discarded all results which τ_s was smaller than 5ps). From the characteristic times τ_s and stretching parameters β, the average characteristic time can be calculated as $\langle\tau\rangle = \tau_s\Gamma(1/\beta)/\beta$. The results on these characteristic times are displayed in Fig. 6.

For Q-values below 1 $Å^{-1}$, the hydrogen motions in the α-relaxation regime of glass-forming systems usually follow the Gaussian prediction [2,4,5,11-13]. At higher Qs, deviations from this Gaussian behavior are found in general. This is also the case of our samples (see Fig. 6). The dotted line there shows the $\tau \sim Q^{-2/\beta}$ dependence corresponding to the Gaussian prediction in stretched exponential intermediate scattering functions. To account for deviations from the Gaussian approximation, the jump diffusion model might be invoked in simple systems where diffusion governs at long times [14-16]. This model considers that an atom remains in a given site for a time, τ_0, where it vibrates around a center of equilibrium. After this time τ_0, it moves rapidly to a new position separated from the previous one by the vector $\vec{l}$. This model was extended to the case of subdiffusive behavior by the anomalous jump diffusion (AJD) model [11,12]. The AJD model introduces stretching in the time-dependent part of the intermediate scattering function. In this framework, the KWW function has a characteristic time given by:

$$\tau_s(Q) = \tau_{s,0}\left[1 + \frac{1}{Q^2 l_0^2}\right]^{\frac{1}{\beta}} \tag{2}$$

where l_0 is the preferred jump length and $\tau_{s,0}$ is the time between jumps [12]. The AJD model was used to describe the Q-dependence of the characteristic times obtained for 250 K for EPrec and for T = 320 and 290 K for the two samples, where the values of the AJD parameters could be determined with better accuracy. The values obtained for the parameters are listed in Table 1. For the two highest temperatures investigated, the uncertainties involved in the values of the characteristic times at high-Q –where the intermediate scattering function decays to a large extent due to vibrations and faster processes, as explained above— prevented a precise independent determination of the two parameters $\tau_{s,0}$ and l_0.

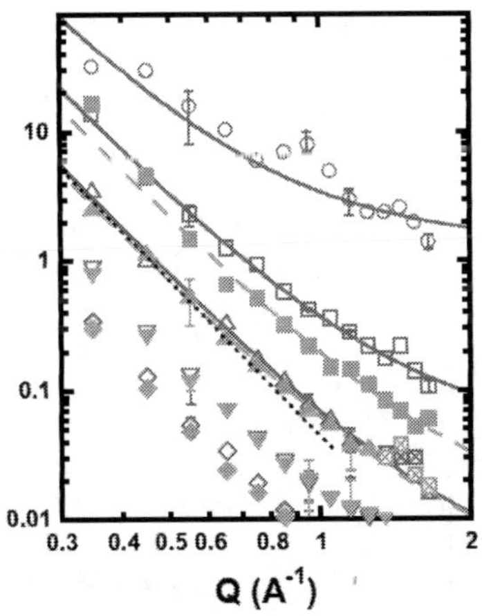

Fig. 6. Average time for H motions obtained for EPrec (unfilled) and ePrec (filled) samples at 250 K (circle), 290 K (square), 320 K (up-triangle), 360 K (down-triangle) and 400 K (diamond), imposing β = 0.5. Lines are fits of Eq. 2 to EPrec (solid lines) and ePrec (dashed lines). The dotted line shows the low-Q Gaussian behavior for the 320 K data. Error bars have been included at selected Q-values (0.55, 0.95, 1.15 and 1.65 Å^{-1}). Results at 320 K and high Q-values represented by crossed squares are subject to very large uncertainties and were not considered in the fit.

Table 1. Values of the parameters involved in the Anomalous Jump Diffusion (AJD) fitting the EPrec and ePrec results described with KWW functions.

Sample	T (K)	$\tau_{s,0}$ (ps)	l_0 (Å)
EPrec	250	678±191	1.3±0.19
	290	21±6	0.71±0.07
	320	1.4±0.4	0.50±0.04
ePrec	250	-	-
	290	4.5±1.2	0.52±0.05
	320	1.5±0.4	0.52±0.04

4 DISCUSSION

Combining "macroscopic" and "microscopic" techniques it is possible to investigate how the composition affects the crystallization capability and the dynamics of copolymer systems in the bulk. The calorimetric studies gave us a direct idea of how the comonomer relative content in the chains can affect the crystallinity of the system. Consistent with the DSC results obtained, WAXS experiments also show directly the difference in crystallization tendency of the samples. This crystallinity can be tuned changing the composition in a copolymer, where one of the components is a semicrystalline polymer (PTHF) and the other one is an amorphous material (PECH). When a c_{ECH} of 27 % was used (EPrec), by calorimetry we have demonstrated that the system become amorphous. On the other hand, when a c_{ECH} of 13% is prepared (ePrec), crystallinity persists in the material. The peaks observed by WAXS at 1.41 and 1.72 Å^{-1} are assignable to the (020) and (110) planes of the monoclinic unit cell of PTHF [17], and should thus correspond to THF comonomer. Thus, we could resolve the pure PTHF crystalline structure in the pattern of ePrec. On the other hand, it is found in the literature that the T_m value for the neat PTHF is around 320 K. In the ePrec sample, the T_m value decreases to 283 K. This is the usual behavior observed for this kind of copolymers, that when the relative content of ECH increases, the T_m value decreases until a composition beyond which the copolymer ends up being amorphous (as it is the case of the EPrec sample here investigated).

The insight provided by QENS data on the H-dynamics shows that the composition, determining the amorphous or semicrystalline morphology, has very little influence on the dynamics at temperatures well above the glass transitions. For experiments done from 320 K up to 400 K, the dynamic behaviour is practically independent of the composition, within the experimental uncertainties. Particularly at temperatures higher than 320 K, the deviations from Gaussian behaviour become very difficult to quantify, due to the fast character of the dynamics. On the contrary, composition strongly affects the decay of the intermediate scattering function at temperatures lower than and slightly above the melting point of the crystalline sample (ePrec) with respect to the amorphous one. At temperatures above but close to the melting point of the crystalline sample (T = 290 K), the atomic motions in the already molten sample are accelerated with respect to the amorphous one (Fig. 4, 5b and 6). This observation is consistent with the lower T_g value observed for the ePrec sample by DSC experiments. These results indicate a lower fragility (in the meaning of Angell [18]) for ePrec.

In terms of the AJD model, we observe longer residence times and larger jump lengths for the EPrec sample than for the molten one (ePrec) at 290 K (see Table 1). This observation could be attributed to a higher degree of heterogeneities in the local atomic displacements induced by the higher amount of ECH monomers in the EPrec sample. We also note that the deviations from Gaussian behaviour observed at 250 K for this sample are very pronounced and characterized by a very large average jump length. At this temperature, the atomic motions in the amorphous portions of the semicrystalline sample ePrec take place within the confined regions imposed by the crystalline regions.

5 CONCLUSIONS

P(THF-*co*-ECH) based copolymers at two different compositions were successfully synthesized. Increasing the relative content of ECH suppresses crystallization as shown by DSC and WAXS experiments. QENS reveals a very similar dynamics at very high temperatures independently of composition, while a strong impact is observed upon cooling down, when approaching the melting point of the sample able to crystallize. In this region, the local motions of the sample that remains in the supercooled state are less Gaussian and slower than those in the sample that crystallizes. This can be attributed to the enhancement of local heterogeneities in the presence of a higher ECH concentration, which could be a determining factor preventing crystallization.

In addition to the interest of this study per se, it will serve as reference for the future investigation of the bulk systems consisting of SCNPs synthesized from these precursor polymers.

We acknowledge the Grant PID2021-123438NB-I00 funded by MCIN/AEI/10.13039/501100011033 and by "ERDF A way of making Europe". We also acknowledge the financial support of Eusko Jaurlaritza, code : IT-1566-22 and from the IKUR Strategy under the collaboration agreement between Ikerbasque Foundation and the Materials Physics Center on behalf of the Department of Education of the Basque Government. The authors would like to express their appreciation and thanks to Dr. Amaia Iturrospe for WAXS measurements.

References

[1] A. Boudenne, L. Ibos, Y. Candau, S. Thomas, *Handbook of multiphase polymer systems* (John Wiley & Sons, 2011)

[2] A. Arbe, J. Rubio, P. Malo de Molina, J. Maiz, J. A. Pomposo, P. Fouquet, S. Prevost, M. Khaneft, J. Colmenero, Journal of Applied Physics **127**, 044305 (2020)

[3] J. Maiz, E. Verde-Sesto, I. Asenjo-Sanz, P. Fouquet, L. Porcar, J. A. Pomposo, P. Malo de Molina, A. Arbe, J. Colmenero, Polymers **13**, 50 (2021)

[4] J. Maiz, E. Verde-Sesto, I. Asenjo-Sanz, P. Malo de Molina, B. Frick, J. A. Pomposo, A. Arbe, J. Colmenero, Polymers **13**, 2316 (2021)

[5] J. Maiz, E. Verde-Sesto, I. Asenjo-Sanz, L. Mangin-Thro, B. Frick, J. A. Pomposo, A. Arbe, J. Colmenero, Macromolecules **55**, 2320 (2022)

[6] M. Rubinstein, R. H. Colby, *Polymer physics* (Oxford university press New York, 2003), Vol. 23

[7] D. Richter, M. Monkenbusch, A. Arbe, J. Colmenero, *Neutron spin echo in polymer systems* (Springer Verlag, Berlin Heidelberg New York, 2005), Vol. 174, Advances in Polymer Sciences

[8] L. Mandelkern, *Crystallization of Polymers: Volume 1: Equilibrium Concepts* (Cambridge University Press, Cambridge, 2002), 2 edn., Vol. 1

[9] L. Mandelkern, *Crystallization of Polymers: Volume 2, Kinetics and Mechanisms* (Cambridge University Press, 2004)

[10] J. Rubio-Cervilla, P. Malo de Molina, B. Robles-Hernández, A. Arbe, A. J. Moreno, A. Alegría, J. Colmenero, J. A. Pomposo, Macromolecular Rapid Communications **40**, 1900046 (2019)

[11] A. Arbe, J. Colmenero, F. Alvarez, M. Monkenbusch, D. Richter, B. Farago, B. Frick, Physical Review Letters **89**, 245701 (2002)

[12] A. Arbe, J. Colmenero, F. Alvarez, M. Monkenbusch, D. Richter, B. Farago, B. Frick, Physical Review E **67**, 051802 (2003)

[13] J. Colmenero, A. Alegria, A. Arbe, B. Frick, Physical Review Letters **69**, 478 (1992)

[14] T. Springer, *Quasielastic neutron scattering for the investigation of diffusive motions in solids and liquids* (Springer-Verlag, Berlin, 1972), Springer Tracts in Modern Physics

[15] K. Singwi, A. Sjölander, Physical Review **119**, 863 (1960)

[16] P. A. Egelstaff, *An Introduction to the Liquid State* (Oxford University Press, New York, 1992)

[17] K. Imada, T. Miyakawa, Y. Chatani, H. Tadokoro, S. Murahashi, Die Makromolekulare Chemie **83**, 113 (1965)

[18] C. A. Angell, Science **267**, 1924 (1995)

Polymers: When S(Q,t) is not what you think ! The role of dynamic RPA

Michael Monkenbusch[1,*]

[1]Jülich Centre for Neutron Science (JCNS), Forschungszentrum Jülich GmbH, Jülich, Germany

Abstract. Investigating a (polymeric) soft-matter system with several components often is aimed at the observation of the motions/mobilities of one of the components. The normal approach to this request in the context of quasielastic neutron scattering (e.g. NSE) is to put contrast on the targeted component, typically by leaving it hydrogenated and have the rest (surrounding, matrix) deuterated. In simple systems with only one molecular species as a homo-polymer melt with a few of the molecules contrasted (e.g. 10% h in 90% d) -provided the h- and d-molecular varieties behave sufficiently equal- this strategy yields a valid single chain (molecule) structure factor S(Q,t)/S(Q), even for a 50/50% mixture. However, if the component of interest and the matrix are of different kind, unexpected distortions of the observed S(Q,t) may occur. Interpretation of such results in terms of the single chain structure factor would then lead to erroneous conclusions. In this contribution the conditions, under which these distortions will occur, are discussed and how dynamic RPA may help to cope with them is explained. A practical method to apply this correction to polymer and similar problems is presented and an experimental verification is discussed [1, 12].

1 Introduction

Neutron scattering investigations of polymers often rely on the use of mixtures of different molecule types. Most common is the mixture of deuterated and protonated species of otherwise identical polymers in order to create contrast in the small-angle scattering (SANS) regime. Further sophistication may comprise the combination of different molecular architectures (e.g. stars, branched, dendrimers ...) or partially labeled chain or -the simplest- different lengths (e.g. molecular weights). While for the interpretation of SANS results it is common practice to use random-phase-approximation (RPA) derived expression to describe the scattering data [2–4], it usually is ignored if the dynamics is inferred from intermediate scattering functions $S(Q,t)$ as obtained by neutron spin-echo spectroscopy (NSE). In SANS investigations the mutual interaction differences as expressed by a Flory-Huggins parameter χ enter the RPA expression for the intensity as in Eq. 1

$$I(Q) = \Delta\rho^2 \frac{1}{[\phi S_1(Q)]^{-1} + [(1-\phi)S_2(Q)]^{-1} - 2\chi} \qquad (1)$$

with the structure functions $S_{1,2}(Q)$ of the two components, ϕ the volume fraction of component 1 and $\Delta\rho$ the scattering length density contrast between components 1 and 2. Depending on the value of χ enhanced critical scattering as precursor of demixing may be observed and may be used to infer χ. But also (independent on χ) different architectures of components 1 and 2 lead to the final form of $I(Q)$ and thus have to be accounted for during the extraction of model parameters, which is realized by Eq. 1 with $\chi = 0$. A simple case for linear flexible polymers with different lengths but $\chi = 0$ is illustrated in figure 1, where some intermediate intensity at low Q is observed

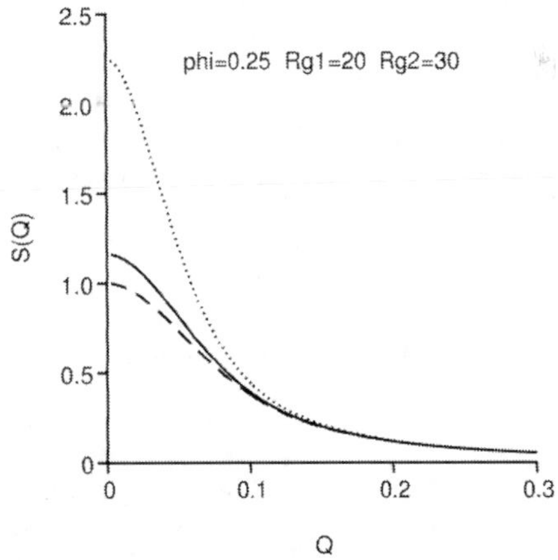

Figure 1. Comparison of the "static" scattering functions of short and long flexible linear polymers (Gaussian coils) and a mixture of both (solid line) as computed using the RPA expression Eq. 1 and $\phi = 0.25$.

in the resulting scattering due to combination of contributions from both components, at larger Q the behaviour of $S_i(Q)$ for both components are the same and coincide. This is true for the whole Q-range if the only difference of components 1 and 2 is the scattering contrast (d-, or h-labeled). In that case the RPA solely gives rise to the standard concentration pre factor $\phi(1-\phi)$. This situation also

*e-mail: m.monkenbusch@fz-juelich.de

was and is given in a larger number of NSE investigations with a mixture of just differently labelled but otherwise equal molecular species. Usually the preparations are also such that $\chi \simeq 0$ may be assumed. In that case there is no further need to observe RPA effects in the observed dynamic intermediate scattering functions. However, let us now contemplate the situation of a chain mixture as depicted in figure 2 more closely. Representative a labelled (minority and short) chain in yellow-green and a longer (majority-matrix) chain in blue are shown. From the point of view of the "labelled" green chain the contrast background is dominated by the blue matrix chains, however, diluted by a finite (small) probability that in any surrounding volume element another green chain can be found (indicated by the little green triangular fillings). Without further specific interactions, i.e. $\chi = 0$ these are randomly occupied and just give rise to the $(1 - \phi)$ term in the intensity pre factor. On the other hand from the point of any one long blue chains the same matrix scattering "dilution" also is present and yields a contrast and thus visibility to the long chains. These effects are described by Eq. 1. At larger Q, where typically most NSE experiments probe the scattering this obviously has little to no influence on the shape of $I(Q)$.

But now let us consider the resulting time dependence of $I(Q,t)$. If both components relax in the same way the resulting time function will be the same and no extra effects would have to be taken into account. However, if the relaxation is significantly different there will be contributions from the labelled (short) chains and from the somewhat visible long "matrix"-chains, then with different time functions. The extreme cases can easily be guessed: if the matrix chain are completely immobile and static there will be no relaxation of the short chains visible since even with mobile short chains the contrast modulation pattern will not change. On the other hand for extremely mobile matrix chains we expect to just "see" the dynamics of the short chains. In between some sort of combination is to be expected. It would be tempting to just enter $S_i(Q,t)$ instead of $S_i(Q)$ into Eq. 1, but as is shown below this is not the right approach.

While in the last decades some approaches on the use of dynamical RPA have been made [5–10] these did not directly yield a way to routinely describe the full resulting intermediate scattering curves as e.g. obtained in NSE experiments and allow model fitting within this framework. In this contribution an approach is described that establishes this.

2 Result for non-interacting 3-component system

To cover most cases of NSE experiments RPA expressions for an $n = 2$ system with an extra matrix component was considered and implemented [1]. For a system of $n + 1$ (interacting) components, a matrix (polymer) and n (different) polymers with different contrasts with respect to the matrix/solvent yields a $n \times n$ matrix problem: In the static formulation this implies:

$$\mathbf{S}(Q)^{-1} = \mathbf{S^0}(Q)^{-1} + \mathbf{v}(Q) \tag{2}$$

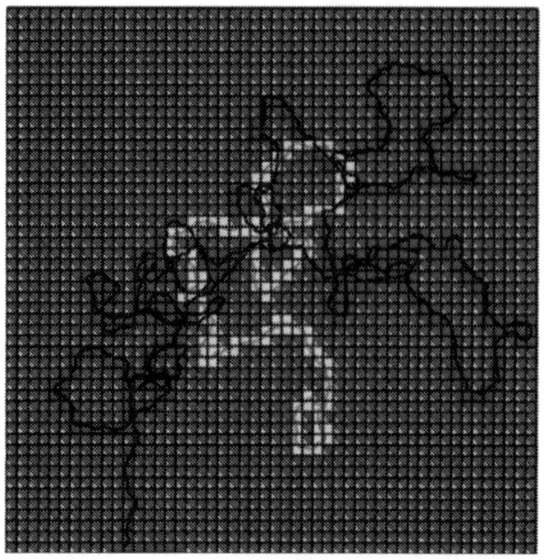

Figure 2. Sketch of the contrast situation -highlighted by the traces of one chain of each species" in a mixture of chains with different scattering length density (indicated by colors green and blue). Occupancy of "blue" and "green" segments of "other" chains simply governed by their volume fractions is indicated by the fraction green triangles in these cells- of chain occupancy (simple RPA) at cells occupied by other chains. Yielding large "matrix" contrast for the minority ($\phi = 0.2$) short, "green" chains and lesser but nonzero contrast for the majority, "blue" chains. The mutually created visibility leads to crosstalk of the relaxation dynamics of bot chains in the scattering signal, which may be described using dynamic RPA.

$\mathbf{S^0}(Q)$ is the matrix of the individual scattering functions and $\mathbf{v}(Q)$ includes the interaction between the components (here incompressibility i.e. mutual exclusion repulsion only). Then the scattering intensity $I(Q)$ from a system with contrast $\mathbf{a} = \{\rho_1 - \rho_c, \rho_2 - \rho_c\}$ between the components and the matrix is

$$I(Q) = \mathbf{a^T S}(Q)\mathbf{a} \tag{3}$$

This is the standard expression used for SANS evaluation. Already decades ago Akcasu and Tombakoglu [11] extended the RPA treatment to include dynamics yielding the Laplace transform of the dynamic scattering matrix $\mathbf{S}(Q,s)$:

$$\mathbf{S}(Q,s) = [\mathbf{1}s + Q^2\mathbf{D}(Q,s)]^{-1}\mathbf{S}(Q) \tag{4}$$

in terms of a frequency dependent diffusion coefficient:

$$\mathbf{D^0}(Q,s) = \frac{1}{Q^2}\left([\mathbf{S^0}(Q,s)\mathbf{S^0}(Q)^{-1}]^{-1} - \mathbf{1}\right) \tag{5}$$

with the relation between the undisturbed free diffusion coefficient $\mathbf{D}^0(Q,s)$ and the RPA influenced $\mathbf{D}(Q,s)$:

$$\begin{aligned}\mathbf{D}(Q,s) \simeq & [\mathbf{1} + Q^2\mathbf{D}^0(Q,s)\mathbf{S}^0(Q)\mathbf{v}'(Q,s)]^{-1}\times \\ & \mathbf{D}^0(Q,s)[\mathbf{1} + \mathbf{S}^0(Q)\mathbf{v}(Q)]\end{aligned} \tag{6}$$

with the incompressibility related interaction term

$$\mathbf{v}' = (\beta/s)[\chi^0_{cc}(Q,s)^{-1} - \chi^0_{cc}(Q,s=0)^{-1}]\mathbf{E}\mathbf{E^T} \tag{7}$$

($\beta = 1/k_B T$ cancels out from the final expression), and $\mathbf{EE^T}$ is a matrix with all elements equal to 1, the index $\underline{cc}$ relates to the "invisible" matrix, most a deuterated "background" compound. The dynamic susceptibility $\chi(Q, s)$ corresponds to the time derivative of $S(Q, t)$ the scattering functions $\chi(Q, s) = -\beta dS(Q, t)/dt = -\beta[sS(Q, s) - S(Q, t = 0^+)]$. With $\mathbf{v}(Q) = (\beta/\chi^0_{cc})\mathbf{EE^T} + \epsilon$ this yields the dynamic RPA result over the full time range using the undisturbed dynamic structure functions $S^0_{ii}(Q, t)$ of the components and the matrix function $S^0_{cc}(Q, t)$ as inputs, where the index c stands for the matrix (background) component. Focussing on the interference effects resulting from incompressibility any small potential interaction parameters $\epsilon_{i,j}$ are ignored. This largely holds for labelled mixtures of h- and d-labelled molecules assembled from the same types of monomers.

Reducing the general treatment to a 3 = (2+1) component system with volume fractions $\phi_1 + \phi_2 + \phi_3 = 1$ (c=3). We use:

$$\hat{S}_i = \phi_i S^0_{ii}(Q) \tag{8}$$

and the Laplace transforms

$$S^0_{ii}(Q, s) = S^0_{ii}(Q)\, F_i(Q, s) \tag{9}$$

as expressed by the Q-dependent, normalized function $f_i = F_i(Q, s)$.

For the polymer type relaxations it is possible to approximate $f(s)$ using:

$$f(s) = \left(\sum_{i=1}^{N} A_i\right)^{-1} \sum_{i=1}^{N} \frac{A_i}{r_i + s} \tag{10}$$

the coefficients A_i and rates r_i have to be derived from fitting the full model functions $S(Q, t)$ to the corresponding sum of exponentials for each Q.

The matrix elements S_{11} of the Laplace transformed scattering matrix $\mathbf{S}$ then is:

$$S_{11} = S_1 \times \frac{\left[f_1 \tilde{f}_2 S_2^2 + f_2 \tilde{f}_1 S_2 S_1 + f_c \tilde{f}_1 S_c S_1 + f_1 (\tilde{f}_2 + \tilde{f}_c) S_c S_2 + S_c^2 \tilde{f}_c f_1\right]}{(S_1 + S_2 + S_c)\left[\tilde{f}_2 S_2 + \tilde{f}_1 S_1 + \tilde{f}_c S_c\right]} \tag{11}$$

with

$$\tilde{f}_i(s) = f_i(s)\, s - 1 = \mathcal{L}\{d(S^0_i(Q, t)/S^0_i(Q))/dt\} \tag{12}$$

For the 1-component labelled $\Delta\rho^2 S_{11}(Q, t)$ represents the observed echo amplitude in an NSE experiment. More complicated labellings require the full $\hat{S}$ matrix, but to discuss the salient features we may resort to just S_{11}. And next the analytically solvable combination of two components with simple exponential relaxations with different rates (time constants) is illuminated in order to highlight some basic properties of the dynamic RPA.

3 Simple exponential combination

For time functions $f_i(t) = \exp(-t\, r_i)$ resulting in $f_i(s) = 1/(s + r_i)$ the inverse Laplace transform of Eq. 11 can be performed analytically yielding Eq. 13. Note that this is different from plainly inserting $S(\cdots, t$ into the SANS RPA expression Eq. 1, which would yield Eq. 15! The proper dynamic RPA result for this generic simple case

$$S(.., t) = \frac{S_1 S_2 \phi(1 - \phi)}{\phi S_1 + (1 - \phi) S_2} e^{-t\, r_{\mathrm{RPA}}} \tag{13}$$

is just another single exponential relaxation decay, however, with a modified relaxation rate r_{RPA}:

$$r_{\mathrm{RPA}} = \frac{r_1 r_2 \left[S_1 \phi + S_2 (1 - \phi)\right]}{r_1 S_1 \phi + r_2 S_2 (1 - \phi)} \tag{14}$$

which is in contrast to the outcome of the **wrong(!)** direct combination of $S_i(Q, t)$ using the SANS expression:

$$S^{static\, RPA}(.., t) = \frac{S_1 S_2 \phi(1 - \phi)}{\phi S_1 e^{t\, r_2} + (1 - \phi) S_2 e^{t\, r1}} \tag{15}$$

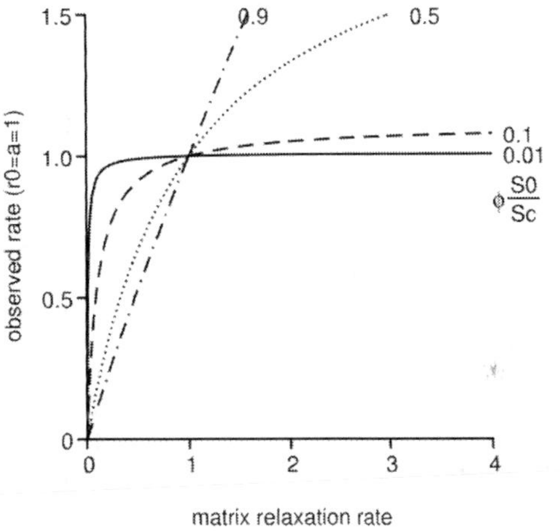

Figure 3. Resulting exponential decay rate for a hypothetical mixture of chains with different simple exponential relaxations. The example shows the dependence of the observed rate r_{RPA} as function of a matrix relaxation rate $r_{\mathrm{matrix}} = r_c = r_2$ for fixed $r_0 = r_1 = 1$ different curves pertain to different ratios $\phi S_0(Q/[(1 - \phi) S_c(Q)]$

Figure 3 illustrates the dependence of the resulting observed relaxation rate r_{RPA} as function of the rate of the second component (here=matrix) for different scattering contributions of the components $\phi S(Q)$ (labelled) and $(1 - \phi) S_c(Q)$ (matrix); for the figure the relaxation rate of the first component is set to 1. The salient feature can directly be extracted: (i) an immovable embedding ($r_c = 0$) in any case also yields a relaxation free resulting scattering, no dynamics is observable, all shown curves meet at (0,0). (ii) in the limit $\phi \to 0$ the observed relaxation rate soon approaches the undisturbed relaxation rate of the labelled component (here $r_1 = 1$). (iii) vice versa for $\phi \to 1$ the rate approaches that of the then dominant matrix. (iv) the crossing of all curves at (1,1) indicates that no change in relaxation (i.e. no distortion of experimental conclusions) is to be expected if the dynamics of both components are equal (very similar).

Note: all changes of the relaxation functions discussed here and throughout this paper only pertain effects that relate to the visibility of molecular movements imposed by the contrast setting in combination with the incompressibility constraint.

This does **not include** any mutual influence on the molecular mobilities as expressed in $S_i^0(Q,t)$. These are subject to modelling or other analysis after the here discussed "distortions" due to their coupling to the observable scattering intensity has been accounted for. The latter is the role of the here discussed dynamic RPA influence!

3.1 Numerics aspects

While the evaluation of Eq. 11 is straightforward the subsequent requirement of an inverse Laplace transform to arrive at $I(Q,t)$ is not. If the single component scattering functions $S(Q,s)$ are just known numerically or as real functions the inversion is impractical. However, if $S(Q,s)$ could be obtained on a complex path as used in the Mellin Laplace inversion

$$\mathcal{L}^{-1}\{f(s)\} = \frac{1}{i\,2\pi}\int_{\xi-i\infty}^{\xi+i\infty} f(s)e^{st}\,ds \qquad (16)$$

the inversion becomes numerically feasible. The required analytic property can be achieved by an intermediate evaluation step that maps the model functions $S_i(Q,t)$ to a sum of simple exponentials, the resulting Laplace transform Eq. 10 than can be inserted into Eq. 11 to yield the required analytic continuation of $S(Q,s)$. The determination of the exp-sum for any given relaxation function, e.g. computed from a suitable polymer scattering model, is automatically done in the software module developed for the dynamic RPA fitting procedure described in ref [1] and used in the following examples. Figure 4 illustrates the N-dependence of achievable accuracy of the N-exp-sum representation of test functions that resemble typical relaxation curves as observed for polymer dynamics over a time span of 6 decades.

Note that limitation in the numeric procedure require the restriction to a minimum value of the individual rates (in order to stay away from nearly divergent $1/(s+a)$ contributions). However, selecting this limit (if it applies) to more than an order of magnitude beyond the longest observed time keeps the associated error of the results low.

4 An experimental verification

The development of a practical implementation of the described dynamics RPA correction was triggered by the observation of "strange" behaviour in the results of an experiment intended to investigate the dynamics of short tracer polymer chains in a matrix of long reptating poleyethylene (PE) chains [12] and in analog way for polyethyleneoxide (PEO) [13]. The data exhibit a leveling off of the relaxation curves at long times which are incompatible with the behaviour of diffusing small molecules with the macroscopic diffusion also established by PFG-NMR.

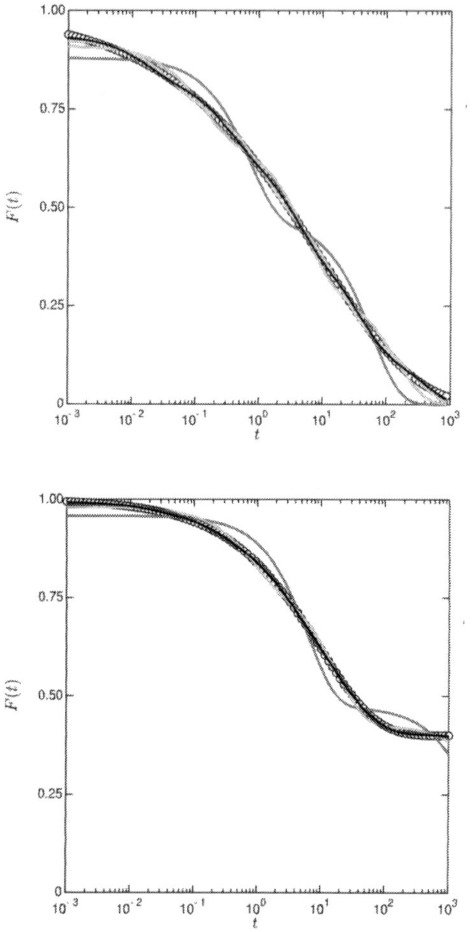

Figure 4. Example sequence of n-exp fits for increasing accuracy requirements (colored lines) as obtained by the automatism used in the software implementation of the dynamic RPA procedure [1]. The text functions (symbols) were: $\exp(-0.3\,t)$ and $0.6\exp(-0.5\,t) + 0.4$. The black lines represent the final result obtained with 5 resp. 6 simple exponentials.

As it turns out the observation is not related to "bad" experiments with excessive background scattering and unsuitable background subtraction or may need other strange explanations but can quantitatively be described within the above described dynamic RPA algorithm. Figure 5 shows a representative example of these data for a concentration $\phi = 0.12$ of the h-labeled tracer polymers in the d-labeled long chain matrix. This concentration yields a clearly visible contribution of the slower long chains in terms of a retardation at longer times more or less halting at a plateau like level. This is in contrast to the expectation for the "pure" diffusion dominated relaxation of the tracers (red lines) which swiftly approaches zero at long observation times.

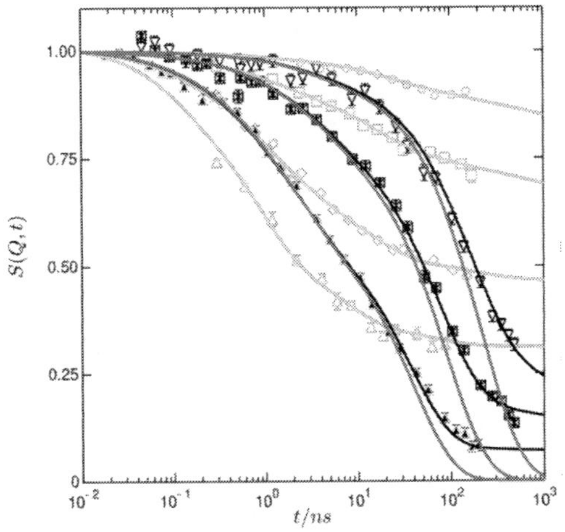

Figure 5. Fits of the scattering function of labelled small PE-tracer molecules (black) consisting of an initial sub-linear diffusion and Rouse dynamics with non-Gaussian effects [12] after including the dynamic RPA influence of the long chain matrix (green symbols and lines). The tracer volume fraction was $\phi = 0.12$, the long chain influence leads to excess residual amplitiudes at large Q as the comparison with the computation without RPA-effects ($\phi \to 0$, red lines) clearly shows. The Q-values for the tracers are: 0.05, 0.078 and 0.12 Å^{-1} and for the long chains: 0.05, 0.077, 0.115 and 0.145 Å^{-1}.

The details of the sub-linear tracer diffusion thus could be extracted from the data by fitting the model function (a product of a diffusion expression and a Rouse mode sum) to the data by observing the dynamic RPA effects that depend on the knowledge of the structure function of the long chains. The details of this treatment are given in ref. [12].

Note that by augmenting the experiment with runs on inversely labelled samples containing a fraction of h-labeled long chains in an otherwise d-labeled systems (e.g. 10% h-long, 78% d-long and 12% d-short) would allow also assessing potential influences of the short chain dilution on the long chains and thus enhance the accuracy of the whole procedure. In the present case the 12% volume fraction is on the limit where such effects would start to significantly influence the results. Again it has to be emphasized that the dynamic RPA treatment only pertains the visibility effects due to the interference of the intermingled incompressible polymers. Any alteration of the chain dynamics due to hindering, topological interaction or added friction by the presence of the "other" polymers still must be modelled into the "undisturbed" $S_i^0(Q,t)$ structure functions; these the are determined (by model refinement) by fitting within the dynamic RPA framework as explained here.

5 Some hypothetical extrapolations

As already mentioned in the introduction a completely immobile matrix would not allow any visible scattering length density variations with time and thus no relaxation in the scattering functions even if the tracer are very mobile. To test this case the fit results of figure 5 have been taken and recomputed but now with a slowing factor of 10^6 for the long chain matrix functions. As figure 6 illustrates this pushes all observed scattering functions close to the static line $S(Q,t)/S(Q,t=0) = 1$.

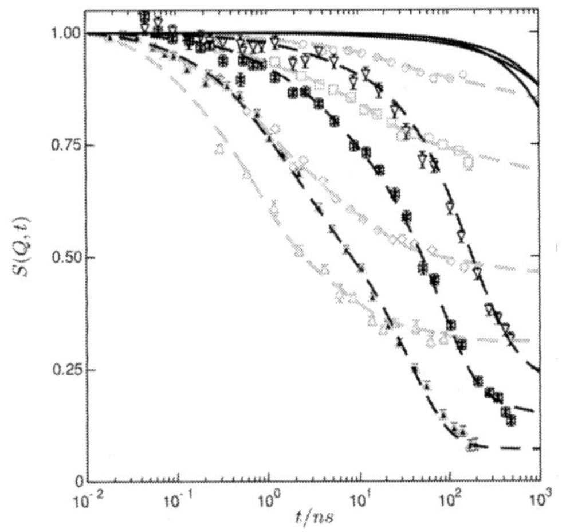

Figure 6. Same data and model (symbols and dashed lines) as in figure 5 but the computed curves (solid black lines) pertain to a slowed down matrix dynamics.

The transition from a more or less close representation of the short chain ($S_1(Q,t)$ to that of the long chain $S_2(Q,t) = S_c(Q,t)$ visibility is illustrated in figure 1 of ref. [1]. Where the plateau visibility steadily increases with increasing concentration ϕ_{tracer}, thereby continuously evolving towards the long chain scattering function for 90%+ concentration. However, note again that this only pertains the visibility, the increased mobility of long chains diluted by a large amount of small tracers is not included and must be modelled into $S_2^0(Q,t)$ of the long chains. If on the other hand a small amount of the long chains are labelled $S_2(Q,t)$ and the rest of long chains $S_c(Q,t)$ and the short chains $S_1(Q,t)$ have the background scattering length density (typically by using deuterated components) the effect on the scattering can be found in figure 2 of ref. [1]. The long chain visibility here is less affected by the short chain presence, still some minor reduction of the observed plateau levels are encountered. But again these curves shown are rather hypothetically since the computations completely ignores the effect of dilution by short chains on the "pure" dynamic structure functions $S_{2,c}^0(Q,t)$ of the long chains (and vice versa). The determination of these influences, which probably would be the goal of an experiment then can be obtained by (model) fitting of the "pure" structure functions $S^0_{1,2,c}(Q,t)$ that enter the dynamic RPA procedure. To serve as tool enabling this in an easy and practically feasible way is the main purpose of the dynamic RPA procedure presented.

6 Conclusion

Interpretation of experiments aiming at the dynamics of a labelled component in a polymer mixture requires special attention if the dynamic properties of the various components are different. Due to the partial visibility of the non-labelled (background, matrix) component in the scattering signal, their different dynamics may modify the characteristic times and/or shapes of the observed relaxation curves. Applying dynamic RPA to the mixture allows to compute this mutual influence. Practical application is made accessible by implementing a procedure that yields the RPA-corrected scattering functions form the "pure" scattering function data (from any kind of modelling). This procedure can be included into the model computation during fitting and thus determine the proper model parameters (model values) of the "pure" scattering functions of labelled components. The Fortran implementation of this procedure is available see ref. [1]. While there is no effect for mixtures of dynamically equal polymers (i.e. the conventional h/d-mixtures used to create contrast in SANS and NSE-experiments) different dynamics as in the example of small diffusing chains in a matrix of reptating slow long chains [12] huge distortions are observed and would lead to erroneous conclusions if the RPA-effect is not properly accounted for. But also in dynamically less different mixtures mutual influence on relaxation rates will occur with possibly worse consequences for the conclusion if their not so obvious influence remains unnoticed.

References

[1] M. Monkenbusch, M. Kruteva, M. Zamponi, L. Willner, I. Hoffman, B. Farago, D. Richter, Journal of Chemical Physics **152**, 054901 (2020)

[2] B. Hammouda, Advances in Polymer Science **106**, 87 (1993)

[3] H. Frielinghaus, Colloid and Polymer Science **299**, 1031 (2021)

[4] Y. Melnichenko, G. Wignall, D. Schwahn, Fluid Phase Equilibria **212**, 209 (2003), 16th European Conference on Thermophysical Properties, Imperial Coll, London, England, Sep 01-04, 2002

[5] A. Akcasu, M. Benmouna, H. Benoit, Polymer **27**, 1935 (1986)

[6] M. Tombakoglu, A. Akcasu, Polymer **33**, 1127 (1992)

[7] A. Akcasu, R. Klein, B. Hammouda, Macromolecules **26**, 4136 (1993)

[8] I. Erukhimovich, Y. Kudryavtsev, European Physical Journal E **11**, 349 (2003)

[9] S. Mochrie, Macromolecules **36**, 5013 (2003)

[10] D. Lumma, S. Mochrie, Macromolecules **34**, 8303 (2001)

[11] A. Akcasu, M. Tombakoglu, Macromolecules **23**, 607 (1990)

[12] M. Zamponi, M. Kruteva, M. Monkenbusch, L. Willner, A. Wischnewski, I. Hoffmann, D. Richter, Physical Review Letters **126**, 187801 (2021)

[13] M. Kruteva, M. Zamponi, I. Hoffmann, J. Allgaier, M. Monkenbusch, D. Richter, Macromolecules **54**, 11384 (2021)

Data reduction strategies at a time-of-flight NSE for a lamellar microemulsion

Olaf Holderer[1,*], *Henrich* Frielinghaus[1], *Piotr* Zolnierczuk[2], *Michael* Ohl[3], and *Michael* Monkenbusch[3]

[1] 1Jülich Centre for Neutron Science (JCNS) at Heinz Maier-Leibnitz Zentrum (MLZ), Forschungszentrum Jülich GmbH, Garching, Germany
[2] SNS - Oak Ridge National Laboratory, Oak Ridge, TN, USA
[3] Jülich Centre for Neutron Science (JCNS), Forschungszentrum Jülich GmbH, Jülich, Germany

Abstract. Neutron spin echo (NSE) spectroscopy provides the ultimate energy resolution in quasi-elastic thermal and cold neutron scattering spectroscopy. A peculiarity of the SNS-NSE, the only NSE spectrometer at a pulsed beam port at the moment, is that the wavelength spread $\delta\lambda/\lambda$ can be chosen during evaluation with an appropriate time channel binning. The Q-resolution can be adjusted in certain limits a posteriori by choosing the appropriate detector binning (as on a continuous source) and time channel binning. This can be exploited for samples with a strongly varying scattering function $S(Q,t)$, e.g. due to Bragg peaks in a crystal or lamellar ordering in microemulsions. The data reduction software DrSpine allows for appropriate slicing and masking for this task. In this contribution the correlation function of microemulsions, thermodynamically stable mixtures of oil, water and surfactant, is measured with NSE on length scales where structural correlations are important, and data reduction strategies varying the Q-resolution by pixel and time channel grouping are discussed. The typical "de Gennes narrowing" or structural narrowing is observed with a relaxation time proportional to *I(Q)*. In these regions of strongly varying intensity it is shown that a too coarse grouping has an influence on the data reduction, with a broadened in Q of the apparent slowing down.

1 Introduction

Microemulsions, i.e. thermodynamically stable mixtures of water, oil and a surfactant, can form different structures such as bicontinuous, lamellar or droplet structures. Neutron scattering is a well suited tool for studying soft matter in terms of its structural and dynamic properties. The bending elasticity or rigidity of microemulsions can be studied with neutron spin-echo spectroscopy (NSE), the highest resolution spectroscopic technique in neutron scattering [1]. The intermediate scattering function determined with NSE pertains to the Fourier transform of the height correlation function and follows approximately a stretched exponential decay for large scattering vectors Q [2]. The influence of the correlation peak on the dynamics has been studied with NSE spectroscopy by looking at bulk bicontinuous microemulsions, but without preferrential ordering, in Ref. [3] and interpreted in terms of hydrodynamic effects [4]. Lamellar microemulsions have been studied with NSE by Mihailescu et al.[5] in detail with added diblock copolymers as an emulsification booster. A single crystal of a lyotropic L_α-phase has been investigated with neutron diffraction by Göcking et al. [6]. A similar oriented lamellar phase has been studied with small angle scattering in order to shed light on electrostatic and steric interactions in such phases [7]. The dynamics of lipid membrane stacks has been studied with NSE in order to observe different fluctuation modes [8, 9], using stacks of lipid bilayers. This allows for a precise measurement of the scattering vector in different directions, in plane and out of plane. Similarly, measurements under grazing incidence conditions allows for a very precise sample geometry for studying microemulsions or lipid membranes [10, 11]. Also neutron reflectometry gives some insight in the structure and dynamics of lamellar stacks by the analysis of the scattering peak shape [12]

Here we want to demonstrate how the time-of-flight mode of a neutron spin echo spectrometer at a spallation source affects the planning and data evaluation of such experiments on structured samples. A more detailed analysis of the deduced bending constants and structural properties of these types of microemulsions can be found e.g. in Refs. [5, 13, 14].

2 Experimental

2.1 Samples

Lamellar microemulsions were prepared with deuterated decane, D_2O and the surfactant $C_{10}E_4$. At surfactant concentrations of $\Psi = 21$ vol% a lamellar phase is formed in such systems [15]. To align the microemulsion in the beam, a cell with quartz lamellae has been used, where the microemulsion is sheard into the gap of 0.2 mm between two lamellae. The lamellae are stacked to fill the full beam. The same cells and samples have been used also in Ref. [16]. The bicontinuous microemulsion which serves as a comparison was prepared with $\Psi = 17$ vol%. Details of the used materials and the bulk microemulsion can be found in Ref. [17].

*e-mail: o.holderer@fz-juelich.de

2.2 Neutron Spin Echo Spectroscopy

Experiments have been carried out at the SNS-NSE at the Oak Ridge National Laboratory [18]. Due to the pulsed neutron source the wavelength of each detected neutron can be tagged by the moderator to detector time-of-flight (TOF). The experiments have been carried out with a wavelength band of 6-9 Å and two scattering angles, for the lamellar sample corresponding to $Q_{center,lam-min}$ = 0.065 $Å^{-1}$ and 0.08 $Å^{-1}$, covering a Q-range from 0.04-0.15 $Å^{-1}$. The bicontinuous microemulsion with different structural parameters has been measured at $Q_{center,lam-min}$ = 0.11 $Å^{-1}$ and 0.14 $Å^{-1}$.

3 Results and Discussion

The microemulsion in the lamellar cell is in an oriented lamellar phase at a temperature of 301 K. The sample cell temperature was set with a copper frame connected to a water thermostat. Aluminum sheets acted as a heat shield in the beam window area. The set point of the thermostat was 305 K allowing for an temperature offset induced by the air conditioning in the SNS-NSE cave. The lower phase boundary to the bicontinuous phase is approximately at 298-299 K.

The inherent time-of-flight mode of the SNS-NSE allows to adjust the wavelength spread during data evaluation by choosing the desired binning over the detector and across the 42 time channels which cover the wavelength band of 3 Å width. Since the sample shows a significant structuring in the Q-range of this experiment with strong correlation peaks, the 32x32 pixel detector was evaluated with different binnings of the detector pixels and time channels. The finest evaluation used the 32x32 detector pixelwise, and the time channels have been grouped into 9 slices. Different detector and time binnings have been tried out and are also presented in the following discussion. Data reduction has been done with the DrSpine software [19]. For the lamellar phase, a detector mask has been applied which covers only a stripe in the middle of the detector, i.e. around $Q_y = 0$. Figure 1 shows the corresponding mask for the full 32x32 pixel detector. Regions marked with 1 are excluded from the evaluation. For other binning during evaluation, the mask has been adapted to cover the same stripe as in Figure 1. A typical detector image of one time bin is shown in Figure 2. The stripe along the x-direction is used for the data treatment of the lamellar phase and corresponds to the region not masked as shown in Figure 1.

The intermediate scattering function for bicontinuous and lamellar phase are shown in Fig. 3. The strong scattering intensity allows for a detailed evaluation of the bicontinuous microemulsion over the whole Q-range measured in this experiment. Since the lamellar phase is strongly structured, the statistics worsens due to the low count rate away from the structure factor peak maximum. The histogrammed intensity as a function of Q as shown in Fig. 4 exhibits for the lamellar phase a correlation peak at Q≃ 0.07 $Å^{-1}$ corresponding to the average distance between surfactant layers, and the second order peak at Q≃ 0.14 $Å^{-1}$. The corresponding length scale $d_{surf} = 2\pi/q$ is 90 Å. The domain size of a bicontinuous microemulsion d=37 Å / Ψ would result in d_{domain}=176 Å and d_{surf} = $d_{domain}/2$ = 88 Å. This indicates that the temperature at the sample might be slightly outside the purely lamellar temperature window with more disorder leading to a larger average distance between lamellae. A perfectly flat lamellar phase would have d=24 Å / Ψ = 114 Å for d_{domain} or 57 Å for d_{surf}. In Figure 4 an attempt has been made for an absolute scaling of the two samples by normalizing the raw scattering data from the SNS-NSE with an absolute measurement of I(Q) vs Q of a bicontinuous microemulsion with similar surfactant contents from a SANS experiment [14]. The histogram of the bicontinuous phase has been fitted with a polynomial of degree 5, the ratio between this fit and the fitted I(Q) of the absolute SANS experiment provides the correction factor for this experiment. The corrected intensity in Figure 4 represents roughly the SANS intensity on an absolute scale.

```
drspine --> mask pix
 > mask pix
 detector pixels:     32 32
        1 2 3 4 5 6 7 8 910111213141516171819202122232425262728293031 32
   1  1 1 1 1 1 1 1 1 1 1 1 1 1 1 1 1 1 1 1 1 1 1 1 1 1 1 1 1 1 1 1 1
   2  1 1 1 1 1 1 1 1 1 1 1 1 1 1 1 1 1 1 1 1 1 1 1 1 1 1 1 1 1 1 1 1
   3  1 1 1 1 1 1 1 1 1 1 1 1 1 1 1 1 1 1 1 1 1 1 1 1 1 1 1 1 1 1 1 1
   4  1 1 1 1 1 1 1 1 1 1 1 1 1 1 1 1 1 1 1 1 1 1 1 1 1 1 1 1 1 1 1 1
   5  1 1 1 1 1 1 1 1 1 1 1 1 1 1 1 1 1 1 1 1 1 1 1 1 1 1 1 1 1 1 1 1
   6  1 1 1 1 1 1 1 1 1 1 1 1 1 1 1 1 1 1 1 1 1 1 1 1 1 1 1 1 1 1 1 1
   7  1 1 1 1 1 1 1 1 1 1 1 1 1 1 1 1 1 1 1 1 1 1 1 1 1 1 1 1 1 1 1 1
   8  1 1 1 1 1 1 1 1 1 1 1 1 1 1 1 1 1 1 1 1 1 1 1 1 1 1 1 1 1 1 1 1
   9  1 1 1 1 1 1 1 1 1 1 1 1 1 1 1 1 1 1 1 1 1 1 1 1 1 1 1 1 1 1 1 1
  10  1 1 1 1 1 1 1 1 1 1 1 1 1 1 1 1 1 1 1 1 1 1 1 1 1 1 1 1 1 1 1 1
  11  1 1 1 1 1 1 1 1 1 1 1 1 1 1 1 1 1 1 1 1 1 1 1 1 1 1 1 1 1 1 1 1
  12  ---------------------------------------------------------------
  13  ---------------------------------------------------------------
  14  ---------------------------------------------------------------
  15  ---------------------------------------------------------------
  16  ---------------------------------------------------------------
  17  ---------------------------------------------------------------
  18  ---------------------------------------------------------------
  19  ---------------------------------------------------------------
  20  ---------------------------------------------------------------
  21  1 1 1 1 1 1 1 1 1 1 1 1 1 1 1 1 1 1 1 1 1 1 1 1 1 1 1 1 1 1 1 1
  22  1 1 1 1 1 1 1 1 1 1 1 1 1 1 1 1 1 1 1 1 1 1 1 1 1 1 1 1 1 1 1 1
  23  1 1 1 1 1 1 1 1 1 1 1 1 1 1 1 1 1 1 1 1 1 1 1 1 1 1 1 1 1 1 1 1
  24  1 1 1 1 1 1 1 1 1 1 1 1 1 1 1 1 1 1 1 1 1 1 1 1 1 1 1 1 1 1 1 1
  25  1 1 1 1 1 1 1 1 1 1 1 1 1 1 1 1 1 1 1 1 1 1 1 1 1 1 1 1 1 1 1 1
  26  1 1 1 1 1 1 1 1 1 1 1 1 1 1 1 1 1 1 1 1 1 1 1 1 1 1 1 1 1 1 1 1
  27  1 1 1 1 1 1 1 1 1 1 1 1 1 1 1 1 1 1 1 1 1 1 1 1 1 1 1 1 1 1 1 1
  28  1 1 1 1 1 1 1 1 1 1 1 1 1 1 1 1 1 1 1 1 1 1 1 1 1 1 1 1 1 1 1 1
  29  1 1 1 1 1 1 1 1 1 1 1 1 1 1 1 1 1 1 1 1 1 1 1 1 1 1 1 1 1 1 1 1
  30  1 1 1 1 1 1 1 1 1 1 1 1 1 1 1 1 1 1 1 1 1 1 1 1 1 1 1 1 1 1 1 1
  31  1 1 1 1 1 1 1 1 1 1 1 1 1 1 1 1 1 1 1 1 1 1 1 1 1 1 1 1 1 1 1 1
  32  1 1 1 1 1 1 1 1 1 1 1 1 1 1 1 1 1 1 1 1 1 1 1 1 1 1 1 1 1 1 1 1
drspine -->
```

Figure 1. Detector mask applied for the lamellar phase. All pixels marked with 1 are excluded from the evaluation. In this representation the main direction of *Q*-variation is horizontal.

Both, the bicontinuous microemulsion and the lamellar microemulsion have been fitted with the amplitude left free, since in microemulsion measurements a proper background subtraction is difficult since the proper oil/water solvent without surfactant can not be prepared. In regions of low intensity (away from the structure factor peak in the lamellar phase or at high Q in the bicontinuous phase), the missing background subtraction requires the amplitude as a fitting parameter. The comparison between the relaxation time τ_0(Q) and the intensity I(Q) is illustrated by figures 4 and 5, both as a function of Q. It shows that $\tau_0(Q) \propto$ I(Q) for the lamellar microemulsion, whereas the bicontinuous microemulsion shows a strong reduction of the relaxation time proportional to Q^{-3} as expected from the height fluctuation model from Zilman and Granek [2]. It has to be pointed out that the relaxation time at the peak position

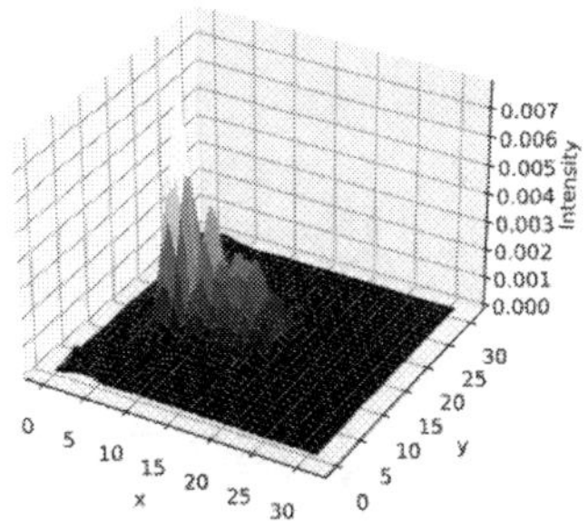

Figure 2. Average intensity at the detector for the oriented lamellar microemulsion with the strong correlation peak. Directions are $x \rightarrow Q_z$ and $y \rightarrow Q_{\text{parallel}}$. The subpeak type modulation is an artifact induced by the analyzer mirror stacking.

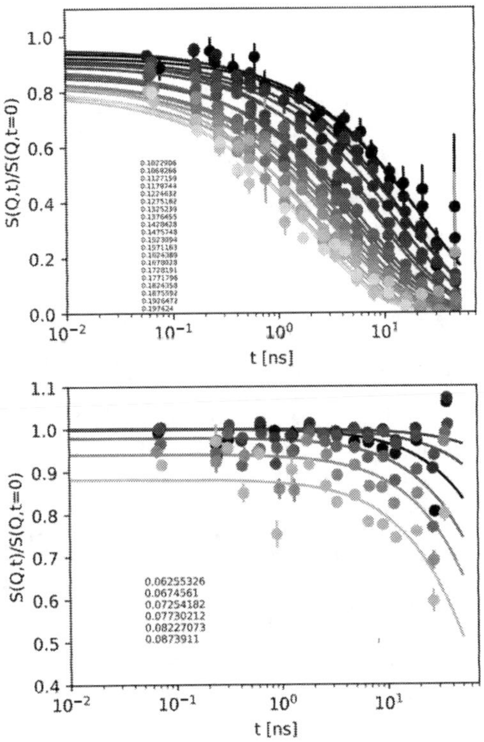

Figure 3. Intermediate scattering function S(q,t)/S(q,0) for a bicontinuous (top) and an oriented lamellar microemulsion (bottom).

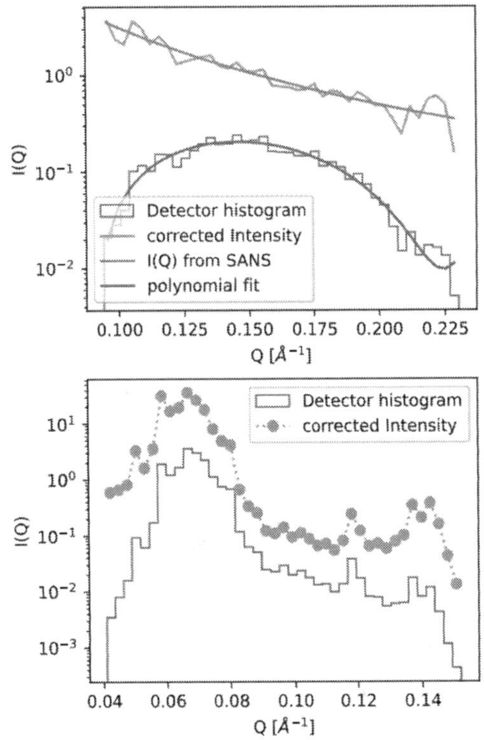

Figure 4. Histogram of q-intensities of the NSE experiment for the bicontinuous (top) and lamellar microemulsion (bottom). From previous SANS experiments on bicontinuous microemulsions the relevant Q-range has been fitted on an absolute scale and used to determine the detector sensitivity.

can not be resolved any more in the current time range of the experiment and represents a basically elastic situation without relaxation in the observation time window. Figure 6 illustrates the effective diffusion coefficient $D = 1/(\tau_0 Q^2)$ with a roughly linear increase for the bicontinuous phase (as expected for the Q^{-3}-dependence) and the strong slowing down in the lamellar phase at the q-position of the correlation peak. The low scattering intensity off the peak position results in a strong scattering of the data above Q=0.1 Å^{-1}. For a more detailed analysis of the bicontinuous phase we refer to fits with the full Zilman-Granek model to such microemulsion systems in Refs [10, 14].

The wavelength band used at the SNS-NSE has a width of 3 Åsplit into 42 time channels for a single pulse. The ΔQ of a time channel is approximately 0.0012-0.0005 Å^{-1} (for the wavelength band of 6-9 Å), the Q-width of a single detector pixel of the 32x32 pixel detector is 0.0026-0.0017 Å^{-1} for the same wavelength band (as inferred from the nse scan). An adequate combination of timechannel binning and detector binning would be 2 time channels for the pixel wise evaluation, 5 time channels for a 2x2 pixel grouping and 10 time channels for the 4x4 pixel grouping evaluation (of the 32x32 pixel detector, where the mask of Fig 1 has been applied). The variation of Q may differ

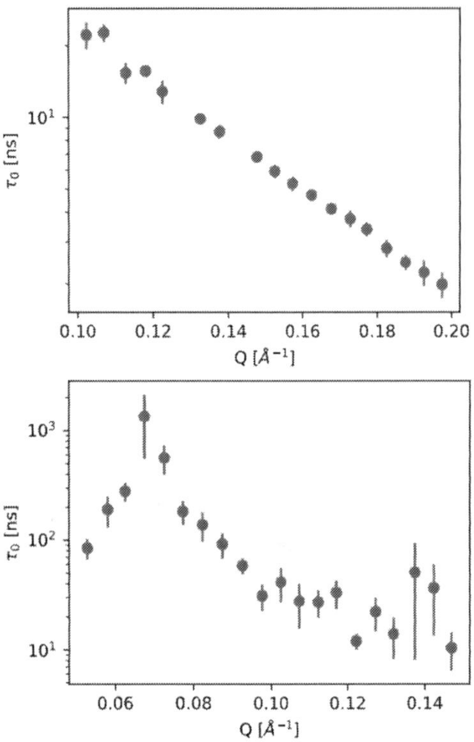

Figure 5. Relaxation time as function of scattering vector for the bicontinuous microemulsion (top) and the ordered lamellar microemulsion (bottom).

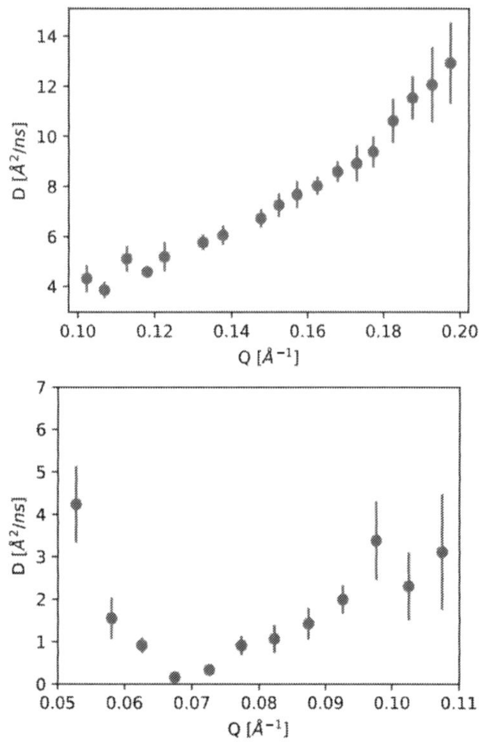

Figure 6. Diffusion constant (i.e. $1/(\tau_0 Q^2)$ for the bicontinuous (top) and lamellar (bottom) microemulsion.

from the minimal possible variation and is determined by the histogramming procedure in the evaluation.

The evaluation has been run in the Q-range of 0.05-0.15 Å^{-1}, with 20, 10 and 5 Q-bins i.e. with a distance between Q's of 0.005 Å^{-1}, 0.01 Å^{-1} and 0.02 Å^{-1}. The Q-variation within the final Q-bins was always about one order of magnitude better than the Q-steps. The limiting factor is therefore not the instrumental capabilities in terms of Q-resolution, but the statistics of the experiment.

The major limiting factor of the Q resolution of the NSE instrument is the beam divergence. It is determined by the size of the neutron guide exit (40 mm x 60 mm), the sample size (30 mm x 30 mm) and the distance between neutron guide and sample (approx. 4 m), leading to about ±0.5° beam divergence. This translates to a ΔQ due to the angular divergence between 0.012 and 0.018 Å^{-1} depending on the wavelength.

The choice of a grouping of 4x4 pixels on the detector matches this divergence quite well.

On the other hand, the lamellar phase acts as a monochromator and affects to some extend the divergence after the scattering process. Especially around the peak position, a fine evaluation has an advantage, also because the intensity variation across the pixel is smaller and better comparable between reference and sample. Figure 7 shows the effect of different binnings and Q collection steps. Around the peak position the statistics is good enough for a single pixel evaluation. Grouping 2x2 pixels together works similarly well. The coarser grouping of 4x4 pixels for the evaluation shows some deviations in the diffusion constant around the peak, although this would be the best matching conditions. Beyond approximately Q=0.1 Å^{-1} the errorbars get too large to compare the different evaluation schemes, it has therefore been omitted. The strong intensity at the peak position dominates the 4x4 grouping evaluation even for pixels centered a bit aside of the peak. Therefore the very slow peak relaxation influences also Q's aside of the peak, if the groupig is coarser.

Omitting the stripe mask for the lamellar phase results in a stronger variation of the resulting relaxation times and diffusion constants due to the added noise of the detector regions with low count rate.

4 Conclusion

The membrane dynamics of a lamellar microemulsion has been studied with the time-of-flight NSE spectrometer, where the Q-resolution can be adjusted by appropriate binning during the data reduction. The structural slowing down has been observed also in this lamellar microemulsion. At Q-values directly beyond the first correlation peak the linear increase of the diffusion constant calculated from the relaxation time shows the height fluctuation signature with a Q^3 dependence of the relaxation rate.

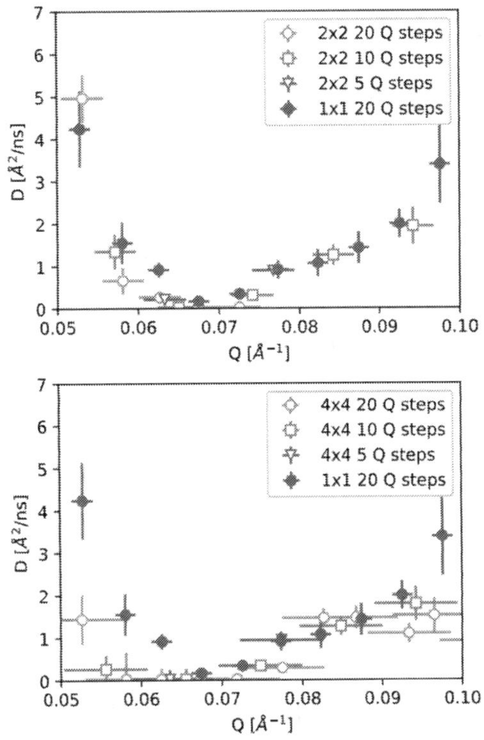

Figure 7. Variation of pixel binning and Q-binning: the most significant difference is the lower diffusion constant if evaluated with a coarser binning, most likely due to contributions from the Bragg peak contributing to more Q-bins. The 2x2 pixel grouping evaluation seems to be a good compromise between statistics and resolution. The dQ errorbars represent the error in Q due to the pixel grouping only.

The Q-resolution provided by the SNS-NSE instrument is mainly governed by the beam divergence in the current experiment. Reducing the pixel binning by grouping 2x2 pixels or using single pixels seems to be advantageous over a too coarse binning. This might be due to subtle resolution effects due to the scattering at the oriented lamellar membrane stack. Only very close to the correlation peak the obtained relaxation time (or diffusion constant) was slightly modified by this effect. The high intensity and slow relaxation at the peak position influences also pixel groups aside of the peak, the more the larger the evaluation pixels are chosen. The limitation is still the intensity per pixel, which prevents a single pixel, single (or two) time channel binning for nominal highest resolution. When conducting such experiments with the presence of correlation peaks on the detector, it has to be taken into account that smaller binnings during evaluation are required, and also that the intensity might vary strongly between the peak region and further away from the peak. The counting time has to be adapted accordingly. These effects of structure factor peaks in the Q-region of the experiment are important not only in such lamellar or structured systems where the dynamics of the system itself is slowed down by the structure factor influence, but also in confined or multi-component materials, where an elastic signal from the matrix might be avoided by chosing the relevant Q-binning. If even sharper Q-resolution is required, then also smaller samples and entrance slits have to be used. Another example might be experiments with large fluctuating magnetic structures such as Skyrmions, where sharp magnetic Bragg peaks are surrounded by weak and slow fluctuations, this high Q resolution might be fully exploited.

Data availabilit statement

The data from the NSE experiments are available from the corresponding author upon reasonable request.

Acknowledgement

This research at Oak Ridge National Laboratory's Spallation Neutron Source was sponsored by the Scientific User Facilities Division, Office of Basic Energy Sciences, US Department of Energy.

References

[1] F. Mezei, in *Neutron spin echo* (Springer, 1980), pp. 1–26
[2] A. Zilman, R. Granek, Physical review letters **77**, 4788 (1996)
[3] O. Holderer, H. Frielinghaus, M. Monkenbusch, J. Allgaier, D. Richter, B. Farago, The European Physical Journal E **22**, 157 (2007)
[4] M. Nonomura, T. Ohta, The Journal of chemical physics **110**, 7516 (1999)
[5] M. Mihailescu, M. Monkenbusch, J. Allgaier, H. Frielinghaus, D. Richter, B. Jakobs, T. Sottmann, Physical Review E **66**, 041504 (2002)
[6] K. Goecking, M. Monkenbusch, EPL (Europhysics Letters) **43**, 135 (1998)
[7] F. Nallet, D. Roux, S. Milner, Journal de Physique **51**, 2333 (1990)
[8] C.L. Armstrong, W. Häußler, T. Seydel, J. Katsaras, M.C. Rheinstädter, Soft matter **10**, 2600 (2014)
[9] M.C. Rheinstädter, W. Häußler, T. Salditt, Physical review letters **97**, 048103 (2006)
[10] H. Frielinghaus, M. Kerscher, O. Holderer, M. Monkenbusch, D. Richter, Physical Review E **85**, 041408 (2012)
[11] S. Jaksch, O. Holderer, M. Gvaramia, M. Ohl, M. Monkenbusch, H. Frielinghaus, Scientific Reports **7**, 4417 (2017)
[12] T. Salditt, C. Münster, U. Mennicke, C. Ollinger, G. Fragneto, Langmuir **19**, 7703 (2003)
[13] M. Mihailescu, M. Monkenbusch, H. Endo, J. Allgaier, G. Gompper, J. Stellbrink, D. Richter, B. Jakobs, T. Sottmann, B. Farago, The Journal of Chemical Physics **115**, 9563 (2001)
[14] O. Holderer, H. Frielinghaus, D. Byelov, M. Monkenbusch, J. Allgaier, D. Richter, The Journal of chemical physics **122**, 094908 (2005)

[15] D. Byelov, H. Frielinghaus, O. Holderer, J. Allgaier, D. Richter, Langmuir **20**, 10433 (2004)
[16] O. Holderer, H. Frielinghaus, D. Byelov, M. Monkenbusch, D. Richter et al., Zeitschrift für Physikalische Chemie **224**, 243 (2010)
[17] F. Lipfert, O. Holderer, H. Frielinghaus, M.S. Appavou, C. Do, M. Ohl, D. Richter, Nanoscale **7**, 2578 (2015)
[18] M. Ohl, M. Monkenbusch, N. Arend, T. Kozielewski, G. Vehres, C. Tiemann, M. Butzek, H. Soltner, U. Giesen, R. Achten et al., Nuclear Instruments and Methods in Physics Research Section A: Accelerators, Spectrometers, Detectors and Associated Equipment **696**, 85 (2012)
[19] P. Zolnierczuk, O. Holderer, S. Pasini, T. Kozielewski, L. Stingaciu, M. Monkenbusch, Journal of applied crystallography **52** (2019)

Microscopic versus Macroscopic Glass Transition(s) in Blends of Industrial Interest

Numera Shafqat[1,2*], *Angel* Alegría[1,3], *Nicolas* Malicki[2], *Séverin* Dronet[2], *Lucile* Mangin-Thro[4], *Bernhard* Frick[4], *Juan* Colmenero[1,3,5], and *Arantxa* Arbe[1]

[1]Centro de Física de Materiales (CFM) (CSIC-UPV/EHU) - Materials Physics Center (MPC), San Sebastián, Spain
[2]Manufacture Francaise des Pneumatiques MICHELIN, Clermont-Ferrand, France
[3]Departamento de Polímeros y Materiales Avanzados: Física, Química y Tecnología, UPV/EHU, San Sebastián, Spain
[4]Institut Laue-Langevin, Grenoble, France
[5]Donostia International Physics Center (DIPC), San Sebastián, Spain

Abstract. We investigate by neutron scattering and calorimetry a mixture of styrene-butadiene rubber (SBR) and a commercial resin. The neat materials present a large dynamic contrast, having SBR a much lower glass-transition temperature than the resin. The focus is to exploit neutron scattering selectivity in an isotopically labelled sample where deuterated SBR is the majority component. This direct insight into the resin atomic motions within the mixture allows determining the 'microscopic' glass transition of the resin in the presence of the a priori much more mobile SBR. This transition takes place in the vicinity of the initial calorimetric glass transition temperature, mainly dictated by the majority component.

1 Introduction

Polymeric mixtures based on styrene-butadiene rubber (SBR) matrices are of high interest in tyre industry. Mixing components with markedly different microscopic mobility (and thereby showing a large difference in their glass-transition temperature (T_g) values) provides a very versatile tool to obtain materials with desired end-use properties. In addition to this applied motivation, the study of mixtures based on systems with strong dynamic asymmetry –i. e., presenting a large difference in the characteristic times of the segmental (α)-relaxation-- has attracted great academic interest in the past years [1].

Frequent experimental observations in polymer blends comprise (i) very broad relaxation responses (much broader than in the neat components) and (ii) distinct characteristic times associated to each of the components, even in thermodynamically miscible blends. It is by now well established that these phenomenological findings can be traced back to two major ingredients: concentration fluctuations which lead to the observed broadening of the measured response functions and self-concentration which causes a distinct dynamical response of the components [1-8]. The self-concentration concept refers to the fact that the local concentration around a given segment is always enhanced with respect to the average (macroscopic) concentration due to chain connectivity. We note that experimental investigations on blends dynamics become difficult due to the need of applying selective techniques allowing to isolate the component dynamics in the mixture, in order to determine their mutual influence upon blending. Therefore, thorough investigations resolving the dynamics of both components are scarce. Most studies involve calorimetric techniques as a first approach to the overall macroscopic characterization of blends' dynamics. In a calorimetry experiment, the effect of blending is observed as a broad feature, leading to a smooth variation of the specific heat over a temperature region between the T_gs of the pure components. If the dynamic asymmetry of the two components is extremely large –as it is the case of e. g. polyethylene oxide (PEO) and poly(methyl methacrylate) (PMMA)– the presence of two distinct contributions to the specific heat step associated to each of the components can be resolved [1,9-11]. However, this is not always possible. In general, an average T_g is determined for the blend from the inflection point of the heat capacity, that is usually considered as the representative glass-transition temperature of the mixture.

The question of the dynamical behaviour in dynamically asymmetric mixtures becomes even more intricate if we consider in addition compositionally asymmetric systems, i. e., mixtures markedly rich in one of the components. In the case of blends where the high-T_g component is the majority one, interesting confinement effects have been reported for the fast minority component. This was the case of the very flexible PEO chains in PMMA [12], poly(vinyl acetate) [13] (PVAc) or polyethersulfone (PES) [14] rigid matrices, or poly(vinyl methyl ether) in polystyrene (PS) [15]. In the present work we face the opposite situation,

* Corresponding author: nshafqat001@ikasle.ehu.eus

considering a mixture where the high-T_g component is the minority one. Our aim is to determine 'microscopically' the vitrification temperature of the minority component, and compare it with the glass transition of the system as determined by calorimetry.

The system here investigated is a binary mixture of industrial interest, composed by SBR as low-T_g component and a commercial resin as high-T_g component. The technique chosen to isolate the dynamics of the latter component and determine its 'microscopic glass transition' is neutron scattering, combined with isotopic (H/D) labelling. This is an ideal tool to provide direct insight into the atomic (in particular, H) motions of a given component in a complex system, if the rest of the material is deuterated. Therefore, to be sensitive to the dynamics of the resin in the mixture, we have investigated a sample where SBR is deuterated. The inspection of the thermal evolution of the scattered intensity has allowed us to discern the 'microscopic effective' T_g of this resin component within the blend. This is compared with the calorimetric results on the same sample. In this way, we can directly compare the 'microscopic' T_g of the resin component as determined by neutron scattering with the 'macroscopic' T_g observed by calorimetry on the whole system. The same kind of study is performed on the pure resin material, as reference.

2 Experimental

Samples were blends of a commercial polylimonene resin (M_n= 374 g/mol and PDI = 2.21) and deuterated SBR (M_n= 56400 g/mol and PDI = 1.40; styrene content: 19%, 1,2-butadiene content: 49.9%) synthesized in the laboratories of Michelin, following a similar procedure as described in Ref. [16].

Differential Scanning Calorimetry (DSC) measurements were carried out on samples of approximately 10 mg placed in aluminum pans using a Q2000 TA instrument. A liquid Nitrogen Cooling System (LNCS) was used with 25 ml/min helium flow rate. Data were acquired during cooling at 3 K/min. Temperature-modulated experiments (MDSC) were performed using a sinusoidal variation of 0.5 K amplitude and 60 s period.

The elastic fixed window scans (EFWS) were carried out by the IN16B backscattering spectrometer at the ILL [17] with a wavelength λ of 6.271Å, covering a scattering vector range $0.2 \leq Q \leq 1.9$Å^{-1}. The energy resolution of IN16B was 0.75μeV. The intensity scattered with energy transfer of 2μeV was also recorded during the heating scans (from 2 to 320 K, with a rate of 2 K/min). This maximum temperature was limited by beam time availability, but is above the glass-transition of the two neat components, being thus enough for the purpose of the current investigation. The blend sample investigated contained a weight fraction of dSBR w_{SBR}=0.80. Experiments on a pure resin sample were conducted as reference.

Whereas the mentioned isotope labelling strongly reduces the incoherent scattering of the deuterated component, it does not completely cancel its contribution, being now of mainly coherent nature. Therefore it is important to determine the Q-dependence of the coherent and incoherent scattering contributions as explained later. This is possible by exploiting polarization analysis. Experiments by the D7 instrument [17] at the ILL allowed accessing the ratio between coherent and incoherent differential scattering cross sections of the blend sample investigated by IN16B. With λ = 4.88 Å, a Q-range from 0.13 to 2.46 Å^{-1} was covered. In addition, we investigated blends with w_{SBR} = 0.50 and w_{SBR} = 0.15, as well as the pure dSBR sample. Experiments were carried out at 300 K. The thicknesses of the samples investigated by neutron scattering were chosen such that a transmission of about 90% was expected. They were filling flat aluminum sample holders and placed at 135° with respect to the incident beam.

3 Results and Discussion

The results on the reversible part of the heat capacity C_p^{rev} corresponding to the neat polymers and the blend with w_{SBR}=0.80 composition are presented in Fig. 1. The glass transition manifests as a jump in this function, and the value of T_g is commonly taken as the inflection point of C_p^{rev}. To give account for the broadening of this transition, the construction illustrated in this figure for the case of the blend results is usually made. For this sample, we deduce a value of T_g = 240 K, with initial and final glass-transition temperature values of T_i = 235 and T_f = 244 K respectively (see the arrows). The temperature-derivative of C_p^{rev} is also included in this figure for the three samples (empty symbols). The glass transition is reflected as a peak in this function, where the position of the maximum corresponds to the inflection point of C_p^{rev} and thus directly gives the value of T_g as usually defined. This function also reflects very clearly the width of the glass transition process and may allow resolving multiple transitions, if present in the sample [18,19]. In Fig. 1 we can observe that the glass transition process in the resin occurs at much higher temperatures than in the dSBR sample. This qualifies the blends of these two systems as dynamically asymmetric mixtures. In addition, we find that the glass-transition process in the resin spans over a much wider temperature interval than in dSBR. This feature can be attributed to the large polydispersity of the resin.

Blends with different compositions were characterized by DSC and the results regarding the location and width of the glass transition are compiled in Fig. 2. As can be seen, the usually invoked Fox equation

$$\frac{1}{T_g} = \frac{\varphi_{SBR}}{T_g^{SBR}} + \frac{(1-\varphi_{SBR})}{T_g^{resin}} \quad (1)$$

where φ_{SBR} is the concentration of the SBR component and T_g^{SBR} and T_g^{resin} the T_g-values of the pure materials, does not describe the concentration dependence of the average glass-transition temperature in these blends. It particularly fails in the range of concentrations in the SBR-poor side. The more flexible equation proposed by Gordon and Taylor [20]:

$$T_g = \frac{(1-\varphi_{SBR})T_g^{resin} + k_{GT}\varphi_{SBR}T_g^{SBR}}{1-\varphi_{SBR}+k_{GT}\varphi_{SBR}} \quad (2)$$

actually describes the results with a much better accuracy. The value deduced for the parameter k_{GT}, that is related with the density and the increment of the isobaric expansivity at T_g of each component, is 2.36.

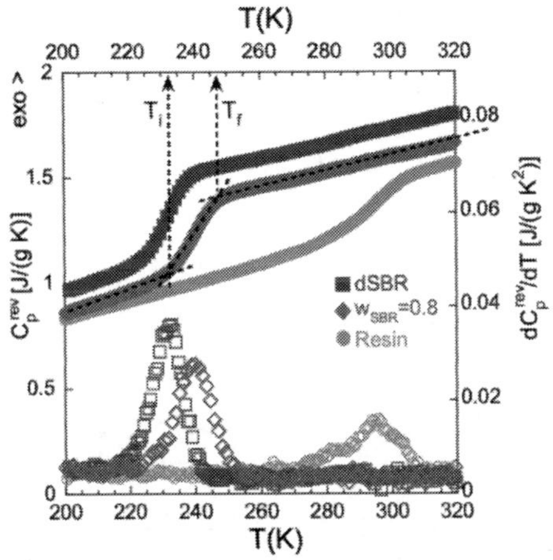

Fig. 1. Results obtained by modulated DSC on the pure components and the blend with 0.80 dSBR weight fraction: reversible part of the specific heat (filled symbols) and its temperature derivative (sign changed, empty symbols). Dotted arrows show, for the blend case, the usual construction to determine the initial and final glass-transition temperature values.

On the other hand, the width of the transition strongly increases with resin concentration. This finding could be attributed to either the persistence of a high degree of dynamic heterogeneity in the mixtures or to the intrinsic pronounced width of the glass-transition process in the resin (or to both factors). We note that at high SBR concentrations –as it is the case on which we are currently interested– the width of the transition is rather small, almost comparable to that observed for the neat dSBR sample.

Now we move to the neutron scattering results. The intensity scattered by a given sample has both, coherent and incoherent contributions [21]. In our investigation we are interested in the incoherent contribution arising mainly from the hydrogens in the protonated component of the blend. Usually this contribution is overwhelming due to the huge incoherent scattering cross section of H, compared with the rest of the scattering cross sections of the nuclei involved in the sample. However, if the composition of the sample is rich in the deuterated component –as it is our case—, since in D the coherent scattering cross section is higher than the incoherent one, the coherent contribution might present non-negligible values in some Q-regions. To determine the Q-range where the intensity scattered is less contaminated by coherent contributions we examined the D7 results. The measured ratio between coherent and incoherent differential scattering cross sections is shown in Fig. 3 for blends of different compositions including the particular one of w_{SBR}=0.80 on which we have focused our study. The figure also shows this magnitude for the pure dSBR blend. For the pure resin –not measured by D7— this ratio is expected to be below 10%. With increasing content of dSBR, the coherent fraction increases. In particular, an important coherent contribution is observed for the blends in the range around $1-1.2$ Å^{-1}, reflecting the structure factor peak associated to the short-range order in these polymeric systems. Even more important, especially for the samples with high dSBR contents, is the pronounced increase of the coherent scattering at low Q-values, due to fluctuations of concentration in the mixtures [6,22]. Thus, if we want to minimize the effects of coherent contributions in our temperature scan results, we have to focus on Q-regions close to the minima: $Q \approx 0.7$ Å^{-1} and above ≈ 1.6 Å^{-1} in the case of the blend with w_{SBR}=0.80 (see Fig. 3).

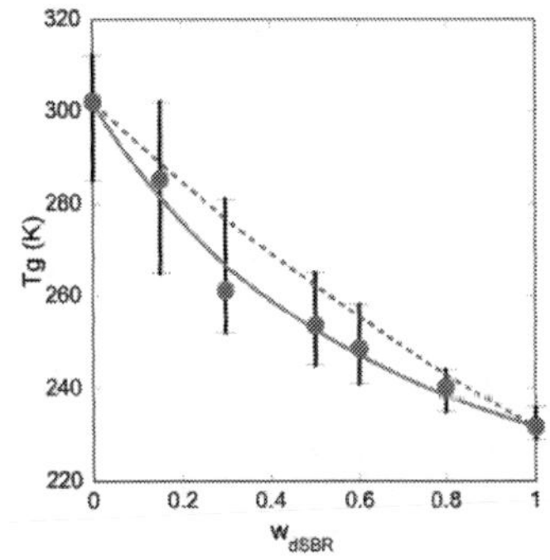

Fig. 2. Composition dependence of the glass-transition temperature T_g (inflection point of the specific heat). Error bars span between the of T_i and T_f determined by DSC. Dashed line shows the Fox prediction, and solid line the Gordon-Taylor description.

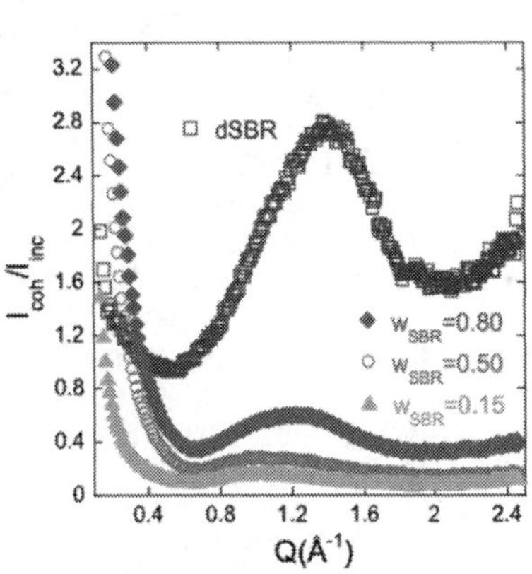

Fig. 3. Ratio between the coherent and incoherent differential scattering cross-sections determined by diffraction with polarization analysis (D7) on the pure dSBR system and blend samples with different compositions.

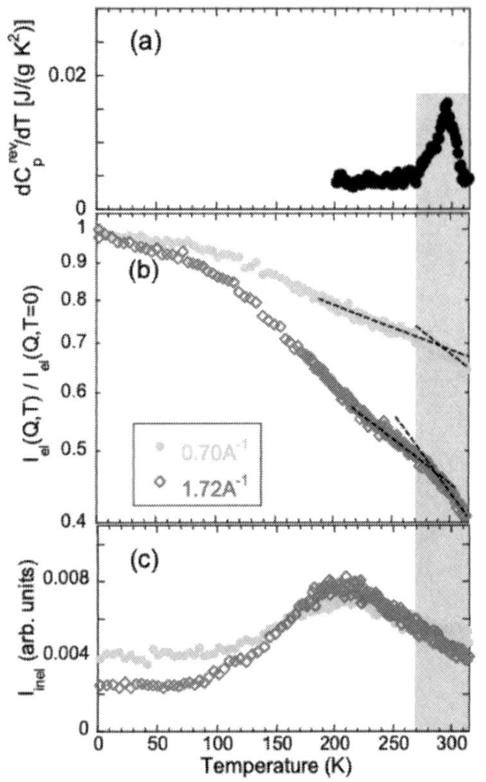

Fig. 4. Temperature dependence of results on the pure resin: (a) temperature derivative of the specific heat; (b) elastic intensity normalized to its extrapolated value at T=0; (c) inelastic intensity at 2μeV. Neutron scattering results correspond to the two Q-values indicated. Red area covers the calorimetric glass-transition region.

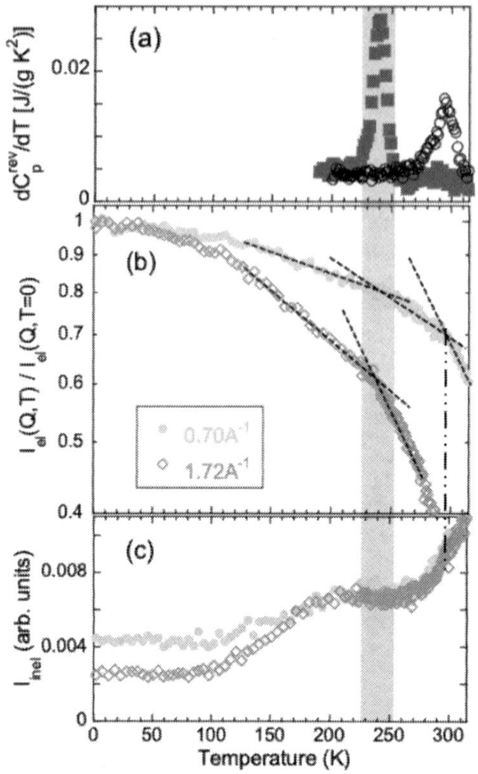

Fig. 5. Temperature dependence of results on the blend with 0.20 resin weight fraction in deuterated SBR: (a) temperature derivative of the specific heat; (b) elastic intensity normalized to its extrapolated value at T=0; (c) inelastic intensity at 2μeV. Neutron scattering results correspond to the two Q-values indicated. Red area covers the calorimetric glass-transition region.

Guided by the D7 results, we have focused our attention on the IN16B data collected at Q = 0.70 Å^{-1} and Q = 1.72 Å^{-1}. They can be seen in Fig. 4 for the case of the pure resin (reference system) and in Fig. 5 for the blend. These results are directly compared with the calorimetric ones, represented in the upper panels of both figures by the derivative of the reversible part of the specific heat. As can be seen in Fig. 4(b) for the neat resin, in the glassy state the elastic intensity shows a decrease that can be attributed to the occurrence of vibrational processes and local motions taking place in a frozen environment, such as methyl group rotations. We note that the characteristic times of these localized processes are expected to be broadly distributed, due to the disorder inherent to the glassy state in addition to the effect of polydispersity in this particular sample. These motions have a clear signature in the inelastic intensity, leading to a pronounced maximum at approx. 200 K. In this temperature range, the average characteristic time of these motions matches the dynamic window explored in these scans. At higher temperatures, we observe an additional abrupt decrease of the elastic intensity for both Q-values investigated. This feature takes place close to the average calorimetric glass transition observed in this material. We note that there is no signature of the glass transition in the inelastic intensity. The dynamic processes leading to the additional decrease of the elastic intensity in this temperature range obviously present associated characteristic times very different from the nanosecond timescale probed by the inelastic scans. We expect that inelastic contributions of the α-relaxation processes (associated to the vitrification phenomenon) with energies ~2μeV would be present at higher temperatures than those here investigated.

In the case of the hydrogens in the resin component *within the blend* –as it is probed by neutrons on the labelled mixture at the two Q-values shown in Fig. 5— we observe that the additional decay of the elastic intensity with respect to the glassy behaviour takes place at a much lower temperature than in the pure resin. In fact, it occurs, within the uncertainties, at the (initial) glass-transition temperature as determined by DSC on the blend sample. This implies that, despite the huge difference in intrinsic mobilities of the two components, the atomic motions of the resin component become 'liquid-like' at the temperature dictated by the fast majority blend component, SBR. Thus, we can say that the minority resin molecules are 'animated' by the surrounding mobile environment imposed by the SBR chains. A similar result has recently been observed on mixtures of SBR and PS-oligomers of 500 g/mol [23].

We note the occurrence of an additional decrease of the elastic intensity in the blend at higher temperatures (approx. 280 K). Actually, it coincides with the calorimetric glass transition of the pure resin. This coincidence could suggest the assignment of this new feature in the elastic scan to the existence of a second glass transition in the blend, attributable to 'dynamically immiscible' resin molecules within the mixture.

However, we observe that this extra-decrease of the elastic intensity corresponding to the resin in the blend occurs simultaneously with an increase of the inelastic intensity of this sample with respect to the pure resin behaviour (compare Fig. 5(c) and 4(c)). This observation suggests that in this temperature range the scattering function of the H-atoms of the resin within the blend has a width resolvable in the IN16B dynamic window, being thus detected as a quasielastic broadening of the signal. This implies an increase of the scattering at the considered energy transfer of 2μeV accompanied by a decrease of the elastic intensity. Thus, the finding of this effect in the neighbourhood of the pure resins' T_g could be just a mere coincidence. Nevertheless, we cannot completely discard that there are still degrees of freedom that were blocked at lower temperatures that become active at the pure resins' T_g. Actually, the apparent Q-insensitivity in the IFWS could also suggest the origin of this high-temperature increase to be due to local rotational motions. This hypothesis is rather speculative, since the peak is out of the range of the temperature scan performed.

4 Conclusions

EFWS provide a very sensitive tool to determine the onset of motions characteristic for the supercooloed liquid state. In the pure resin, we have found that the microscopic fingerprint of the glass transition revealed by the EFWS coincides very well with the glass transition temperature as determined by calorimetry. When blended with a polymer with a much lower glass-transition temperature, the resin shows its 'microscopic' glass transition signature close to the initial glass transition temperature of the whole sample as determined by calorimetry. These experiments provide the microscopic proof of 'animation' effects on the molecular motions of rigid molecules by the surrounding mobile environment imposed by the majority blend component in a mixture of industrial interest. Whether this effect is particular for systems where the slow component molecules are much smaller in size and highly polydisperse than the fast component is a question that shall be elucidated in future works.

We acknowledge the Grant PID2021-123438NB-I00 funded by MCIN/AEI/10.13039/501100011033 and by "ERDF A way of making Europe". We also acknowledge financial support of Eusko Jaurlaritza, code: and IT1566-22, as well as from the IKUR Strategy under the collaboration agreement between Ikerbasque Foundation and the Materials Physics Center on behalf of the Department of Education of the Basque Government.

References

1. J. Colmenero, A. Arbe, Soft Matter **3**, 1474-1485 (2007)
2. A. Zetsche, E.W. Fischer, Acta Polymerica **45**, 168-175 (1991)
3. G. Katana, E.W. Fischer, T. Hack, V. Abetz, F. Kremer, Macromolecules **28**, 2714-2722 (1995)
4. T.P. Lodge, T.C.B. McLeish, Macromolecules **33**, 5278-5284 (2000)
5. S. Shenogin, R. Kant, R.H. Colby, S.K. Kumar, Macromolecules **40**, 5767-5775 (2007)
6. T. Gambino, N. Shafqat, A. Alegría, N. Malicki, S. Dronet, A. Radulescu, K. Nemkovski, A. Arbe, J. Colmenero, Macromolecules **53**, 7020-7160 (2020)
7. T. Gambino, A. Alegría, A. Arbe, J. Colmenero, N. Malicki, S. Dronet, Polymer **187**, 122051 (2020)
8. N. Shafqat, A. Alegría, A. Arbe, N. Malicki, S. Dronet, L. Porcar, J. Colmenero, Macromolecules **55**, 7614-7625 (2022)
9. Y. Miwa, K. Usami, K. Yamamoto, M. Sakaguchi, M. Sakai, S. Shimada, Macromolecules **38**, 2355-2361 (2005)
10. D. Herrera, J.C. Zamora, A. Bello, M. Grimau, E. Laredo, A.J. Muller, T.P. Lodge, Macromolecules **38**, 5109-5117 (2005)
11. S. Arrese-Igor, A. Alegría, J. Colmenero, Macromolecules **43** 6406 (2010)
12. A.C. Genix, A. Arbe, F. Alvarez, J. Colmenero, L. Willner, D. Richter, Phys. Rev. E **72**, 031808 (2007)
13. M. Tyagi, A. Arbe, J. Colmenero, B. Frick, R. Stewart, Macromolecules **39**, 3007-3018 (2006)
14. A.C. Genix, A. Arbe, S. Arrese-Igor, J. Colmenero, D. Richter, B. Frick, P.P. Deen, J. Chem. Phys. **128**, 184901 (2008)
15. C. Lorthioir, A. Alegría, J. Colmenero, Phys. Rev. E **68**, 031805 (2003)
16. A. Bouty, L. Petitjean, J. Chatard, R. Matmour, C. Degrandcourt, R. Schweins, F. Meneau, P. Kwasniewski, F. Boue, M. Couty, J. Jestin, Faraday Discuss. **186**, 325–343 (2016)
17. A. Arbe, M. Appel, J. Colmenero, B. Frick, J. Maiz, L. Mangin-Thro, A. Mendia, J. Ollivier, J.A. Pomposo, N. Shafqat, Institut Laue-Langevin ILL (2019) ; https://doi.ill.fr/10.5291/ILL-DATA.9-11-1901
18. D. Bhowmick, J.A. Pomposo, F. Juranyi, V. García-Sakai, M. Zamponi, Y. Su, A. Arbe, J. Colmenero, Macromolecules **47**, 304-315 (2014)
19. D. Bhowmick, J.A. Pomposo, F. Juranyi, V. García-Sakai, M. Zamponi, Y. Su, A. Arbe, J. Colmenero, Macromolecules **47**, 3005-3016 (2014)
20. M. Gordon, J. S. Taylor, J. Appl. Chem. **2**, 493 -500 (1952)
21. S.W. Lovesey, *Theory of Neutron Scattering from Condensed Matter*, Clarendon Press: Oxford (1997)

22. J.S. Higgins, H.C. Benoit, *Polymers and Neutron Scattering*, University Press: Oxford (1997)
23. N. Shafqat, A. Alegría, N. Malicki, S. Dronet, F. Natali, L. Mangin-Thro, L. Porcar, A. Arbe, J. Colmenero (in progress)

Intracellular molecular dynamics studied by neutron scattering

Giuseppe Zaccai[1,2*], *Dominique* Madern[1] and *Bruno* Franzetti[1]

[1]Univ. Grenoble Alpes, CNRS, CEA, IBS, F-38000 Grenoble, France
[2]ILL, F-38042 Grenoble, France

Abstract. Incoherent neutron scattering experiments have produced important insights into intracellular molecular dynamics *in vivo*. Selected results highlight the role of water dynamics in cancer and brain cells, as well as cellular adaptation through the evolution of appropriate molecular dynamics, in order to respond to environmental challenges.

1 Introduction: dynamics, why?

Dynamics defines biological molecules through the forces that maintain their structures and allow appropriate functional motions. These forces have been studied by calorimetry, which revealed the thermodynamics of structural stability and importance of solvent interactions in the folding of a protein to achieve its active architecture [1]. They form the basis of force fields in molecular dynamics simulations [2].

The study of proteins from extremophiles, organisms that thrive under, for example, extreme temperature, pressure, or salinity, revealed the evolutionary adaptation of dynamics to maintain structure and activity under violent physical-chemical conditions. In Fig. 1, is shown the example of malate dehydrogenase, an enzyme of the Krebs cycle, from a halophile living in high salinity in the Dead Sea [3] and a hyperthermophile living at temperatures up to 95°C, in oil fields [4].

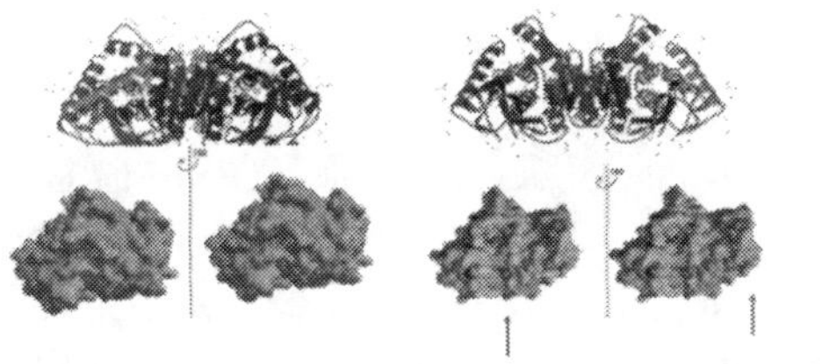

Fig. 1. Ribbon and surface structural representations of malate dehydrogenase from *Haloarcula marismortui* (on the Left) and *Archeoglobus fulgidus* (on the Right).

Note how, in spite of non-identical amino-acid composition, the two enzymes have maintained a similar tertiary structure throughout evolution. The adaptation to their respective physiological environments is clearly illustrated in the surface representation of the exploded monomers. The surface of the halophile enzyme is predominantly red, corresponding to an enrichment in negatively charged aspartic and glutamic acid residues, involved in stabilising the structure through solvent interactions with salt ions. In contrast, the surface of the hyperthermophile enzyme displays both positive (in blue) and negative charges (red), placed in such a way that in the formation of the dimer, by 180° rotation of one monomer over the other, the places indicated by red and blue arrows overlap for strong +/- ionic bonds to maintain stability at high temperature.

2 QENS and neutron incoherent elastic window scattering temperature scans

Quasi-elastic neutron scattering (QENS) is a classical approach to the study of molecular dynamics diffusive processes. The method has been applied successfully to measurements of intracellular molecular dynamics on the pico- to nanosecond time scale on back-scattering spectrometers at ILL, Juelich, ISIS and ORNL [5-11]. The Fourier transform of a diffusive process is a Lorentzian function, from which can be derived a diffusion coefficient and relaxation time. Cell contents being heterogeneous, the experimental data are the sum of scattering from several diffusive processes. A way to tackle the complexity is clearly explained in the pioneering works of Marques et al. on intracellular water dynamics in cancer cells [8, 10]. Three components were considered: (i) very slow motions from the largest organelles and cyto-skeleton and global motions of the macromolecules—represented by a Delta function (elastic line); (ii) slow diffusion of the intracellular water molecules (reorientations mediated by hydrogen bonds) for both the cytoplasmic and hydration water—defined by two Lorentzian functions; (iii) internal localized motions comprising conformational rearrangements from the biomolecules and lipid motions, such as the lateral diffusion of phospholipids and the movement of cholesterol within the membranes—ascribed to a broader Lorentzian.

The landmark 1989 paper by Doster et al. [12], introducing the neutron incoherent scattering elastic window temperature scan on IN13 [15] to study

* Corresponding author: zaccai@ill.fr

myoglobin, opened up new opportunities in experimental molecular dynamics in biology. A few years after, a similar shaped plot was obtained for bacteriorhodopsin in its natural membrane environment, suggesting it represented a general property of macromolecular dynamics [13]. The scan from [12] is shown in Fig. 2, together with data from a study of myoglobin in a trehalose glass [14].

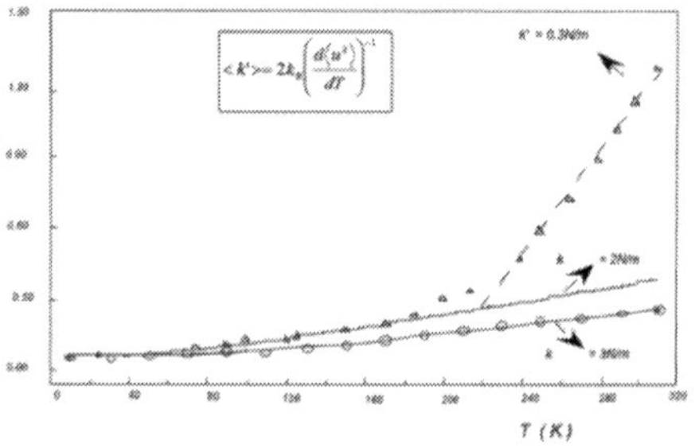

Fig. 2. Elastic window temperature scans on IN13 (8 μeV resolution) at the Institut Laue Langevin (ILL) [15] of myoglobin in a hydrated powder (triangles) from [12], and in a trehalose glass (circles) from [14].

The scans are plots of the mean square displacement (msd), $<u^2>$, calculated from the elastic scattering (see [12-14]), as a function of absolute temperature. The hydrated powder msd values follow a straight line until a break at about $T = 200$ K, where the structure *softens* and the msd climb steeply with temperature. For myoglobin trapped in a trehalose glass, the initial straight line subsists until just below 320 K, the highest temperature measured. A quasi-harmonic interpretation of the scans has been proposed by Zaccai [16]. The linear dependence of the msd at low temperature, extrapolating to zero at $T = 0$, is indicative of a harmonic regime, described by the equation in the inset in Fig. 2, where k' is a force constant. The quasi-harmonic analysis applies the harmonic equation to the portion of the plot above the break at ~200 K, with $<k'>$, now, a mean effective force constant. As shown in Fig. 3, $<k'>$ can be related to free energy configurations (such as the *double-well*) that have been used by authors to interpret elastic window temperature scans [12, 17, 18].

In qualitative terms, the msd informs on the *flexibility* of the macromolecular structure and the mean effective force constant informs on its *resilience* [16-18].

The $<k'>$ values derived from the different portions of the scans are shown in Fig. 2. Interestingly, they are of the same order as values in our macroscopic world, like the stiffness of a new tennis ball, ~1 N/m, or that of a soft rubber ball, ~0.1 N/m.

3 The length-time window and incoherent scattering from H_2O

The energy resolution and Q range of a neutron spectrometer delimit the length-time window seen by elastic incoherent neutron scattering [29]. The IN13 Italian-French back-scattering spectrometer at ILL is specifically dedicated to biology [15]. In particular, the 8 μeV resolution and large Q range delimit a window of ~ 1 Å / 100 ps. Neutron scattering is strongly dominated by the incoherent cross-section of H atoms. On the one hand, atoms bound to macromolecular groups, in proteins, for example, will move well inside the window. The square of the diffusion length of H_2O at 20 °C, on the other hand, is 25 Å^2, well outside the IN13 window. An elastic window scan on IN13, therefore, would filter out most of the incoherent scattering from bulk water. This was an important realisation. Formerly, neutron experiments on biological molecules were performed in D_2O, to avoid the signal from being swamped by the incoherent scattering of H_2O. There is an isotope effect, however, and macromolecular dynamics is significantly different in H_2O and D_2O as measured, for example, in the halophilic malate dehydrogenase [19]. The IN13 length/time window permitted to extend the studies of biological macromolecules to solutions in H_2O—most relevant to physiological conditions.

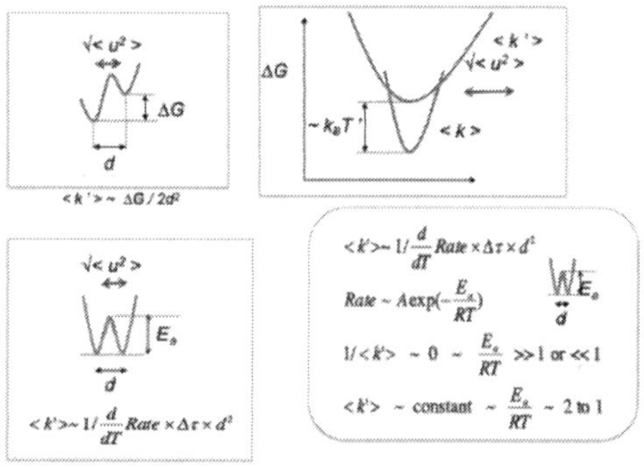

Fig. 3. The two-well model (top left) is from [12], the top right one is from [17]. The model on the bottom, in which two wells of the same free energy are separated by an activation energy barrier, would be appropriate, for example, to methyl group rotations

4 Live cell samples

The elastic window neutron scattering method is not site-specific but provides mean values over the motions of all H atoms in the sample. From this point of view a protein is already a complex entity in which the mean is over H atoms bound to different groups moving in different environments, as well as in strongly associated water molecules. It was obviously interesting to go one step further and to examine live cells. But would the measured mean values (msd, $<k'>$) contain useful information for biology?

In an exploratory experiment, neutron scattering from live cells of bacteria adapted to life at different temperatures was measured on IN13 [20]. In order to avoid freezing or over-heating the samples, the temperature range of the scans was adjusted to between a few degrees C to below 40 °C. The msd *versus* T plots for four bacterial species are shown in Fig. 4.

The plot for the psychrophile was particularly encouraging that the scans gave meaningful information. It revealed a break in the line above 20 °C from a softening of the macromolecular structures due to denaturation at this *high* temperature from the bacterium's view-point!

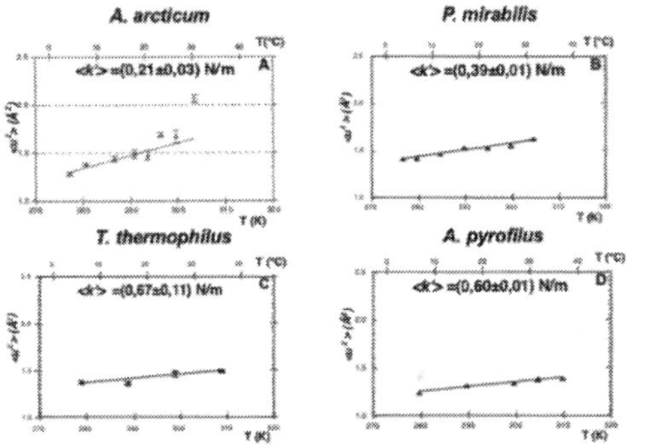

Fig. 4. From [20]. Elastic window temperature scans of *Aquaspirillum arcticum* (a psychrophile, adapted to 4 °C in glacial waters), *Proteus mirabilis* (a mesophile adapted to body temperature), *Thermus thermophilus* (a thermophile, adapted to 65 °C) and a hyperthermophile, *Aquifex pyrofilus* (85 °C).

Cell content calculations proposed that the signal was dominated by the protein fraction in the cell (the proteome). The increasing macromolecular resilience with physiological temperature, suggested the dominance of the enthalpic term in the mean free energy of protein stabilisation, *i.e.*, stronger internal bonds maintained the folded structures to higher temperature. The question then arises why is psychrophile macromolecular resilience low, when a mesophile resilience, for example, would surely maintain structural stability above freezing? In order to achieve functional activity a macromolecule not only needs to be stable but also to have the appropriate *flexibility*. Both resilience and msd have to be adapted to the temperature. At psychrophile temperatures a mesophile resilience would lead to low msd values, incompatible with activity, and similarly for the mesophile if the resilience were as high as that of the thermophiles.

5 Live cells

The elastic window neutron QENS and elastic window neutron scattering experiments on live cells (partially reviewed in [21]) include studies on water mobility in Dead Sea archaea, in which a slow component was discovered, hypothetically associated with ion-binding structures [5], in *E. coli*, indicating about 90% of cytoplasmic water behaves similarly to bulk water, 10% being associated in macromolecular hydration shells [6], in *Shewanella oneidensis* at ambient and high pressure [7], in cancer [8, 10, 11, 15] and brain cells [23], as well as on intracellular macromolecular dynamics in red-blood cells and archaea (see below).

In their extensive study of intracellular water as a drug target in cancer cells [8, 10] and in the transition to malignancy [11], Marques et al. observed an increased intracellular plasticity. The combined results corroborated the key role of intracellular water dynamics in cell proliferation and differentiation. With respect to drug action, perturbation of the water's dynamical profile led to inhibition of cell growth and death. In the case of normal-to-malignant transformation, results suggested that water dynamics were closely associated with the onset of carcinogenesis and invasiveness.

Intracellular water dynamics in breast cancer cells, untreated and treated with the drug paclitaxel, was compared on TOSCA at ISIS, UK [22]. Paclitaxel is known to 'freeze' the cell-line in mitosis and, in this work also, the data supported the hypothesis that water mobility plays an important role in the cell cycle and response to chemotherapy drugs.

Diffusion Magnetic Resonance Imagery (dMRI) is used effectively to diagnose ischemic strokes and brain tumours by measuring water mobility in the brain. A combined dMRI/QENS quantified water dynamics in brain cells from the μm/ms scale of dMRI down to the atomic Å/ps scale of QENS [23]—important data for the understanding of water diffusion in the brain in order to improve MD simulations and assist medical diagnostics.

Comparative experiments on red blood cells of different animals revealed an adaptation of proteome molecular dynamics to body temperature. Remarkable because, in contrast to the psychrophile, mesophile and thermophile bacteria, the temperature range is small (~10 K) [24].

Studies of halophilic archaea showed how proteome molecular dynamics responded to temperature [25] and salt stress [26] in the organisms. The elastic window temperature scans turned out to be good indicators of protein functionality and cell viability.

Neutron scattering experiments at high hydrostatic pressure (HHP) on deep ocean archaea are particularly challenging. First results from elastic window scattering revealed molecular dynamics response mechanisms to HHP of hydration water and the proteome [27].

It is appropriate to conclude with a molecular dynamics study of *Mycoplasma genitalium* [28], which opens the way towards whole-cell simulations to link with experiment. QENS and elastic window neutron scattering data are already here to contribute when the times comes!

This work was supported by the French National Research Agency in the framework of the *Investissements d'Avenir* program (ANR-15-IDEX-02) through the funding of the "Origin of Life" project of the Univ. Grenoble-Alpes, and AlloSpace, ANR-21-CE44-0034-01). IBS acknowledges integration into the Interdisciplinary Research Institute of Grenoble (IRIG, CEA).

References

1. N. R. Zaccai, I. Serdyuk, J. Zaccai, Methods in Molecular Biophysics: Structure, Dynamics, Function for Biology and Medicine (Cambridge University Press, UK, 2017)
2. L. Meinhold, J.C. Smith, A. Kitao, A.H. Zewail., Proc. Natl. Acad. Sci. U. S. A. **104** (44), 17261 (2007)

3. A. Irimia, C. Ebel, D. Madern, S. B. Richard, L. W. Cosenza, G. Zaccai, F.M. Vellieux, J Mol Biol. **326** (3), 859 (2003)
4. A. Irimia, F.M. Vellieux, D. Madern, G. Zaccai, A. Karshikoff, G. Tibbelin, R. Ladenstein, T. Lien, N. K. Birkeland, J Mol Biol. **335** (1), 343 (2004)
5. M. Tehei, B. Franzetti, K. Wood, F. Gabel, E. Fabiani, M. M. Jasnin, M. Zamponi, D. Oesterhelt, G. Zaccai, M. Ginzburg, B.Z. Ginzburg, Proc Natl Acad Sci U S A, **104**, 766 (2007)
6. M. Jasnin, M. Moulin, M. Haertlein, G. Zaccai, M. Tehei, EMBO Rep, **9**, 543 (2008)
7. F. Foglia, R. Hazael, G. G. Simeoni, M-S Appavou, M. Moulin, M. Haertlein, V. T. Forsyth, T. Seydel, I. Daniel, F. Meersman, P. F. McMillan, Sci. Rep. **6**,18862 (2016)
8. M. P. M. Marques, A. L. M. Batista de Carvalho, V. G. Sakai, L. Hatter, and L.A. E. Batista de Carvalho, Phys. Chem. Chem. Phys. **19** (4), 2702 (2017)
9. E. Mamontov, Sci. Rep. **8**(1), 4190 (2018)
10. M. P. M. Marques, A. L. M. Batista de Carvalho, A. P. Mamede, S. Rudic´, A.Dopplapudi, V. Garcia Sakai, and L. A. E. Batista de Carvalho, Int. Rev. Phys. Chem. **39** (1), 67 (2020)
11. M. P. M. Marques, A. L. M. Batista de Carvalho, A. P. Mamede, A. Dopplapudi, V. Garcıa Sakai, L. A. E. Batista de Carvalho, Struct. Dyn. **7**, 054701 (2020)
12. W. Doster, S. Cusack, W. Petry, Nature **337**, 754 (1989)
13. M. Ferrand, A. J. Dianoux, W. Petry, G. Zaccai, Proc Natl Acad Sci U S A **90**, 9668 (1993)
14. L. Cordone, M. Ferrand, E. Vitrano, G. Zaccai, Biophys. J. **76**, 1043 (1999)
15. https://www.ill.eu/users/instruments/instruments-list/in13/description/instrument-layout
16. G. Zaccai, Science **288**, 1604 (2000)
17. D. J. Bicout, G. Zaccai, Biophys. J. **80**, 1115 (2001)
18. Y.Miao, L. Hong, Z. Yi, J.C. Smith. Zaccai, Eur. Phys. J. E. **36** (7), 72 (2013)
19. M. Tehei, D. Madern, C. Pfister, G. Zaccai, Proc. Natl. Acád. Sci. U S A **98**, 14356 (2001)
20. M. Tehei, B. Franzetti, D. Madern, M. Ginzburg, B.Z. Ginzburg, M.T. Giudici-Orticoni, M. Bruschi, G. Zaccai, EMBO Rep. **5**, 66 (2004)
21. G. Zaccai, BBA - General Subjects **1864**, 129475 (2020)
22. M.L. Martins, A.B. Dinitzen, E. Mamontov, S. Rudić, J.E.M. Pereira, R. Hartmann-Petersen, Kenneth W. Herwig, Heloisa N. Bordallo, Sci. Rep. **9** (1), 8704 (2019)
23. F. Natali, C. Dolce, J. Peters, C. Stelletta, B. Demé, J. Ollivier, M. Boehm, G. Leduc, I. Piazza, A. Cupane, E. L. Barbier, J. R. Soc. Interface **16** (157), 20190186 (2019)
24. A. M. Stadler, C J Garvey, A. Bocahut, S. Sacquin-Mora, I. Digel, G J Schneider, F. Natali, G M Artmann, G Zaccai, European Biophysics Journal with Biophysics Letters **40**, 157 (2011)
25. V. Marty, M. Jasnin, E. Fabiani, P. Vauclare, F. Gabel, M. Trapp, J. Peters, G. Zaccai, B. Franzetti, J. Royal Soc. Interface **10** (82), 20130003 (2013)
26. P. Vauclare, V. Marty, E. Fabiani, N. Martinez, M. Jasnin, F. Gabel, J. Peters, G. Zaccai, B. Franzetti, Extremophiles. **19** (6), 1099 (2015)
27. N. Martinez, G. Michoud, A. Cario, J. Ollivier, B. Franzetti, M. Jebbar, P. Oger, J. Peters, Sci. Rep. **6**, 32816 (2016)
28. I. Yu, T. Mori, T. Ando, R. Harada, J. Jung, Y. Sugitai, et al., eLife **5**, e:19274 (2016)
29. G. Zaccai, J. Non-Cryst. Solids **357**, 615 (2011)

https://doi.org/10.1051/epjconf/202227201010

Controlling terahertz sound propagation: some preliminary Inelastic X-Ray Scattering result

Scott T. Lynch[1], *Alessio* De Francesco[2,*], *Luisa* Scaccia[3], and *Alessandro* Cunsolo[1]

[1]Department of Physics, University of Wisconsin at Madison, 1150 University Avenue, Madison, WI, USA
[2]CNR-IOM & INSIDE@ILL c/o Operative Group in Grenoble (OGG), F-38042 and Institut Laue Langevin, Grenoble, France
[3]Dipartimento di Economia e Diritto, Università di Macerata, Via Crescimbeni 20, 62100 Macerata, Italy

Abstract. The control of sound propagation in materials via the design of their elastic properties is an exciting task at the forefront of Condensed Matter. It becomes especially compelling at terahertz frequencies, where phonons are the primary conveyors of heat flow. Despite the increasing focus on this goal, this field of research is still in its infancy. To achieve a few advances in this field, we performed several Inelastic X-Ray Scattering (IXS) measurements on elementary systems as dilute suspensions of nanoparticles (NPs) in liquids. We found that nanoparticles can effectively impact the sound propagation of the hosting liquid. We also explored the possibility of shaping terahertz sound propagation in a liquid upon confinement on quasi-unidimensional cavities. These results are here reviewed and discussed, and future research directions are finally outlined.

1 Introduction

The control of sound propagation through the design of a material's structure is a long-term objective of a relevant portion of Condensed Matter physicists. In particular, achieving this goal at terahertz frequencies, where phonons are the leading heat carriers in insulators, will pave new avenues for developing novel thermal devices [1]. Furthermore, it will promote a transformation of the core activity of researchers: from the mere observation and formalization of a physical phenomenon - sound propagation - to its active steering via the engineering of novel artificial materials. Especially promising appears the capability of structuring nanoscale structures to function as the so-called phononic crystals (PCs). These devices are ideal candidates for implementing acoustic manipulation thanks to their mass and elastic modulus distributions with periodicity suited to impact the propagation of sound waves of specific frequencies, slowing it down, reflecting, or even stopping them [2]. Overall, PCs proved to be critical to the development of novel thermal devices, such as thermal diodes [3], thermoelectrics [4], and thermocrystals [5], and applications as relevant as energy harvesting [6], and management of thermal properties [7].

A prerogative of these devices is the ability to forbid acoustic propagation in some frequency bands, customarily referred to as phononic gaps, representing the mechanical equivalent of electronic and photonic bandgaps in semiconductors and photonic crystals, respectively. While in photonic crystals, band gaps originate from periodic variations of the refractive index, in PCs, they stem from the periodic modulation of density and elastic modulus. Inelastic X-Ray (IXS) [8, 9] and Neutron Scattering (INS) [10, 11] are the best-suited techniques to characterize the propagation of terahertz sound mode, via their direct access to the spectrum of density fluctuations in disordered systems or the phonon spectrum from crystalline materials. An alternative pathway to control terahertz acoustic propagation is to reduce acoustic lifetime by controlling the amount of disorder in the system. To understand this point, one may recognize that at sufficiently short distances and timescales - as well as at low temperatures – the dominant contribution to sound damping derives from either: 1) the inherent disorder of the liquid structure, *i.e.* the coexistence of a finite distribution of first neighboring atomic distances, and 2) the orientational component of the disorder, arising from the random orientation of locally ordered microdomains, similar to that of polycrystalline materials [12, 13]. The quasi-continuous distribution of local lattice parameters prevents density fluctuations from propagating with a well-defined sound velocity but rather with a range of propagation velocities; hence, the dominant sound mode spreads over a continuous superposition of eigenmodes. The mutual interference between these modes substantially damps acoustic propagation, causing a visible broadening of the collective excitations in the measured inelastic spectrum. These arguments suggest that the structural disorder impacts sound propagation, highlighting the possibility of manipulating the acoustic properties of a fluid by including randomly arranged structural heterogeneities. Indeed, the immersion of nanoparticles can substantially affect the fluid ability to support terahertz collective modes propagation [14]. Despite the undoubted scientific relevance, *IXS* and *INS* studies on these systems are still sporadic, partially owing to the challenging interpretation of the often unstructured

*e-mail: defrance@ill.fr

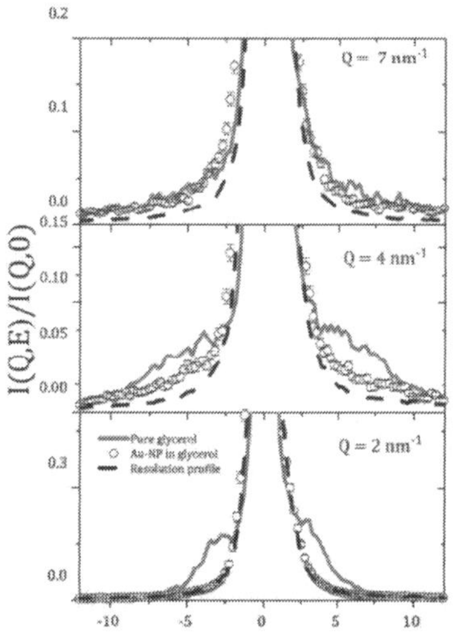

Figure 1. Selected *IXS* spectral shapes measured on a dilute suspension of Au nanorods in water at the Q values indicated. Data are from Ref. 14. Experimental spectra (open circles) are compared with their counterparts in the pure solvent (grey line) and the resolution function. All spectral profiles are normalized to their maxima for the sake of comparison.

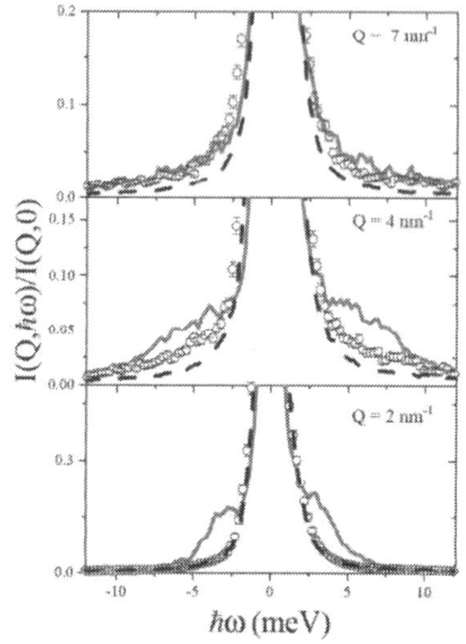

Figure 2. As in figure 1, but referring to a gold nanoparticle (AuNP) suspension in glycerol. Data are again from Ref. 14.

spectral profile, especially considering the lack of firm theoretical predictions of the spectral shape of complex materials. In several recent works, we showed how some of these inherent difficulties could be successfully addressed by using Bayesian inferential methods [15–17] enabling hypothesis tests, which might establish *e.g.* the number of inelastic modes most likely to appear in the spectrum, or even the most plausible analytical function describing measured lineshapes on a sound probabilistic basis. This paper briefly reviews some relevant results emerging from these studies.

2 The damping effect of nanoparticles in immersion

In a colloidal suspension, the sound damping mechanism strongly depends on the wavevector of the excited acoustic waves [14, 18–22] which, in a scattering experiment, coincides with the exchanged wavevector Q. At Q values well beyond the hydrodynamic limit, yet smaller than those probed by *IXS*, i.e., in the so-called multiple scattering regime, immersed colloids behave as point scattering sources for the density waves impinging on them. The interference between repeatedly scattered waves causes mutual dephasing, which substantially reduces the lifetime of the dominant acoustic excitations. Although a theoretical description is available to predict the behavior of collective modes when acoustic wavelength $2\pi/Q$ vastly exceeds the colloid diameter, much less is known about the opposite regime, which best describes the cases investigated in all *IXS* measurements. In these situations, it is reasonable to anticipate that elastic heterogeneity plays a significant role. Indeed, the mismatch between the elastic constants of the immersed colloids and the hosting medium originates sound speed fluctuations, which ultimately reduce the coherence of the acoustic mode, shortening its lifetime. A more subtle effect of this heterogeneity arises from the different expansion coefficients of the colloid and the hosting fluid. When a compression-rarefaction (acoustic) wave propagates at the colloid interface, the unmatched expansions of the colloid and the surrounding fluid cause interfacial shear stresses which enhance the system's resistance to sound propagation. Several of our previous measurements [14, 21, 23] suggested that one can significantly enhance sound damping via the immersion of even sparse amounts of nanoparticles (NPs). Indeed, this strategy visibly changes the high-frequency acoustic damping probed, e.g., by *IXS* methods [8, 9]. This damping effect can make barely visible inelastic features arising from sound propagation, as shown by figures, 1 and 2, which respectively refer to the case of NP suspensions in water and glycerol. In both plots, the *IXS* spectrum of the suspension (open circles) is compared to the one of the pure solvent (thick grey line) and the instrumental resolution profile (dot-dashed line). For comparison, all spectral profiles are normalized to the respective maxima. It readily appears that the side shoulders present in the lineshape of the pure liquid medium disappear upon NP immersion.

3 the Bayesian modeling of the lineshape

At this stage, the cause of the inelastic shoulders' disappearance in the suspension spectra, as evidenced by figures. 1 and 2 is still unclear. [15, 16]. In particular, it is

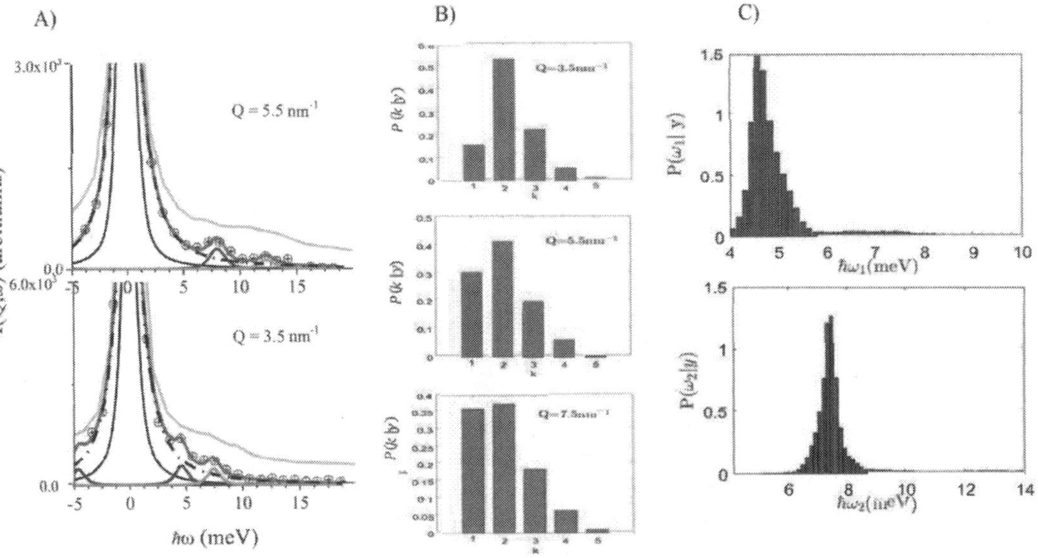

Figure 3. Panel A: the *IXS* spectra from a dilute suspension of Au nanoparticles in water measured at two representative Q's are displayed along with their best fit lineshape and individual spectral components. Panel B: the posterior distribution drawn by the Bayesian inferential algorithm (see text) for the number of inelastic components contributing to the spectral shape is reported for three Q values. Panel C: the posteriors of the inelastic shift of the two phonon modes are displayed as obtained from the Bayesian lineshape analysis of the $Q = 3.5$ nm^{-1} spectrum.

still questionable whether this effect is due to the increase in the damping (width) of inelastic mode, its attenuation (relative amplitude decrease), or the combined effects. Detailed lineshape modeling is the only option to unravel this query properly. A difficulty hampering this endeavor is the lack of a firm analytical prediction for the lineshape in the dynamic ($Q\omega$) range of interest which suggests the use of suitable phenomenological *ansatz*. The latter entails, for instance, the choice of the most plausible number of spectral modes, their shape, their possible coupling with multiple timescale relaxation phenomena, and so forth. Within the approximation eventually adopted, the Q-dependence of shape parameters is determined *a posteriori* only by best-fitting the model to the spectra measured at various Q's. This result is usually achieved through a conventional chi-square minimization routine, whose outcome strongly depends on the *ansatz* adopted. Also, the hybrid, solid-liquid character of a liquid suspension adds further complexity to the interpretation of the spectral shape. To sort out this puzzling scenario, a minimal requirement is to limit the inherent bias of the adopted modeling while opting for an evidence-based analysis of experimental results. The strategy adopted should clarify:

(1) how accurately competitive models account for the measured scattering profiles; (2) how investigators should cope with the opposite risks of "confirmation bias" and data over-interpretation (3) how to circumvent common shortcomings of chi square-based approaches as the convergence to local, rather than global, minima, and, most importantly, (4) how to implement minimally invasive external constraints without wasting information from prior knowledge (if any) of the problem at hand, and (5) how to efficiently handle the limited statistical accuracy of experimental results by enhancing the inferential power of data analysis [24].

Bayesian inferential methods successfully address all these concerns. Furthermore, when applied to the analysis of spectroscopic results, these methods can enable probabilistic hypothesis tests involving, for instance, the number of inelastic modes most likely to appear in the spectrum or the number of relaxation phenomena coupling with them. These unknown numbers can be treated as parameters in the inference procedure. Even more interesting is the possibility of submitting to a hypothesis test the number of relaxations associated with the observed dynamics and the analytic expression that best describes the available data. A very instructive example, not discussed in this short review because it refers to a time scale different from the one tackled here, is provided in Ref. [25]. The inferential analysis performed in this work estimates the number of relaxation channels affecting the time correlation function of an NP aqueous solution, also unraveling the physical nature of such decay channels. Of course, the choice of the model involves a narrow range of options, in this case, limited to combinations of simple exponentials, stretched exponentials, mixtures of stretched and unstretched exponentials, or a simple Kohlrausch–Williams–Watts function [26]. Bayesian methods have already been successfully exploited to extract accurate structure factor amplitudes from a powder diffraction pattern with strongly overlapping Bragg peaks [27]. In other studies [14, 17, 22, 28], they were employed to determine how many excitations are present in an inelastic neutron or X-Ray scattering spectrum. Another application is the determination of the number of diffusive processes affecting the density correlation function in the frequency [29, 30] or in the time domain [25]. The inferential method uses a numerical algorithm based on a Markov Chain Monte Carlo (MCMC) [31] routine with reversible jump (RJ) [32] steps. This numerical routine eventually delivers the joint posterior probability distribution, or simply the posterior, of each model parameter, which represents the best guess on the parameter probability distribution given the experimental evidence and the *a priori* knowledge of the physical problem at hand. The maximum of such a posterior can be naturally interpreted as the most plausible, or best-fitting, value of the parameter involved, provided the posterior profile is sharply peaked, well-shaped, and unimodal. The posterior shape bears direct insight into the uncertainty associated with the best-fitting value. Further details on the Bayesian approach are given in Refs. [17, 25, 33, 34]. The typical outcome of this inferential analysis, when applied to the modeling of the *IXS* spectra from a nanoparticle suspension, is summarized by figure 3, which displays results discussed in Ref. 14. In Panel A of such a figure, we compare the *IXS* spectra from an aqueous suspension of gold NPs with the corresponding best-fitting model lineshapes and their individual spectral components. Panel B shows that the relative probability of the most plausible model option ($k = 2$) decreases upon Q increase, while the probability of the alternative model option ($k = 1$) instead increases. In Ref. 14, the two dominant modes corresponding to the most plausible model option are assigned to the longitudinal (LA) and transverse acoustic (TA) phonons propagating in the Au NP interior. This interpretation primarily rests on two main observations: 1) the maximum values of the dispersion curves of the modes - roughly attained at the edge of the first Brillouin zone - are consistent with the lower and higher frequency maxima of the Au phonon Density of States. Furthermore, the LA dispersion is compatible with the one reported in the literature for liquid gold [35] and 2) The width of these phonon excitations is far narrower than the typical inelastic linewidth of a generic liquid, particularly the one of the hosting liquid (water). In summary, figure 3 indicates that the intra-NP phonons are the only inelastic excitations identifiable by the model in the measured spectral shape of the suspension, despite the NPs being present in it in a very sparse amount (about 0.1 % in volume). This evidence unveils the nearly complete vanishing of collective modes in the hosting liquid. Again, it seems still unclear the phenomenology underlying this disappearance. The results of the Bayesian analysis of our previous IXS measurements [36, 37] reported in figure 4 shed light on this aspect. In particular, Panels A and B of figure 4 display respectively the Q-dependence of the acoustic damping Γ and the ratio A_{in}/A_{tot} between the inelastic component and the total amplitude of the spectrum. Open circles and dots refer to a dilute Au NP suspension in glycerol and to pure glycerol respectively. Panels A and B shows that for $Q \leq 2.5$ nm^{-1} the sound mode in the suspension has smaller amplitude, while for 2.5 nm$^{-1} \leq Q \leq 4.5$ nm^{-1} it has larger damping. A quantitative assessment of the relative lifetime of the longitudinal acoustic mode came from a comparison between its dominant frequency Ω and damping coefficient Γ via the ratio $R = \Omega/\Gamma$. From a physical point of view, the parameter R measures the oscillations experienced by the sound wave before being completely damped off. In Panel C of figure 4, this ratio is reported as a function of Q as obtained from the lineshape modeling of our *IXS* measurements in Ref. 37 on aqueous suspensions of vitreous silica (v-SiO_2) NPs of various sizes and concentrations, either neutral, positively, or negatively charged. The trends shown suggest that the immersion of v-SiO_2-NPs causes the longitudinal mode of the liquid to approach the critical damping condition $\Omega_2/\Gamma_2 = 1$, i.e., visibly reduces its relative lifetime. However, based on the plot, it seems still unclear if and to what extent this damping effect depends on the v-SiO_2 NP characteristics or their relative concentration.

4 New excitations in the dynamics of a suspension

In a series of *IXS* measurements on a dilute (< 1 % volume concentration) suspension of Au NP in glycerol [21] we gain evidence for a low-frequency inelastic mode never observed in the previous investigations on pure glycerol [38–41]. The complete characterization of this spectral mode largely owes to the discussed Bayesian inferential approach, as its presence was often partially hidden by the Lorentzian wings of the instrumental resolution profile. Two relevant aspects of this additional mode, i.e., its spectral fingerprints and dispersive behavior, are illustrated in the two plots of figure 5. This additional spectral feature was, of course, connected with the immersed NPs, and resembled the one previously reported by Brillouin Light Scattering investigations of the hydrodynamic spectrum from colloidal suspensions [18–20]. We assigned this spectral mode to Stoneley waves [42] propagating on the external surface of immersed NPs. Stoneley waves are surface vibrations that primarily propagate at the interface between a liquid and a solid medium as their amplitude rapidly vanishes away from such a boundary. Its propagation velocity is lower than the sound velocities of the solid and liquid media.

As illustrated in figure 5, this low-frequency mode (red squares) coexists with the longitudinal mode of glycerol (open circles) and exhibits a distinctive dispersion behavior characterized by a slow linear Q-increase over an impressively large range of distances $\approx Q^{-1}$, which extends down to mesoscopic scales, while showing no evidence for the Q oscillations typically arising from the interaction with the local arrangement of the liquid. Most importantly, its propagation speed, derived from the slope of its linear Q-trend, is about 520 m/s, i.e., much lower than the sound velocities of glycerol and gold, as to be expected for a Stoneley wave. The cascade of *IXS* spectra in the left panel of figure 5 shows that the spectral contribution of

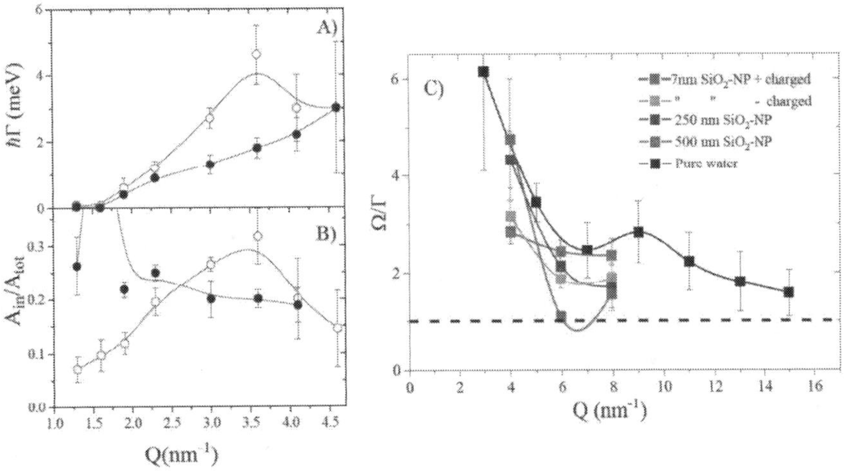

Figure 4. Panel A: The damping coefficient of the collective mode in the spectrum of an Au-NP suspension in glycerol (open circles) is compared with its counterpart in pure glycerol (black dots) as derived from Ref. 36. Solid lines through the data are spline included for visualization purposes. Panel B: The relative amplitude of the inelastic components of the *IXS* spectra, referring to the same measurement and with the same symbols as in panel A. Panel C: The ratio Ω/Γ representing the relative lifetime of the inelastic mode is reported as a function of *Q*, as obtained from the *IXS* measurements on amorphous SiO_2 NPs of various sizes, either neutral, positively, or negatively charged, as indicated in the legend. Data are from Ref. 37. The dashed horizontal line represents the critical damping condition.

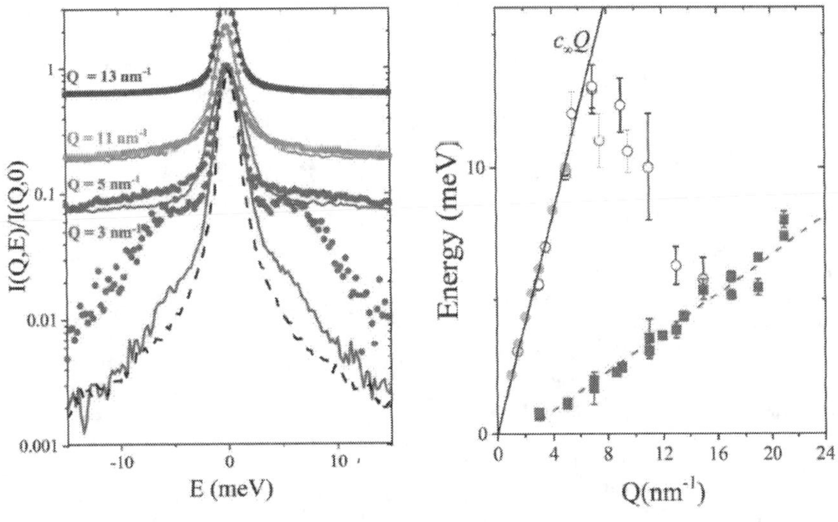

Figure 5. Left panel: *IXS* spectra measured at the indicated *Q* values in glycerol (dots) and on a dilute Au NP suspension in glycerol (solid lines of the corresponding color). All lineshapes are normalized to the respective intensity maxima and vertically offset for clarity. Right panel: *Q* dependence of the longitudinal acoustic frequency in both glycerol and pure suspension (open circles) is compared with the low-frequency dispersive branch (red squares) ascribed to Stoneley waves (see text). The solid straight line represents the high-frequency linear dispersion of glycerol (corresponding to a 3100 m/s sound speed) as derived from Ref. 38. The dashed red line is the linear best fit of the low-frequency dispersion branch and has a slope of 520 m/s. Gray dots represent the sound dispersion of pure glycerol reported by Sette and collaborators [39].

this Stoneley mode, albeit somewhat elusive at first sight, consists in a seemingly quasielastic broadening of the central peak in the spectrum. Such a broadening emerges by comparing the scattering profiles of the solution with the one of the pure substrate.

5 The anisotropic dynamics of water under highly directional confinement

Steering acoustic propagation in isotropic systems as fluids is far from trivial as density waves propagating in these systems are much shorter-lived than their counterparts in crystalline materials, and their propagation has no preferential directions. The achievement of this task likewise entails containing the liquid inside highly directional, nanometer-sized cavities. Carbon nanotubes (CNTs) seem ideally suited to this scope as they can act as quasi-unidimensional flow channels for the liquid confined in their interior. In particular, confinement in CNTs is known to drastically modify the physical properties of embedded water [43] as its dipolar configuration [44, 45] and orientational relaxation [46]. Furthermore, when trapped inside CNTs, water exhibits a non-trivial temperature evolution [47, 48]. In this scenario, it seems especially interesting the hypothesis that a highly directional confinement of a liquid may cause an "isotropy breach" of its sound propagation properties. To clarify this point, we conducted an *IXS* study on hydrated multi-walled aligned carbon nanotubes "forest-grown" on a Si substrate [22]. We conducted our measurements with the exchanged wavevector in either the basal or the axial directions of CNTs, these scattering geometries being here referred to as $\boldsymbol{Q}_{\perp}$ and $\boldsymbol{Q}_{\parallel}$, respectively.

Such a measurement demonstrates that the water dynamics under this quasi-1D confinement is strongly anisotropic. In particular, we observed that the $\boldsymbol{Q}_{\perp}$ *IXS* spectrum of hydrated CNTs is dominated by a low-frequency inelastic feature resembling the well-known transverse acoustic mode of water. At the same time, no signatures of collective dynamics appear in the $\boldsymbol{Q}_{\parallel}$ spectrum. This finding suggests that aligned CNTs visibly affect the propagation of terahertz acoustic modes of water embedded in their interior by acting either as "polarization filters" ($\boldsymbol{Q}_{\perp}$ geometry) or "acoustic dampers" ($\boldsymbol{Q}_{\parallel}$ geometry). This finding might inspire the use of these devices to efficiently shape terahertz collective excitations, which is deemed to impact the emerging domain of terahertz phononics. The polarization filtering ability of hydrated CNT is schematically illustrated in figure 6. Panel A displays a Transmission Electron Microscopy image of the forest-grown CNT array we used in Ref. 22 while showing the CNT axis and the perpendicular direction of the wavevector ($\boldsymbol{Q}_{\perp}$ geometry). Panel B compares the sound dispersion of the only inelastic mode surviving in the $\boldsymbol{Q}_{\perp}$ spectrum of the hydrated sample and its counterpart in bulk water. The overall agreement between the two curves indicates that orienting the hydrated sample in the $\boldsymbol{Q}_{\perp}$ scattering geometry selectively enables the propagation of the transverse acoustic mode of water. We finally stress that IXS measurements of the phonon spectrum are not the only characterization tool of the phonon response of a system, a valuable alternative being Inelastic Neutron Scattering. This technique has the undoubted advantage of a substantially narrower and sharper energy resolution profile, as required for a detailed study of low-frequency collective modes gathering in the quasielastic portion of the spectrum. Aside from the direct measurement of the phonon spectrum, neutron scattering can shed insight into the phonon Density of States, possibly revealing its signatures of a non-trivial phonon response. In this perspective, it is worth mentioning recent joint Inelastic Neutron Scattering (INS) measurements and Density Functional Theory ab initio computations on NP superlattices evidencing the increasing visibility of the inter-NP phonon dynamics at low temperatures [50].

6 Perspectives

The results discussed in this brief review are intended as very preliminary steps in a field of research still in its infancy, which essentially consists of the shaping of terahertz sound propagation via the design of the nanoscale structure. The extension of the domain of sound propagation engineering to this uncharted high-frequency territory appears especially compelling, because terahertz phonons are the leading conveyors of heat in insulators, and their control is crucial to implement heat flow management. Perhaps the most encouraging aspect of the whole body of results discussed here is the evidence that visible changes in the sound propagation can be simply achieved by slightly perturbing its internal structure through the immersion of a sparse amount of nanoparticles. Yet these results are only preliminary, and there is still a long distance keeping us apart from the ultimate goal of terahertz sound control. As mentioned in the introductory section, to pursue this objective seems particularly promising the arrangement of nanoparticles in superlattices functioning as the so-called phononic crystals. The density and elastic moduli distribution of these devices has periodicity that interferes directly with sound propagation at some frequency, potentially enabling its full shaping. Any advance in this field critically depends on the joint use of world-class nanotechnology to produce novel samples and spectroscopic methods to characterize their terahertz phonon behavior. Our group has recently become very active in this new field, and this will hopefully soon enable us to share this excitement with the broad community of Condensed Matter physicists.

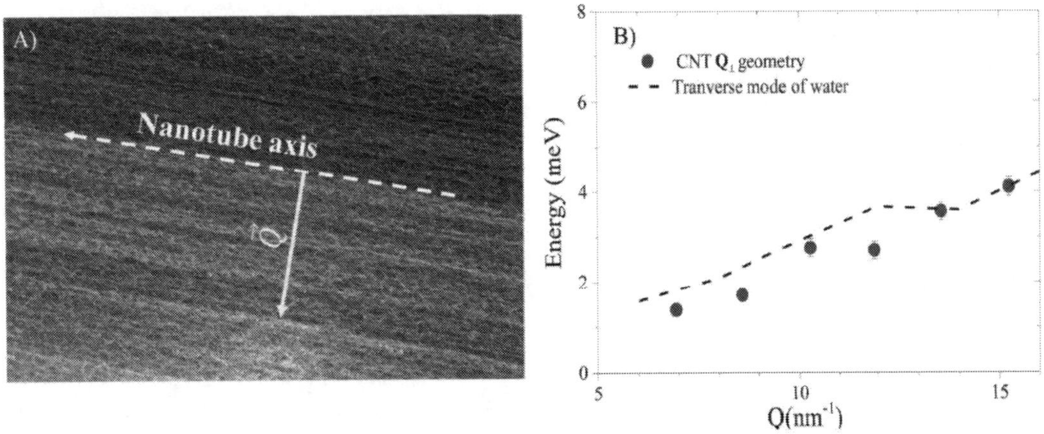

Figure 6. Panel A: Transmission Electron Microscopy image of the forest-grown carbon nanotube (CNT) array used in the *IXS* measurements in Ref. 22, in which the direction of the exchanged momentum and the CNT axis are specified. Panel B: The dispersion curve derived from the inelastic shift of the low-frequency mode in the $Q_\perp$ *IXS* spectra of the hydtrated sample is compared with the Q dispersion of the transverse mode of pure water as derived from Ref. 49 (dashed line).

References

[1] M. Maldovan, Nature **503**, 209 (2013)
[2] Z. Liu, X. Zhang, Y. Mao, Y. Zhu, Z. Yang, C.T. Chan, P. Sheng, science **289**, 1734 (2000)
[3] C.W. Chang, D. Okawa, A. Majumdar, A. Zettl, Science **314**, 1121 (2006)
[4] T. Harman, P. Taylor, M. Walsh, B. LaForge, science **297**, 2229 (2002)
[5] M. Maldovan, Physical review letters **110**, 025902 (2013)
[6] H. Lv, X. Tian, M.Y. Wang, D. Li, Applied Physics Letters **102**, 034103 (2013)
[7] P.E. Hopkins, C.M. Reinke, M.F. Su, R.H. Olsson III, E.A. Shaner, Z.C. Leseman, J.R. Serrano, L.M. Phinney, I. El-Kady, Nano letters **11**, 107 (2011)
[8] A. Cunsolo, *The THz dynamics of liquids probed by inelastic X-ray scattering* (World Scientific, 2021)
[9] S.K. Sinha, Journal of Physics: Condensed Matter **13**, 7511 (2001)
[10] S.W. Lovesey, *Theory of Neutron Scattering from Condensed Matter. Vol. 1: Nuclear Scattering* (Oxford, 1984)
[11] G.L. Squires, *Introduction to the theory of thermal neutron scattering* (Courier Corporation, 1996)
[12] V.M. Giordano, G. Monaco, Proceedings of the National Academy of Sciences **107**, 21985 (2010)
[13] V.M. Giordano, G. Monaco, Physical Review B **84**, 052201 (2011)
[14] A. De Francesco, L. Scaccia, M. Maccarini, F. Formisano, Y. Zhang, O. Gang, D. Nykypanchuk, A.H. Said, B.M. Leu, A. Alatas et al., ACS Nano **12**, 8867 (2018)
[15] A. Gelman, J.B. Carlin, H.S. Stern, D.B. Dunson, A. Vehtari, D.B. Rubin, *Bayesian data analysis* (Chapman and Hall/CRC, 2013)
[16] D. Sivia, J. Skilling, *Data analysis: a Bayesian tutorial* (OUP Oxford, 2006)
[17] A. De Francesco, E. Guarini, U. Bafile, F. Formisano, L. Scaccia, Physical Review E **94**, 023305 (2016)
[18] J. Liu, L. Ye, D. Weitz, P. Sheng, Physical Review Letters **65**, 2602 (1990)
[19] R. Penciu, G. Fytas, E. Economou, W. Steffen, S. Yannopoulos, Physical Review Letters **85**, 4622 (2000)
[20] L. Ye, J. Liu, P. Sheng, D. Weitz, Physical Review E **48**, 2805 (1993)
[21] A. De Francesco, L. Scaccia, F. Formisano, E. Guarini, U. Bafile, M. Maccarini, A. Alatas, Y.Q. Cai, D. Nykypanchuk, A. Cunsolo, Physical Review E **102**, 022601 (2020)
[22] A. De Francesco, L. Scaccia, F. Formisano, M. Maccarini, F. De Luca, A. Parmentier, A. Alatas, A. Suvorov, Y.Q. Cai, R. Li et al., Physical Review B **101**, 054306 (2020)
[23] A.D. De Francesco, L. Scaccia, F. Formisano, E. Guarini, U. Bafile, M. Maccarini, A. Alatas, Y.Q. Cai, A. Cunsolo, Nanomaterials **10**, 860 (2020)
[24] D.V. Lindley, Teach. Stat. **15**, 22 (1993)
[25] A. De Francesco, L. Scaccia, R.B. Lennox, E. Guarini, U. Bafile, P. Falus, M. Maccarini, Phys. Rev. E **99**, 052504 (2019)
[26] G. William, D.C. Watts, Transac. Faraday Soc. **66**, 80 (1970)
[27] D. Sivia, W. David, Acta Crystallogr. A **50**, 703 (1994)
[28] A. De Francesco, U. Bafile, A. Cunsolo, L. Scaccia, E. Guarini, Scientific Reports **11**, 1 (2021)
[29] D. Sivia, W. Hamilton, G. Smith, Physica B **173**, 121 (1991)

https://doi.org/10.1051/epjconf/202227201010

[30] L. Pardo, M. Rovira-Esteva, S. Busch, M. Ruiz-Martin, J.L. Tamarit, T. Unruh, arXiv preprint arXiv:0907.3711 (2009)

[31] W.R. Gilks, S. Richardson, D.J. Spiegelhalter, *Markov chain Monte Carlo in practice* (Chapman & Hall/CRC, 1996)

[32] P.J. Green, Biometrika **82**, 711 (1995)

[33] A. De Francesco, A. Cunsolo, L. Scaccia, in *Inelastic X-Ray Scattering and X-Ray Powder Diffraction Applications*, edited by A. Cunsolo, M.K.K.D. Franco, F. Yokaichiya (IntechOpen, 2020), chap. 2, p. 26

[34] A. De Francesco, L. Scaccia, M. Maccarini, F. Formisano, E. Guarini, U. Bafile, A. Cunsolo, Materials **12**, 2914 (2019)

[35] E. Guarini, S. Bellissima, U. Bafile, E. Farhi, A. De Francesco, F. Formisano, F. Barocchi, Physical Review E **95**, 012141 (2017)

[36] A. De Francesco, F. Formisano, L. Scaccia, E. Guarini, U. Bafile, M. Maccarini, D. Nykypanchuck, A. Suvorov, Y.Q. Cai, S.T. Lynch et al., Nanomaterials **12**, 2401 (2022)

[37] A. De Francesco, L. Scaccia, F. Formisano, E. Guarini, U. Bafile, M. Maccarini, Y. Zhang, D. Nykypanchuck, A. Alatas, A. Cunsolo, Scientific Reports **11**, 1 (2021)

[38] A. Cunsolo, B. Leu, A. Said, Y. Cai, The Journal of chemical physics **134**, 184502 (2011)

[39] F. Sette, M.H. Krisch, C. Masciovecchio, G. Ruocco, G. Monaco, Science **280**, 1550 (1998)

[40] C. Masciovecchio, G. Baldi, S. Caponi, L. Comez, S. Di Fonzo, D. Fioretto, A. Fontana, A. Gessini, S.C. Santucci, F. Sette et al., Physical Review Letters **97**, 035501 (2006)

[41] A. Giugni, A. Cunsolo, Journal of Physics: Condensed Matter **18**, 889 (2006)

[42] R. Stoneley, Proceedings of the Royal Society of London. Series A, Containing Papers of a Mathematical and Physical Character **106**, 416 (1924)

[43] B. Mukherjee, P.K. Maiti, C. Dasgupta, A. Sood, ACS nano **4**, 985 (2010)

[44] J. Köfinger, G. Hummer, C. Dellago, Proceedings of the National Academy of Sciences **105**, 13218 (2008)

[45] C. Cametti, F. De Luca, A. Parmentier, The Journal of Chemical Physics **137**, 094908 (2012)

[46] B. Mukherjee, P.K. Maiti, C. Dasgupta, A.K. Sood, Acs Nano **2**, 1189 (2008)

[47] A.I. Kolesnikov, J.M. Zanotti, C.K. Loong, P. Thiyagarajan, A.P. Moravsky, R.O. Loutfy, C.J. Burnham, Physical review letters **93**, 035503 (2004)

[48] G. Briganti, G. Rogati, A. Parmentier, M. Maccarini, F. De Luca, Scientific reports **7**, 1 (2017)

[49] A. Cunsolo, C. Kodituwakku, F. Bencivenga, M. Frontzek, B. Leu, A. Said, Physical Review B **85**, 174305 (2012)

[50] N. Yazdani, M. Jansen, D. Bozyigit, W.M. Lin, S. Volk, O. Yarema, M. Yarema, F. Juranyi, S.D. Huber, V. Wood, Nature communications **10**, 1 (2019)

Understanding the coherent dynamic structure factor of liquid water measured by neutron spectroscopy with polarization analysis: a Molecular Dynamics simulations study

Fernando Alvarez[1,2], *Arantxa* Arbe[1], and *Juan* Colmenero[1,2,3*]

[1]Centro de Física de Materiales (CFM) (CSIC-UPV/EHU) - Materials Physics Center (MPC), San Sebastián, Spain
[2]Departamento de Polímeros y Materiales Avanzados: Física, Química y Tecnología, UPV/EHU, San Sebastián, Spain
[3]Donostia International Physics Center (DIPC), San Sebastián, Spain

Abstract. This work is focused on atomistic molecular dynamics (MD) simulations of water carried out at 300 K. The main goal is to better understand the experimental results of the coherent dynamic structure factor $S(Q,\nu)$ of D_2O that were obtained by means of neutron scattering with polarization analysis and previously reported by us [A. Arbe et al. Phys. Rev. Res. **2**, 022015 (2020)]. From the simulations, we have calculated the coherent dynamic structure factor in the time domain $S(Q,t)$ as well as its self- and distinct-contributions. We have also calculated $S(Q,t)$ corresponding to a H_2O sample. The main results obtained are: (i) The Q-independent relaxation process identified in $S(Q,\nu)$ in the mesoscopic range (Q^0-mode) is the responsible of the restructuring of the hydrogen bond (HB) network at times shorter than that corresponding to the molecular diffusion; (ii) the vibrational contribution identified at high frequency in $S(Q,\nu)$ corresponds to a hydrodynamic-like mode propagating in an elastic medium (fixed HB bonding pattern); (iii) in the crossover range from mesoscopic to intermolecular scales, diffusion also progressively contributes to the decay of density fluctuations; (iv) MD-simulations suggest that it would be basically impossible to measure $S(Q,\nu)$ of H_2O in the mesoscopic range with the current neutron scattering capabilities.

1 Introduction

Nowadays there is not any doubt that water is a liquid that plays a prominent role in nature. Moreover, water is also one of the most interesting liquids concerning its physical and chemical properties. It is believed the peculiar behaviour of water is mainly due to the fact that water is an associated liquid, showing a network of hydrogen bonds (HB) that breaks and reforms continuously.

The molecular dynamics in general of liquid water has been deeply investigated by means of many different techniques, including neutron scattering. In particular, the collective dynamics has been previously investigated by inelastic neutron scattering (INS) [1-5], quasielastic neutron scattering (QENS) [6], inelastic X-ray scattering (IXS) [5, 7-9] and molecular dynamics (MD) simulations [10-14]. However, in spite of this effort, the collective dynamics of water is still poorly understood, in particular, at the mesoscale, i.e., at length scales larger than the intermolecular ones but not yet in those corresponding to the hydrodynamic range. The coherent structure factor $S(Q,\nu)$ (ν being the frequency) measured by IXS was usually described in terms of a viscoelastic model that involves a 'structural' relaxation process, usually called 'α-relaxation'. However, due to the current resolution of this technique ($\gtrsim$ 1.5 meV) and to the shape of the resolution function, a direct observation of this relaxation process has remained elusive. In principle, QENS is the ideal technique for directly observing the 'structural' collective relaxation of water in a wide Q-range, due to the available high energy resolution. However, the problem is that the measured intensity always contains a combination of coherent (collective) and incoherent (self) contributions. In samples like water with a high content of hydrogen atoms, it is generally assumed that replacing hydrogen atoms by deuterons (i.e., measuring D_2O) the measured neutron intensity would be basically coherent. Although this is true for D_2O in the Q-range of the first maximum of the static structure factor $S(Q)$, this is not the case at the mesoscale. In that Q-range ($0.1\text{Å}^{-1} \lesssim Q \lesssim 0.7\text{Å}^{-1}$) it has been demonstrated that the incoherent intensity is of the order of the coherent one [15]. The only way to unambiguously separate coherent and incoherent neutron scattering contributions is by using scattering of polarized neutron beams and analysing the polarization of the scattered neutrons. However, this technique has also many technical difficulties, in particular, for neutron spectroscopy with sub-meV resolution. The first measurements of quasielastic neutron scattering with polarization analysis on D_2O, in a wide Q-range from mesoscopic to intermolecular scales, were recently reported by us in Ref. [15]. They were obtained by means of LET direct geometry time of flight TOF spectrometer at the ISIS Neutron and Muon Source,

* juan.colmenero@ehu.eus

Oxfordshire, UK. In that paper, it was clearly proved the separation between coherent and incoherent contributions to the measured scattering. The obtained coherent and incoherent data were analyzed in terms of the imaginary part of the corresponding susceptibility, $\chi''(Q,\nu)$. In general, $\chi''(Q,\nu)$ shows a low-frequency relaxational-like process and a high frequency vibrational-like contribution. Fig. 1 displays the Q-dependence of the relaxation times obtained by fitting the low-frequency peak of $\chi''(Q,\nu)$ by a simple Lorentzian-like expression --equivalent to a simple exponential in the time domain. As can be seen, the incoherent relaxation time follows the expected behaviour for a diffusion-like Q-dependence ($\tau\sim Q^{-2}$) in the low-Q range. However, the coherent relaxation time hardly depends on Q (within the experimental uncertainties) in the low-Q mesoscopic range. This implies a non-diffusive decay of density fluctuations in such a range.

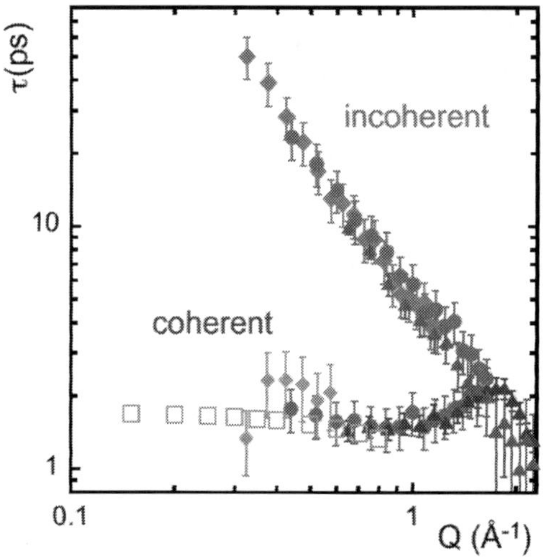

Fig. 1. Characteristic relaxation times (blue symbols: coherent; red symbols: incoherent) obtained by the procedure described in the text from the neutron scattering data with polarization analysis corresponding to a D_2O sample at 300 K [15]. Different solid symbols correspond to data obtained with different incident neutron wavelengths. Open blue squares correspond to the coherent relaxation times obtained from the MD-simulations described in the text.

In this short paper, we will focus on MD-simulations results trying to contribute to the understanding of the experimental results mentioned above. In particular, we would like to answer the following questions: (i) why the coherent relaxation in the mesoscopic range is not diffusive? (ii) what is the physical origin of such a process? On the other hand, the experimental results we are dealing with correspond to D_2O. We note that we tried to measure the coherent contribution in the mesoscale of a H_2O sample without success. Then, the questions emerging are: (iii) what should be observed if we were able to measure separately the coherent scattering on H_2O sample? and (iv) are this kind of measurements possible with the current neutron scattering capabilities?

After the Introduction (Section 1), the paper is organized as follows. First (Section 2) we will give a summary of the details of MD-simulations carried out. In Section 3 we will describe the relevant neutron scattering magnitudes to be calculated from the MD-simulations atomic trajectories. After that, in Section 4, we will describe and discuss the main results obtained. Finally, the main conclusions will be summarized in Section 5.

2 Molecular Dynamics Simulations

MD-simulations described in this work were carried out by means of the facilities implemented in the GROMACS module. The simulated cubic cell constructed was large enough (100555 water molecules) to facilitate the calculation of collective properties in the mesoscopic range. After the equilibration procedure and the needed NPT (fixed number of molecules, pressure value and temperature) dynamics, the side of the equilibrated cubic cell at 300K and 1 atm was $L = 144.56$ Å. The production runs were carried out under the canonical ensemble, NVT, conditions (fixed number of molecules, cell side value and temperature). The water model used was that of TIP4P-EW. In order to evaluate the electrostatic-like terms of the potential, we used the Ewald algorithm. Different production runs were performed allowing to extend the simulation time up to 10ns. The simulated cell allowed us to calculate the coherent dynamic structure factor, $S(Q,t)$, for Q-values in the range $0.1 \lesssim Q \lesssim Q_{max} \approx 2$ Å^{-1} where Q_{max} means the Q-value corresponding to the first maximum of the static structure factor of D_2O. More details about the simulation procedure can be found in Ref. [14].

3 Calculation of Magnitudes Measured in Neutron Scattering Experiments

From the atomic trajectories obtained in the MD-simulation runs we have calculated the partial dynamic structure factors defined as:

$$a_{\alpha\beta}(Q,t) = \int_0^\infty 4\pi r^2 G_{\alpha\beta}(r,t)\frac{\sin(Qr)}{Qr}dr \qquad (1)$$

where we have assumed isotropic behaviour. We note that in the case of water, $\alpha,\beta \in \{\mathrm{H},\mathrm{O}\}$. $G_{\alpha\beta}(r,t)$ in expression (1) is the van Hove correlation function that in the classical limit and for an isotropic sample is defined as

$$G_{\alpha\beta}(r,t) = \langle \frac{1}{N}\sum_{i\alpha,j\beta}^{N_\alpha,N_\beta} \delta\{r - |\vec{r}_{i\alpha} - \vec{r}_{j\beta}|\}\rangle \qquad (2)$$

where r is the radial distance from a given atom and $\vec{r}_{i\alpha}\,(\vec{r}_{j\beta})$ is the position of the $i^{th}(j^{th})$ atom of type $\alpha(\beta)$. $N_\alpha(N_\beta)$ is the number of nuclei of kind $\alpha(\beta)$ and N is the total number of atoms. The brackets in (2) mean the ensemble average that considers a large number of atomic configurations.

In neutron scattering experiments, the partial dynamic structure factors are weighted with the neutron

https://doi.org/10.1051/epjconf/202227201011

scattering lengths of the involved isotopes $\bar{b}_\alpha, \bar{b}_\beta$. Thereby, the coherent dynamic structure factor in the time domain, as should be experimentally accessed, is obtained as:

$$S(Q,t) = \sum_{\alpha,\beta} \bar{b}_\alpha \bar{b}_\beta a_{\alpha\beta}(Q,t) \quad (3)$$

For t=0, this expression gives the static structure factor $S(Q)$. In the case of water ($\alpha, \beta \in \{\mathrm{H}, \mathrm{O}\}$) expression (3) reads as:

$$S(Q,t) = \bar{b}_O^2 a_{OO}(Q,t) + \bar{b}_H^2 a_{HH}(Q,t) + 2\bar{b}_O \bar{b}_H a_{OH}(Q,t) \quad (4)$$

where $\bar{b}_O$=5.803fm and $\bar{b}_H$=-3.741fm.

As it has already been mentioned, the MD-simulations were carried out with the model corresponding to light water (H_2O). Then, in order to calculate $S(Q,t)$ for deuterated water (D_2O) we have just used the same partial dynamic structure factors in expression (4) but changing $\bar{b}_H$ by $\bar{b}_D$=6.671fm. We have shown that this approximation works rather well [14]. By considering a shift factor of 0.74 in the frequency scale of the simulations, an excellent agreement was found between simulated data of the imaginary part of the susceptibility for D_2O and those measured by neutron scattering with polarization analysis on an actual D_2O sample [14].

It is worthy of remark that the van Hove correlation function $G_{\alpha\beta}(r,t)$ contains the so-called self-part of the van Hove correlation function. This is obtained restricting the correlations included in expression (2) to those relating the positions of a single particle of type $\alpha(\beta)$ at different times. As consequence, the partial dynamic structure factor, $a_{\alpha\beta}(Q,t)$, for $\alpha = \beta$ can also be decomposed in 'distinct' and 'self' parts. For example, $a_{\alpha\alpha}(Q,t) = a_{\alpha\alpha}^{self}(Q,t) + a_{\alpha\alpha}^{dist}(Q,t)$. Then, for water, $S(Q,t)$ given by expression (4) can be rearranged as

$$S(Q,t) = S_{self}(Q,t) + S_{dist}(Q,t) \quad (5)$$

where, for light water, $S_{self}(Q,t) = \bar{b}_O^2 a_{OO}^{self}(Q,t) + \bar{b}_H^2 a_{HH}^{self}(Q,t)$ and $S_{dist}(Q,t) = \bar{b}_O^2 a_{OO}^{dist}(Q,t) + \bar{b}_H^2 a_{HH}^{dist}(Q,t) + 2\bar{b}_O \bar{b}_H a_{OH}(Q,t)$. In the framework of our approximation, for heavy water it would be the same but replacing $\bar{b}_H$ by $\bar{b}_D$ as it was mentioned above.

4 Results and Discussion

First of all, we would like to point out that the MD-simulations considered in this work were previously validated by direct comparison with the neutron scattering results [$S(Q)$ and the imaginary part of the susceptibility $\chi''(Q,\nu)$] as it has been mentioned above and described in a previous publication [14]. Before focusing on the different partial dynamic structure factors, $a_{\alpha\beta}(Q,t)$, Fig. 2 shows both, coherent and incoherent dynamic structure factors, as they are calculated from the simulations for Q-values in the range $0.1\text{Å}^{-1} \lesssim Q \lesssim 0.5\,\text{Å}^{-1}$. For the incoherent contribution we observe the expected behaviour in this low-Q regime: only one time-decay function (basically exponential), which characteristic relaxation time, τ, strongly depends on Q ($\tau \sim Q^{-2}$). The coherent dynamic structure factor is very different. It shows a two-step decay, being the second one basically Q-independent in the low-Q range considered in the plot. In Ref. [14] we demonstrated that the first --Q-dependent-- decay can be explained in terms of a longitudinal hydrodynamic-like component. We note that similar results were recently reported by Sciortino et al. [13] from massive MD-simulations of water but corresponding to the supercooled regime of water (T~250K) and to a rather low Q-range ($0.03\text{Å}^{-1} \le Q \le 0.3\,\text{Å}^{-1}$). Now we obtain similar results but for liquid water (300K) and for a larger Q-range. Even though these differences, in the rest of the manuscript we will follow the terminology introduced in Ref. [13] and we will call the Q-independent relaxation process Q^o-mode. We note that the experimental relaxation time also hardly depends on Q in the low Q-range (see Fig. 1). Following the procedure carried out for obtaining the relaxation times reported in Fig. 1 that was described in the introduction, we have fitted the low-Q time-domain coherent simulation data by the expression:

$$S(Q,t) = S(Q)\{(1 - C(Q))S_{hydro}(Q,t) + C(Q)\exp(-t/\tau_c)\} \quad (6)$$

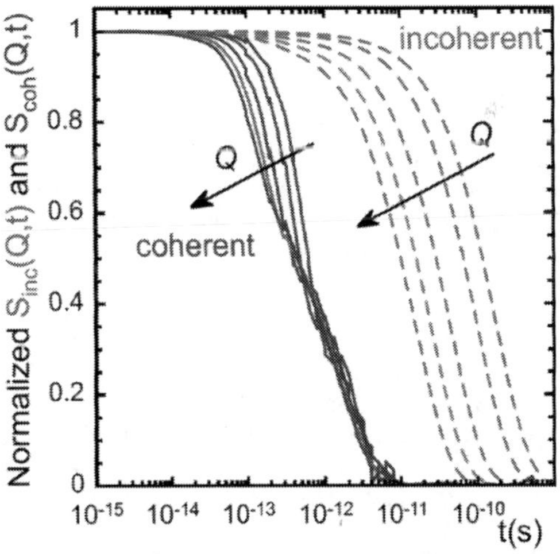

Fig. 2. Normalized coherent and incoherent dynamic structure factors at different Q-values [$Q = 0.15\,\text{Å}^{-1}$; $Q = 0.2\,\text{Å}^{-1}$; $Q = 0.3\,\text{Å}^{-1}$; $Q = 0.5\,\text{Å}^{-1}$] calculated from the simulations for D_2O at 300K.

Since for water the specific heat ratio ($\gamma = C_p/C_v$) is close to one, the hydrodynamic equation reduces to:

$$S_{hydro}(Q,t) = e^{-\Gamma Q^2 t}\left[\cos(cQt) + \frac{\Gamma Q}{c}\sin(cQt)\right]$$

Here, Γ is the acoustic attenuation coefficient and c the adiabatic sound velocity. Both parameters were allowed to vary with Q (molecular hydrodynamics approximation). The values obtained for the coherent relaxation time τ_c in the Q-range from Q=0.15 Å^{-1} until Q=0.8 Å^{-1} are shown in Fig.1. They nicely agree with the experimental times obtained by a similar procedure but in the frequency domain to fit the susceptibility $\chi''(Q,\nu)$ obtained by neutron scattering with polarization analysis [15]. This agreement gives additional support to our MD-simulations.

Now, taking advantage of MD-simulations, we can try to understand why the relaxation process observed does not depend on Q in the low-Q mesoscopic range. To do that, we have calculated from the simulated trajectories the self and distinct parts of the partial dynamic structure factors $a_{\alpha\beta}(Q,t)$ $(\alpha,\beta \in \{\mathrm{H,O}\})$. They are shown in Fig. 3 for a representative low-Q value Q = 0.2 Å^{-1}. As can be seen in the figure, the distinct part of both oxygen and hydrogen atoms are negative for this Q-value. The short-time range ($t \lesssim$ 3 ps) of the hydrogen-hydrogen distinct part shows clear signatures of two processes. They are also evident in the oxygen-hydrogen cross correlations that are positive. If we magnify the function corresponding to the oxygen-oxygen distinct correlations, we realize that these short-time processes are also there. On the other hand, both oxygen-oxygen and hydrogen-hydrogen correlations show a long-time decay to zero. This long-time decay is the only process in the self-part of both oxygen and hydrogen atoms, which are positive. Doing the same calculations but for other values in the low-Q range, we realized that the long-time decay depends on Q as a diffusive process. Interestingly enough, this long-time decay is the same –apart from the sign— for both self- and distinct contributions. Thereby, they cancel each other when we make the summations to calculate $a_{HH}(0.2,t)$ and $a_{OO}(0.2,t)$. These two partial dynamic structure factors are represented in Fig. 4. This figure also contains the partial dynamic structure factor corresponding to oxygen-hydrogen correlations that, obviously, only have distinct component. As can be seen, the three partial dynamic structure factors show the same behaviour apart from the amplitude: (i) they decay to zero at about 10^{-11}s; (ii) they show a first decay, that can be identified with a hydrodynamic-like longitudinal mode, and a second relaxation decay; the amplitude of the $a_{HH}(0.2,t)$ is the same as that corresponding to $2a_{OH}(0.2,t)$. The fact that all these partial dynamic structure factors display the same behaviour tells us that all atoms are participating in a similar way in the two processes involved –something that could be expected taking into account the low-Q range considered. Based on these results, we understand why the coherent dynamic structure factor $S(Q,t)$ at the mesoscale does not show any signature of diffusive processes. The reason is that the diffusive component of the self and distinct parts of the partial dynamic structure factor cancel each other.

Starting from the partial dynamic structure factors shown in Fig. 4 for Q = 0.2 Å^{-1} and taking into account the suitable coherent scattering lengths ($\bar{b}_O$=5.803fm, $\bar{b}_H$=-3.741fm and $\bar{b}_D$=6.671fm) we can calculate the dynamic structure factor $S(Q,t)$ [eq. (3)]. In Fig. 5 we show the behaviour obtained for heavy water (D_2O) and

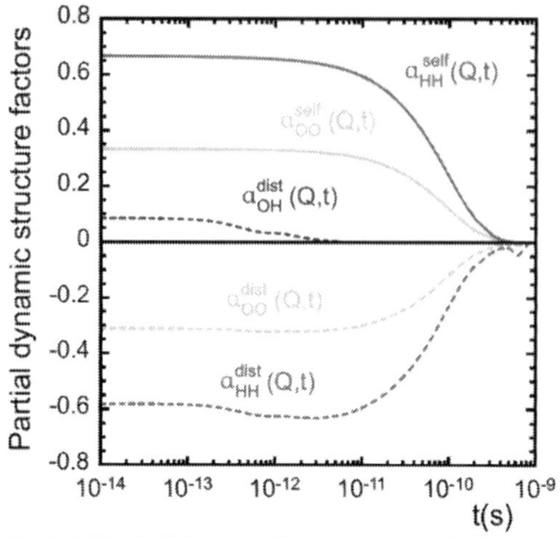

Fig. 3. Self and distinct contributions to the partial dynamic structure factors $a_{\alpha\beta}(Q,t)$ $(\alpha,\beta \in \{\mathrm{H,O}\})$ calculated from the simulations for Q = 0.2 Å^{-1}: $a^{self}_{OO}(0.2,t)$; $a^{dist}_{OO}(0.2,t)$; $a^{self}_{HH}(0.2,t)$; $a^{dist}_{HH}(0.2,t)$ and $a^{dist}_{OH}(0.2,t)$. Hatched area indicates the diffusion range.

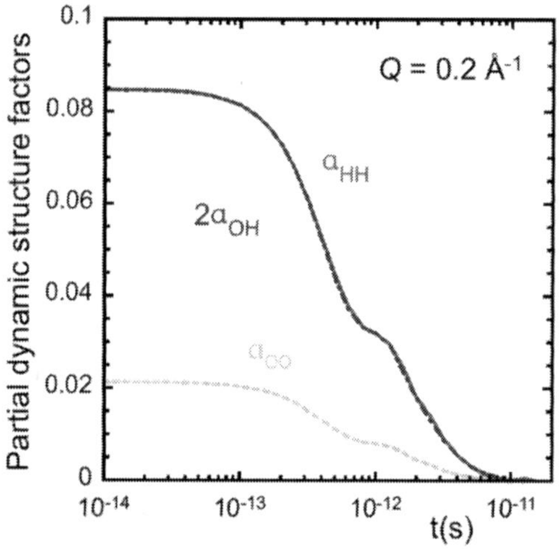

Fig. 4. Partial dynamic structure factors $a_{\alpha\beta}(Q,t)$ $(\alpha,\beta \in \{\mathrm{H,O}\})$ calculated from the simulations for Q = 0.2 Å^{-1}: $a_{HH}(0.2,t)$ (continuous red line); $a_{OO}(0.2,t)$ (dashed green line); $2a_{OH}(0.2,t)$ (dashed-dotted blue line).

light water (H_2O). The intensity of $S(Q,t)$ for Q = 0.2 Å^{-1} and $t \cong 10^{-14}$s results to be in the case of D_2O about 125 times larger than that corresponding to H_2O. We note that in the low-Q thermodynamic limit ($Q \longrightarrow$ 0), the coherent intensity measured in a neutron experiment on a molecular fluid can be expressed as

$$I_{coh}(Q) = n^2 \left[\sum_j \bar{b}_j\right]^2 k_B T \beta_T$$

Where n is the molecular number density and the summation runs over all the nuclei of a molecule. k_B is the Boltzmann constant and β_T the isothermal compressibility, which is basically the same for heavy and light water (see, e.g. [16]). Then, considering the values of the coherent scattering lengths about mentioned, we obtain that $I_{coh}^{D_2O}/I_{coh}^{HO} \simeq 130$ in rather good agreement with the value of about 125 obtained from the simulations (see above). This, not only gives additional support to our simulations, but also tells us that it is basically impossible to measure *S(Q,t)* of H_2O by neutron scattering, even using methods based on polarization analysis.

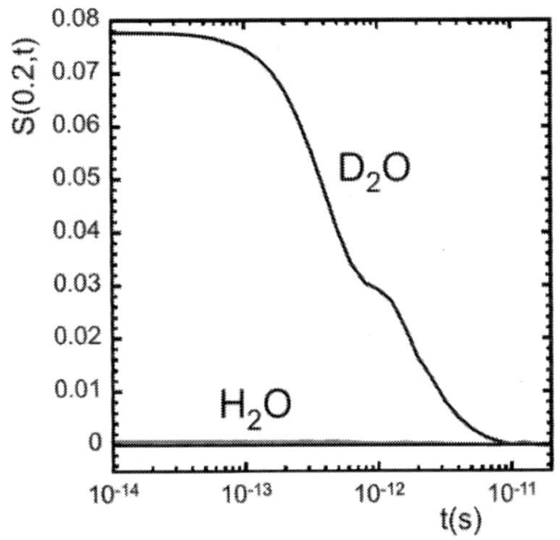

Fig. 5. Coherent dynamic structure factor $S(Q,t)$ corresponding to neutron scattering measurements calculated from the simulations for both D_2O and H_2O at $Q = 0.2$ Å^{-1}.

Now we can ask what is the physical origin of the two-steps decay shown by the partial dynamic structure factors and *S(Q,t)* as well. Since in the mesoscopic scale the diffusion process takes place at times longer than τ_c (see Figs. 1 and 2) this implies that, in the time range of the fast relaxation decay ($\tau_c \approx$2ps) the water molecules are trapped in the hydrogen bond (HB) network. Thereby, the decay of the density fluctuations by this process has to be related with the restructuring process of the HB network, likely mediated by a sequence of elementary bond-switch processes. We note that experimental results, corresponding to different temperatures in the range 280K-350K, deliver an activation energy of ≈4 Kcal/mol for τ_c [15]. This energy is larger than the values usually associated with the enthalpy for breaking a hydrogen bond in water (≈2 Kcal/mol) [17]. It is worth mentioning that signatures of this HB network restructuring processes have also been observed in the short-time behaviour of single-particle correlations as incoherent neutron scattering of H_2O [6, 15] or single water molecule reorientation [18]. On the other hand, at times shorter than τ_c (the HB-network restructuring time) the HB network has, thereby, a fixed bonding pattern. Then, the hydrodynamic-like modes mentioned above [first decay of *S(Q,t)*] are propagating in an elastic medium. Since the hydrodynamic-like decay of *S(Q,t)* depends on *Q* (see Fig. 2) there will be a low *Q*-value at which the vibrational process takes place on a time scale comparable with the network restructuring. Only at *Q*-values below this crossover-*Q*, the hydrodynamic predictions are expected to properly model the decay of density fluctuations in the system. From our MD-simulation results we can estimate (from extrapolation) a value of the crossover *Q* in the range 0.03 -- 0.1 Å^{-1}, which was not directly covered by the simulations [14].

Now, once we have clarified why the diffusion processes do not play any role in the relaxation of density fluctuations in the mesoscopic *Q*-range and what is the physical origin of the remaining short-time relaxation process, the question is in which *Q*-range the diffusion processes start to be an effective mechanism for the decay of density fluctuations. To address this question we have calculated the self and distinct part of *S(Q,t)*, at different *Q*-values from the mesoscopic to the intermolecular range, for both D_2O and H_2O, and the static part of these magnitudes as well. The results obtained for D_2O are shown in Figs. 6 and 7. Fig. 6 displays the self-part of *S(Q,t)* and the distinct part but with the sign changed to facilitate the comparison. The resulting *S(Q,t)* is also shown in the same scale. As it was previously discussed, at $Q = 0.2$ Å^{-1} the diffusive contribution to self and distinct parts is, basically, the same but with opposite sign. Thereby, they cancel each other. Fig. 6 shows that the same happens at $Q = 0.5$ Å^{-1} and $Q = 0.7$ Å^{-1}, although, as can be seen, the characteristic time of the long-time diffusive part decreases as expected ($\tau_d \sim Q^{-2}$). However, for $Q >$ 0.7 Å^{-1}, the absolute value of the distinct part diminishes and, as consequence, (i) the diffusive contribution to *S(Q,t)* starts to be more and more important as soon as *Q* increases (see the case of $Q =$ 1.2 Å^{-1} in Fig. 6 (d)), and (ii) this contribution merges with the fast process identified at low *Q*. Thereby, *S(Q,t)* cannot be described anymore by only a hydrodynamic-like term plus a single-exponential fast process [eq. (6)]. In Ref. [15] we proposed a "convolution model" for describing the relaxation contribution to the decay of density fluctuations in this crossover *Q*-range from mesoscopic to intermolecular scales. In that framework, the relaxational contribution can be expressed as a convolution product:

$$S_R(Q,t) = S_c(Q,t) \cdot S_d(Q,t) \qquad (7)$$

where $S_d(Q,t) = exp[-t/\tau_d]$ is a diffusive process ($\tau_d \propto Q^{-2}$) and $S_c(Q,t)$ is identified with the Q^0-mode above mentioned. $S_c(Q,t)$ is parametrized as:

$$S_c(Q,t) = A(Q) + [1 - A(Q)]exp[-t/\tau_c^0] \qquad (8)$$

with amplitude $[1 - A(Q)]$ and *Q*-independent relaxation time, τ_c^0. We note that when $A(Q) \sim 0$ and $\tau_d \gg \tau_c^0$ expression (7) reduces to the simple exponential used in the low *Q*-range, where the Q^0-mode dominates the relaxation decay of density fluctuations. As soon as *Q* increases, atomic diffusion enters in the game and this kind of processes dominate in the *Q*-range of the maximum of *S(Q)*. The idea behind the convolution model is that in the crossover range there are two channels for the relaxation decay of density fluctuations. In this framework, these two channels are considered independent. As it was reported in refs. [14, 15] the convolution model nicely describes both, experimental (frequency domain) and MD-simulations (time domain) results in the crossover range above mentioned.

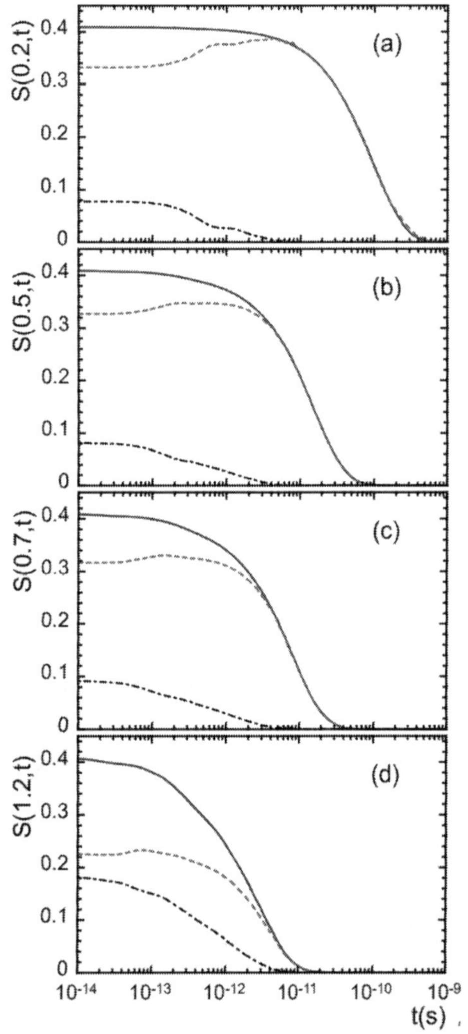

Fig. 6. Coherent dynamic structure factor *S(Q,t)* (dashed-dotted lines) corresponding to D_2O, its self part $S_{self}(Q,t)$ (blue continuous line), and the distinct part with opposite sign $-S_{dist}(Q,t)$ (red dashed line) for Q = 0.2 $Å^{-1}$(a); Q = 0.5 $Å^{-1}$(b); Q = 0.7 $Å^{-1}$(c) and Q = 1.2 $Å^{-1}$(d).

We note that, according to the previous discussion, the diffusive contribution to *S(Q,t)* starts to increase just when *S(Q)* starts to increase as well towards the first maximum of the static structure factor. Fig. 7 shows that at $Q \approx 1.6$ $Å^{-1}$ the amplitude of the distinct part is going to zero [$S(1.6Å^{-1}) \approx S_{self}(1.6Å^{-1})$]. For $Q \gtrsim 1.6$ $Å^{-1}$ and until about $Q \approx 2.9$ $Å^{-1}$, the distinct part becomes positive and, thereby, the diffusive contributions of self and distinct parts, instead of being cancelled they are reinforced. Then, in this *Q*-range, that corresponds to the first maximum of *S(Q)*, diffusive processes dominate the relaxational part of *S(Q,t)* as it has been mentioned above.

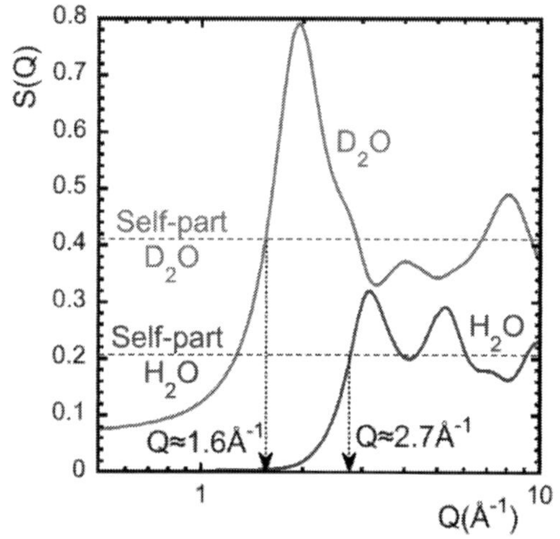

Fig. 7. Static structure factor *S(Q)* calculated from the simulations for both D_2O and H_2O. The corresponding self part is indicated by the dotted line in each case. The vertical arrows mark the *Q*-value at which the self-part equals *S(Q)* in each case. At this *Q* -value, the corresponding distinct part is zero, thereby crossing over from negative to positive values.

Until now we have discussed the behavior of both *S(Q,t)* and *S(Q)* corresponding to D_2O in the crossover range from the mesoscale to inter-molecular length scales [first maximum of *S(Q)*]. Now the question emerging is: what would be the behaviour observed in the case of H_2O? We remind that the atomic correlations are the same because we are using the same MD-simulations for calculating *S(Q,t)* and *S(Q)* corresponding to D_2O and H_2O. Only the scattering lengths are different. We remind that for $Q \approx 0.2$ $Å^{-1}$ and $t \sim 10^{-14}$s the amplitude of *S(Q,t)* corresponding to H_2O was more than 125 times smaller than the equivalent for D_2O (see Fig. 5). On the other hand, Fig. 7 shows that *S(Q)* is basically constant until about $Q \sim 1.6$ $Å^{-1}$ in the case of H_2O. Then, according to the results previously discussed for D_2O, we would expect that in the case of H_2O: (i) the diffusive part of both, self and distinct contribution to *S(Q,t)* cancel each other until $Q \sim 1.6$ $Å^{-1}$; and (ii) the amplitude of *S(Q,t)* will be very low until the *Q*-range of the first maximum of *S(Q)* that occurs at $Q \sim 3$ $Å^{-1}$. This is the behaviour that we obtain. Fig. 8 shows the self and distinct contributions to *S(Q,t)* corresponding to different *Q*-values (Q = 0.5 $Å^{-1}$; 1.0 $Å^{-1}$; 1.6 $Å^{-1}$; 2.4 $Å^{-1}$) together with the resulting *S(Q,t)*. As can be seen in Fig. 8 (a), for Q = 0.5 $Å^{-1}$, 1.0 $Å^{-1}$ and 1.6 $Å^{-1}$, the diffusive part of the self and distinct contributions cancels and the remaining short-time differences are very low. As consequence, the amplitude of the resulting *S(Q,t)* is very low as well. Only for *Q*-values in the range of the first maximum of

S(Q) as $Q = 2.4$ Å^{-1}, represented in Fig. 8 (b), the self and distinct parts are clearly different and thereby *S(Q,t)* becomes 'visible'. These MD-simulation results suggest that the only possibility of measuring the coherent dynamic structure factor *S(Q,t)* of H_2O by neutron scattering –even using polarization analysis techniques-- would be in the *Q*-range of the first maximum of *S(Q)*, which for H_2O takes place at about 3 Å^{-1}. However, in this *Q*-range the decay of the density fluctuations would be dominated by diffusive processes. Then, it would be basically impossible to reach experimental evidence by neutron scattering of the so-called Q^0 – mode in the case of H_2O.

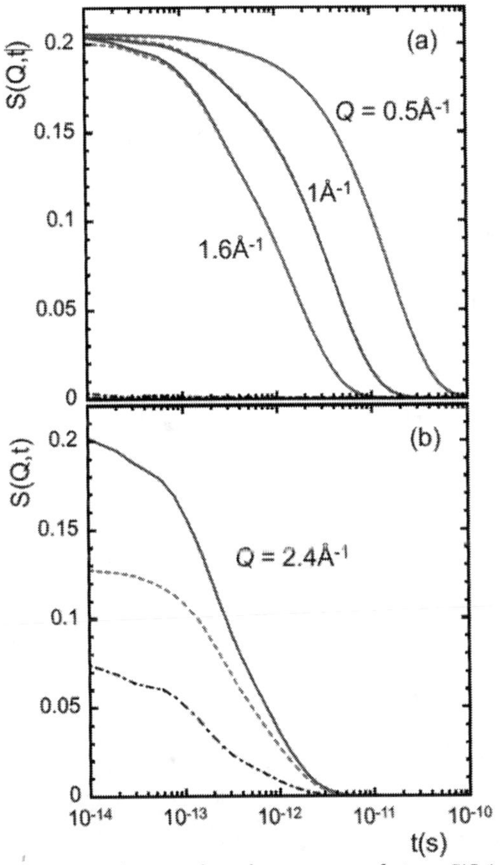

Fig. 8. Coherent dynamic structure factor *S(Q,t)* corresponding to H_2O (black dashed-dotted line), its self part $S_{self}(Q,t)$ (blue continuous line), and the distinct part with opposite sign -$S_{dist}(Q,t)$ (red dashed line) for *Q*-values: 0.5 Å^{-1}; 1 Å^{-1}; and 1.6 Å^{-1} in (a) and for 2.4 Å^{-1} in (b). Note that in the case of 0.5 Å^{-1}; 1 Å^{-1} and 1.6 Å^{-1}, $S_{self}(Q,t)$ and -$S_{dist}(Q,t)$ basically coincide, and the amplitude of *S(Q,t)* is almost zero in the scale of the figure.

5 Summary and Conclusions

In this paper we have first given a short overview of the behavior of the collective density fluctuations in liquid water (300K) in the mesoscale and in the crossover range towards intermolecular scales. The work was focussed on MD-simulations results that were carried out in order to better understand the first neutron scattering measurements of the coherent dynamic structure factor of liquid water (D_2O) with neutron polarization analysis that were previously performed by us [15]. After the validation of the MD-simulations, we have exploited the advantages of this method to calculate the neutron scattering magnitudes in the time domain and in a broader frequency range than the experimental counterpart. We have also been able to separate the self and distinct contributions to the coherent dynamic structure factor as well as the different atomic pair-correlations involved. We have also calculated the neutron scattering dynamic coherent structure factor corresponding to a fully protonated H_2O sample. The main conclusions obtained can be summarized as follows:

-In the mesoscopic range $(0.1\ \text{Å}^{-1} \lesssim Q \lesssim 0.7\text{Å}^{-1})$ the diffusive contribution to the self- and distinct-part of the coherent dynamic structure factor cancel each other and thereby the decay of the density fluctuations take place only by hydrodynamic-like modes and a basically *Q*-independent relaxation process (Q^0 – mode).

-In the time scale of the Q^0-mode, the water molecules are trapped in the HB-network (diffusion is not possible). Then, the physical nature of this relaxation mode has to be related with the restructuring of the HB-network, likely mediated by a sequence of elementary bond-switch processes.

-At times shorter than the time scale of the Q^0-mode the HB-network has to show a fixed bonding pattern. Thereby, the hydrodynamic modes are propagating in an elastic medium.

-The density fluctuations will decay only by hydrodynamic modes in the *Q*-range at which the time scale of the Q^0-mode is shorter than that characterizing the hydrodynamic mode decay, which is *Q*-dependent. It can be estimated that this crossover-*Q* should be in the range 0.03 Å^{-1}- 0.1 Å^{-1}.

-In the crossover range from mesoscopic to intermolecular scales $(0.7\ \text{Å}^{-1} \lesssim Q \lesssim 2\text{Å}^{-1})$, the diffusion processes do not cancel and they progressively also contribute to the decay of density fluctuations. A convolution model for the relaxational part of the decay of density fluctuations seems to be a rather good approximation in this crossover range.

-The MD-simulations results show that the amplitude of the intensity measured by neutron scattering in a H_2O sample, is extremely low in the *Q*-range from ~0.7 Å^{-1} to ~2 Å^{-1}. This suggests that it would be basically impossible to measure this coherent dynamic structure factor in this *Q*-range with the current neutron scattering capabilities.

We acknowledge the Grant PID2021-123438NB-I00 funded by MCIN/AEI/10.13039/501100011033 and by "ERDF A way of making Europe". We also acknowledge financial support of Eusko Jaurlaritza, code: and IT1566-22, as well as from the IKUR Strategy under the collaboration agreement between Ikerbasque Foundation and the Materials Physics Center on behalf of the Department of Education of the Basque Government.

References

1. J. Texeira, M. C. Bellisent-Funel, S. H. Chen, B. Dorner, Phys. Rev. Lett. **54**, 2681 (1985)
2. C. Petrillo, F. Sacchetti, B. Dorner, J. B. Suck, Phys. Rev. E **62**, 3611 (2000)
3. F. Sacchetti, J. B. Suck, C. Petrillo, B. Dorner, Phys. Rev. E **69**, 061203 (2004)
4. P. Bosi, F. Dupré, F. Menzinger, F. Sacchetti, M. C. Spinelli, Lett. Nuovo Cimento **21**, 436 (1978)
5. A. Cunsolo, C. N. Kodituwakku, F. Bencivenga, M. Frontzek, B. M. Leu, A. H. Said, Phys. Rev. B **85**, 174305 (2012)
6. A. Arbe, P. Malo de Molina, F. Alvarez, B. Frick, J. Colmenero, Phys. Rev. Lett. **117**, 185501 (2016)
7. G. Monaco, A. Cunsolo, G. Ruocco, F. Sette, Phys. Rev. **E** 60, 5505 (1999)
8. F. Sette, G. Ruocco, M. Krisch, C. Masciovecchio, R. Verbeni, U. Bergmann, Phys. Rev. Lett. **77**, 83 (1996)
9. E. Pontecorvo, M. Krisch, A. Cunsolo, G. Monaco, A. Mermet, R, Verbeni, F. Sette, G. Ruocco, Phys. Rev. E. **71**, 011501 (2005)
10. A. Rahman, F. H. Stillinger, Phys. Rev. A **10**, 368 (1974)
11. F. Sciortino, S. Sastry, J. Chem. Phys. **100**, 3881 (1994)
12. M. Sampoli, G. Ruocco, F. Sette, Phys. Rev. Lett. **79**, 1678 (1997)
13. P. H. Handle, L. Rovigatti, F. Sciortino, Phys. Rev. Lett. **122**, 175501 (2019)
14. F. Alvarez, A. Arbe, J. Colmenero, J. Chem. Phys. **155**, 244509 (2021)
15. A. Arbe, G. J. Nilsen, J. R. Stewart, F. Alvarez, V. García-Sakai, J. Colmenero, Phys. Rev. Res. **2**, 022015 (2020)
16. S. Bhagavantam, S. Sundara Rama Rao, Nature **140**, 1099 (1937)
17. Kevin, A. T. Silverstein, A. D. J. Haymet, Ken A. Dill, J. Am. Chem. Soc. **122**, 8037 (2000)
18. J. Qvist, H. Schober, B. Halle, J. Chem. Phys. **134**, 144508 (2011)

Revisiting the modeling of quasielastic neutron scattering from bulk water

Martin H. Petersen[1,*], *Mark T. F.* Telling[2,**], *Gerald* Kneller[3,4,***] and *Heloisa N.* Bordallo[1,5,****]

[1]Niels Bohr Institute, University of Copenhagen, Universitetsparken 5, 2100 Copenhagen, Denmark
[2]STFC ISIS Facility Rutherford Appleton Laboratory, Harwell Campus, Oxfordshire, OX11 0QX, UK
[3]Centre de Biophysique Moléculaire, CNRS and Université d'Orléans, Rue Charles Sadron, 45071 Orléans, France
[4]Synchrotron Soleil, L'Orme de Merisiers, 91192 Gif-sur-Yvette, France
[5]European Spallation Source ERIC Box 176, SE-221 00, Lund, Sweden

Abstract. Quasi-elastic neutron scattering (QENS) from bulk-water at 300 K, measured on the IRIS backscattering neutron spectrometer (ISIS, UK), is interpreted using the jump diffusion model (JDM), a "minimalistic" multi-timescale relaxation model (MRM) and molecular dynamics simulations (MD). In the case of MRM data analysis is performed in the time domain, where the relaxation of the intermediate scattering function is described by a stretched Mittag-Leffler function, $E_\alpha(-(|t|/\tau)^\alpha)$. This function displays an asymptotic power law decay and contains the exponential relaxation function as a special case ($\alpha = 1$). To further compare the two approaches, MD simulations of bulk water were performed using the SPCE force field and the resulting MD trajectories analysed using the nMoldyn software. We show that both JDM and MRM accurately describe the diffusion of bulk water observed by QENS at all length scales, and confirm that MD simulations do not fully describe the quantum effects of jump diffusion.

1 Introduction

Quasi-elastic neutron scattering (QENS) is a powerful spectroscopic technique used to explore diffusive motions of atoms on length scale between 1 Å and 100 Å and times scales between 10^{-9}s and 10^{-13}s. Because of the dominant incoherent scattering response from hydrogen atoms, QENS studies explore, in particular, the diffusive motions of individual hydrogen atoms and are thus ideal to study water dynamics. Considering the study of bulk-water, the QENS spectra cannot be completely described by a small-step diffusion model for free diffusion. This happens because the diffusion in liquids takes place in discrete diffusive jumps [1] and a corresponding model for simple liquids was formulated by Chudley, using both Schofield's semiclassical approximation [2] and Van Hove's formalism of correlation functions [3]. Following these ideas, Teixeira *et al.* [4] extended the model for water, including vibrational and rotational degrees of freedom. As with any "spatial motion model" [5], the resulting intermediate scattering function is composed of several Lorentzians, related to different motion types. The ability to rationalize QENS data in terms of a relatively small number of parameters makes this approach widely adopted in QENS data analysis for a variety of other systems [6, 7, 8, 9, 10, 11]. Although successful, the method described in [4] requires prior information of the system, since it is essential to attribute each Lorentzians needed to represent the measured dynamics to a specific motion. A good example here is the confined water in clay minerals, where different populations of water molecules coexist [7, 10]. To avoid such *a priori* assumptions, alternative approaches for interpreting QENS data are urgently needed. Having that in mind, a "minimal model approach", where the multi-scale relaxation of the water molecules can be described without any prior information [12], has been recently proposed and successfully used to analyse the dynamics of confined water molecules in proteins and clay minerals [13, 14, 15]. This said analysis shows that the new method can capture the full dynamics of the confined water with the benefit that it rationalizes extensive sets of QENS data in terms of a minimal number of parameters. Here we apply both methods to analyse bulk water, H_2O, at 300K.

To further understand the differences and similarities between the two approaches, a complementary technique is needed. Molecular dynamics (MD) simulation is the method of choice since it provides detailed information about the dynamics of molecules and atoms of complex systems within the length and time scale probed by QENS. Considering the system of bulk-water, we can then compare the motion of water molecules seen by MD trajectories with those probed experimentally and, as a "side effect", verify the simulations [16]. For this purpose, MD simulations have been performed for bulk-water at 300K using the LAMMPS Molecular Dynamics Simulator with the Velocity-Verlet algorithm as its numerical solver. To directly compare the measured QENS spectra to the MD results, the conversion of the trajectories and velocities to either $S(Q, \omega)$ or its Fourier transformed

*e-mail: martinhp27@gmail.com
**e-mail: mark.telling@stfc.ac.uk
***e-mail: gerald.kneller@cnrs.fr
****e-mail: bordallo@nbi.ku.dk

$F(Q,t)$ was carried out using the program package nMoldyn [17, 18, 19, 20].

2 Neutron scattering functions

2.1 Dynamic structure factor

In a standard QENS experiment from hydrogen rich systems one measures the dynamic structure factor [4, 6],

$$S(\mathbf{Q},\omega) = \frac{1}{2\pi}\int_{-\infty}^{+\infty} dt\, e^{-i\omega t} F(\mathbf{Q},t), \tag{1}$$

which is the time Fourier transform of the intermediate scattering function containing within it information about the structural dynamics of the system under consideration; $|\mathbf{Q}| = Q$ being the magnitude of the scattering vector and $\hbar\omega$ the energy transferred between the neutron and the nucleus.

Defining $\hat{\mathbf{R}}_j(t)$ as the time-dependent position operator of scattering atom j, we have

$$F(\mathbf{Q},t) \approx \frac{1}{N}\sum_{j=1}^{N}\left\langle e^{-i\mathbf{Q}\cdot\hat{\mathbf{R}}_j(0)} e^{i\mathbf{Q}\cdot\hat{\mathbf{R}}_j(t)}\right\rangle, \tag{2}$$

where $\langle\ldots\rangle$ denotes a quantum ensemble average summed over all N hydrogen atoms. We note in this context that $\hbar\omega$ is the energy transfer from the neutron to the sample.

The intermediate scattering function and the dynamic structure factor fulfill the symmetry relations

$$F(\mathbf{Q},-t) = F(-\mathbf{Q}, t + i\beta\hbar), \tag{3}$$
$$S(\mathbf{Q},\omega) = e^{\beta\hbar\omega} S(-\mathbf{Q},-\omega), \tag{4}$$

where $\beta = 1/k_B T$ is the inverse Boltzmann temperature. Equation (4) is the well-known detailed-balance relation.

2.2 Generic form of $F(\mathbf{Q},t)$

For the further considerations we write the intermediate scattering function in the generic form [12, 13, 14],

$$F(\mathbf{Q},t) = (1 - EIS\,F(\mathbf{Q}))\phi(\mathbf{Q},t) + EIS\,F(\mathbf{Q}) \tag{5}$$

where $(1 - EIS\,F(\mathbf{Q}))$ and $EIS\,F(\mathbf{Q})$ are, respectively, the quasi-elastic and elastic amplitudes and $\phi(\mathbf{Q},t)$ is an atom-averaged normalized self-correlation function describing the relaxation dynamics of the hydrogen atoms,

$$\phi(\mathbf{Q},t) = \frac{1}{N_H}\frac{\sum_\alpha \left\langle \delta\bar{\rho}_\alpha^\dagger(\mathbf{Q},0)\delta\bar{\rho}_\alpha(\mathbf{Q},t)\right\rangle}{\sum_\alpha \left\langle \delta\bar{\rho}_\alpha^\dagger(\mathbf{Q},0)\delta\bar{\rho}_\alpha(\mathbf{Q},0)\right\rangle}. \tag{6}$$

Here the sum $\sum_\alpha$ runs over all hydrogen atoms, where N_H is their total number and $\delta\bar{\rho}_\alpha(Q,t) = \bar{\rho}_\alpha(Q,t) - \langle\bar{\rho}_\alpha(Q)\rangle$ is the deviation of the spatially Fourier-transformed single particle density of atom α, $\bar{\rho}_\alpha(\mathbf{Q},t) = e^{i\mathbf{Q}\cdot\hat{\mathbf{R}}_j(t)}$, with respect to its mean value. By construction, $\phi(\mathbf{Q},t)$ has the same symmetry properties as the intermediate scattering function and must satisfy $\phi(\mathbf{Q},0) = 1$ and $\lim_{t\to\infty}\phi(\mathbf{Q},t) = 0$.

Since we consider here scattering from an isotropic sample, $S(\mathbf{Q},\omega) = S(Q,\omega)$ and $F(\mathbf{Q},t) = F(Q,t)$, and $Q \equiv |\mathbf{Q}|$.

2.3 Semiclassical approximation

We now introduce the symmetrized and normalized relaxation function [14],

$$\phi^{(+)}(Q,t) = \frac{\phi(Q, t + i\beta\hbar/2)}{\phi(Q, i\beta\hbar/2)}, \tag{7}$$

which is real and symmetric in time, and the corresponding intermediate scattering function,

$$F^{(+)}(Q,t) = (1 - EIS\,F(Q))\phi^{(+)}(Q,t) + EIS\,F(Q). \tag{8}$$

The choice of using $\phi^{(+)}(Q,t)$ instead of $\phi(Q,t)$ is motivated by Schofield's semi-classical correction [2], which consists of relating the shifted relaxation function with its classical counterpart [14],

$$\phi^{(+)}(Q,t) \approx \phi^{(cl)}(Q,t), \tag{9}$$

and is valid up to the first order in $\hbar$, giving some justification for analyzing QENS data with classical diffusion models [2, 13].

3 QENS models

3.1 Jump diffusion model (JDM)

To obtain the half-width at half-maximum (HWHM), $\gamma(Q)$, of the Lorenztian, $\mathcal{L}(Q,\omega)$, which represents the broadened energy distribution that results from neutron–nucleus collisions and corresponds to the population statistics of one relaxation process, the QENS spectra were fitted using the following function,

$$S(Q,\omega) = [EIS\,F(Q)\delta(\omega) + (1 - EIS\,F(Q))\mathcal{L}(Q,\omega)] \circledast R(Q,\omega), \tag{10}$$

where $R(Q,\omega)$ denotes the resolution function of the instrument and defines the observation time of the spectrometer. The symbol $\mathcal{L}(Q,\omega)$ stands for a Lorentzian function,

$$\mathcal{L}(Q,\omega) = \frac{1}{\pi}\frac{\gamma(Q)}{\gamma(Q)^2 + \omega^2}. \tag{11}$$

The observed QENS spectra essentially arise from the translational diffusion of the molecules and we use here the well-established jump diffusion model (JDM) [1] to describe the experimental data, with the Chudley-Elliot relaxation rate

$$\gamma_{CE}(Q) = \frac{D_t Q^2}{1 + D_t Q^2 \tau_0} \tag{12}$$

where D_t is the translation diffusion coefficient and τ_0 the residence time, i.e. the time between the discrete jumps of the hydrogen atoms.

The fit with a single Lorentzian, $\mathcal{L}(Q,\omega)$ corresponds to choosing the relaxation function in the generic form (5) of the intermediate scattering function to be an exponentially decaying function for $t > 0$,

$$\phi^{(+)}_{JDM}(Q,t) = \exp(-\gamma(Q)|t|), \tag{13}$$

which is symmetrized by choosing the absolute value of t as argument.

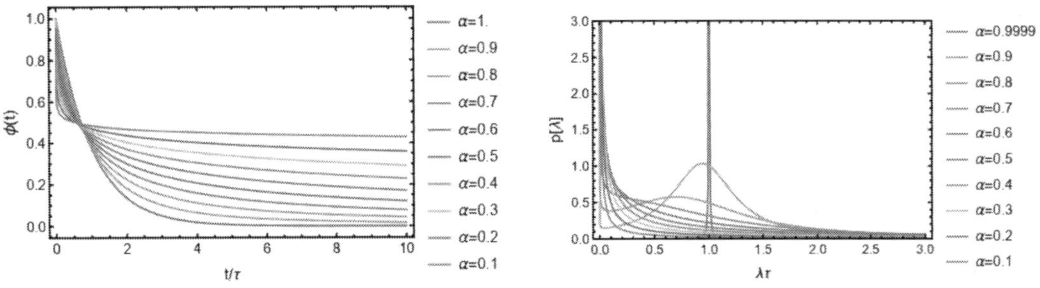

Figure 1. Left panel: Model relaxation function, $E_\alpha(-(|t|/\tau)^\alpha)$, for $0 < \alpha \leq 1$. **Right panel:** The corresponding relaxation rate spectra defined by Eq. (18).

3.2 "Minimalistic" multi-scale relaxation model

The QENS data were also analyzed in the time domain using a "minimalistic" multi-timescale relaxation model (MRM) [12, 13, 14]. The intermediate scattering function required for such analysis was obtained by Fast Fourier transforming the experimental QENS spectra. In the fitting procedure the generic form of the intermediate scatting function, Eq. 8, is assumed to have a (symmetrized) relaxation function of the form of a stretched Mittag-Leffler (ML) function [12, 13, 14] (the Q-dependence is omitted here),

$$\phi_{\mathrm{ML}}^{(+)}(t) \equiv E_\alpha(-(|t|/\tau)^\alpha) \quad (0 < \alpha \leq 1, \tau > 0), \quad (14)$$

which can be considered a generalization of the exponential function and admit $EIS\, F(Q) > 0$. We note that,

$$E_\alpha(z) = \sum_{k=0}^{\infty} \frac{z^k}{\Gamma(1 + k\alpha)} \quad (15)$$

is an entire function in the complex plane [21], where $\Gamma(z)$ is the generalized factorial [22], and that the exponential function is retrieved in the special case $\alpha = 1$. Here α sets the form of the relaxation function and τ its time scale. The left panel of Fig. 1 shows the model relaxation function for various values of α. For $0 < \alpha < 1$ the stretched ML function takes a scale-invariant power law form,

$$\phi_{\mathrm{ML}}^{(+)}(t) \overset{t \gg \tau}{\sim} \frac{(t/\tau)^{-\alpha}}{\Gamma(1-\alpha)}, \quad (16)$$

and since $\lim_{z \to 0} \Gamma(z) = \infty$, this long time tail will vanish for $\alpha \to 1$, i.e for exponential relaxation. The Q-dependent parameters in the above model are thus α, determining the form of the relaxation function, τ, setting the time scale, and the EISF [23].

Important for the following discussions is that the stretched ML function can be written as a continuous superposition of exponentially decaying functions,

$$\phi_{\mathrm{ML}}^{(+)}(t) = \int_0^{\infty} d\lambda\, p_{\alpha,\tau}(\lambda) e^{-\lambda t} \qquad (t \geq 0) \quad (17)$$

where the relaxation rate spectrum, $p_{\alpha,\tau}(\lambda)$, has the form [24],

$$p_{\alpha,\tau}(\lambda) = \frac{\sin(\pi\alpha)}{\pi\lambda\left((\lambda\tau)^{-\alpha} + (\lambda\tau)^{\alpha} + 2\cos(\pi\alpha)\right)}. \quad (18)$$

The above relaxation rate spectrum fulfills the conditions of a probability distribution, which are positivity, $p_{\alpha,\tau}(\lambda) \geq 0$, and normalization, $\int_{-\infty}^{+\infty} d\lambda\, p_{\alpha,\tau}(\lambda) = 1$. The right panel of Fig. 1 shows the relaxation rate spectrum for different values of α on a dimensionless scale, where $\tau = 1$. We note here that all moments $\langle \lambda^k \rangle \equiv \int_{-\infty}^{+\infty} d\lambda\, \lambda^k p_{\alpha,\tau}(\lambda)$ diverge for $k \geq 1$, which follows from the fact that, according to Eq. (14), $\langle \lambda^k \rangle = \partial_t^k \phi_{\mathrm{ML}}^{(+)}(t)\big|_{t=0}$, and that $\phi_{\mathrm{ML}}^{(+)}(t)$ is not differentiable at $t = 0$. The median, $\lambda_{1/2}$, and which is defined through $\int_0^{\lambda_{1/2}} d\lambda\, p_{\alpha,\tau}(\lambda) = 1/2$, has, however, the particularly simple form,

$$\lambda_{1/2} = 1/\tau, \quad (19)$$

which will be used in the following.

4 Experimental and computational methods

4.1 Experimental details and data reduction

Data on bulk-water were collected using IRIS, a cold neutron indirect backscattering spectrometer installed at the ISIS Spallation Neutron and Muon source. The spectrometer provides an intrinsic energy resolution of $17.5\,\mu$eV (FWHM), corresponding to an upper experimental observation of $\approx$ 200 ps for those scattered neutrons energy analysed at $\lambda = 6.3$Å [8]. The bulk-water sample was mounted in a cylindrical sample holder and measurements were conducted at $300K$.

The measured dynamic structure factor is inevitably resolution broadened, which can be mathematically expressed through the convolution integral

$$S_m(Q, \omega) = \int_{-\infty}^{+\infty} d\omega'\, R(Q, \omega - \omega') S(Q, \omega), \quad (20)$$

where the index "m" stands for "measured". A vanadium standard, that can be considered a completely elastic and isotropic material, is used to define the intrinsic resolution function of the spectrometer, $R(Q, \omega)$, and calibrate the measured data. Such intrinsic broadening is accounted for using the reduction and analysis program Mantid [25]. After calibration and background subtraction, using data collected from an empty sample container, the measured QENS spectra were transformed into the dynamic structure factor, $S(Q,\omega)$, and the spectrum grouped to obtain 9 constant scattering vectors in the range $0.55\text{Å}^{-1} \leq Q \leq 1.85\text{Å}^{-1}$.

4.2 MD simulation and analysis

A cubic box of 1728 water molecules, with a side length of $3.72nm$, was simulated at $300K$ using the SPCE force field [26] and applying periodic boundary conditions. Coulomb interactions in presence of periodic boundary conditions were treated by the particle-particle-particle-mesh (PPPM) method [27] and the Lennard-Jones part of the potential was cut at $1.4\,nm$. Prior to the production run, the system was energy minimized and subsequently equilibrated. The production run was performed for 1 ns, with a time step of 1 fs and the Berendsen thermostat [28] was used to keep the system at the desired temperature of $300K$. During the production run, the positions and velocities of the atoms were saved every $0.01\,ps$ and the SHAKE algorithm [29] was used to constrain the interatomic distances within the water molecules. The resulting MD trajectories were analysed using the nMoldyn software package [17, 18, 19, 20] which allowed direct comparison between the simulation results and the experimental data. Note that for this project some of the nMoldyn algorithms were rewritten in Python3 [30], and the physical quantities computed using Computerome [31].

5 Results and Discussion

5.1 Data analysis of the QENS spectra

The measured QENS spectra were analyzed

1. in the frequency domain, using the jump diffusion model Eq. (10), and
2. in the time domain, using the multiscale relaxation function (14) for the symmetrized relaxation function with the generic form (5) of the intermediate scattering function.

The details will be described separately in the following.

5.1.1 Data analysis with the jump diffusion model

The measured QENS spectra were first analysed using Eq. (10) and the Mantid software package [25]. Here a single Lorentzian relaxation is assumed and fast vibrational motions accounted for by adding a linear background term. As a result only a few priors were needed to analyse the data. The result of the fit and obtained parameters, are shown in Figs. 2a and 2b. The translational diffusion coefficient, obtained by fitting $\gamma(Q)$ using Eq. 12, was found to be $D_t = 2.38 \cdot 10^{-9} m^2/s$ and the residence time $\tau_0 = 0.61ps$, corresponding to a jump length of about 1 Å. These values are in agreement with the literature [4, 6]. The quality of the fit can be further verified by inspecting Fig. 3.

5.1.2 Data analysis with the multi-scale relaxation model

Here the QENS spectra are analyzed in the time domain, using the generic form given by Eq. (8) of the intermediate scattering function with the model relaxation function described by Eq. ((14). For the analysis three steps are performed. For more details we refer the reader to the original work describing the model [13]:

1. **Symmetrization the QENS spectra.** Based on the detailed balance relation, Eq (4), the QENS spectra are symmetrized with respect to ω,

$$S^{(+)}(Q,\omega) = \frac{e^{-\beta\hbar\omega/2} S(Q,\omega)}{\int_{-\infty}^{+\infty} d\omega\, e^{-\beta\hbar\omega/2} S(Q,\omega)} \quad (21)$$

which guarantees that $\int_{-\infty}^{+\infty} d\omega\, S^{(+)}(Q,\omega) = F^{(+)}(Q,0) = 1$. By defining and using $S^{(+)}(Q,\omega)$ we make sure that our spectra are symmetric and fulfill the detail balance relation, making it possible to easily Fourier transform the spectra and describe it using semi-classical models [13].

2. **Compute and deconvolve the intermediate scattering function.** Assuming that the resolution function, Eq. 20, is symmetric in ω, the "measured" intermediate scattering function is computed using a discrete approximation of the inverse Fourier transform $F_m^{(+)}(Q,t) = \int_{-\infty}^{+\infty} d\omega\, e^{i\omega t} S_m^{(+)}(Q,\omega)$. By writing $F_m^{(+)}(Q,t) = R(Q,t) S^{(+)}(Q,t)$ according to the convolution theorem of the Fourier transform, we obtain,

$$F^{(+)}(Q,t) = F_m^{(+)}(Q,t)/R(Q,t) \quad \text{for} \quad |t| < t_c. \quad (22)$$

The cutoff time is determined by the resolution ΔE (FWHM) of the instrument, writing $t_c = \hbar/2\Delta E$. An explicit illustration can be found in Ref. [23].

3. **Parameter fit.** Once the experimental data was reduced following the steps described above, the deconvolved $F^{(+)}(Q,t)$ was fitted using Eq. (8) using the IMinuit package [32]. This approach uses a least-square fitting routine to find the set of parameters which minimize the χ^2-value [32].

The result of fitting $F^{(+)}(Q,t)$ with Eq. (14) is shown at selected Q-value Fig. 2c. It can be observed that the model follows the experimental data quite well. From the fit parameters, shown in Fig. 2d, we note that the evolution of $\alpha(Q)$ accounts for a certain heterogeneity of the local environment of the diffusing water molecules, which becomes visible on increasingly larger spatial scales where $\alpha(Q) \approx 0.9$. Besides, the fact that local motions are faster than more delocalized ones is clearly reflected by the decay of the relaxation time scale $\tau(Q)$ with Q.

5.1.3 Comparing the results obtained in frequency and time domain

We start the comparison between the results of the QENS data analyses with the two models by resuming that the

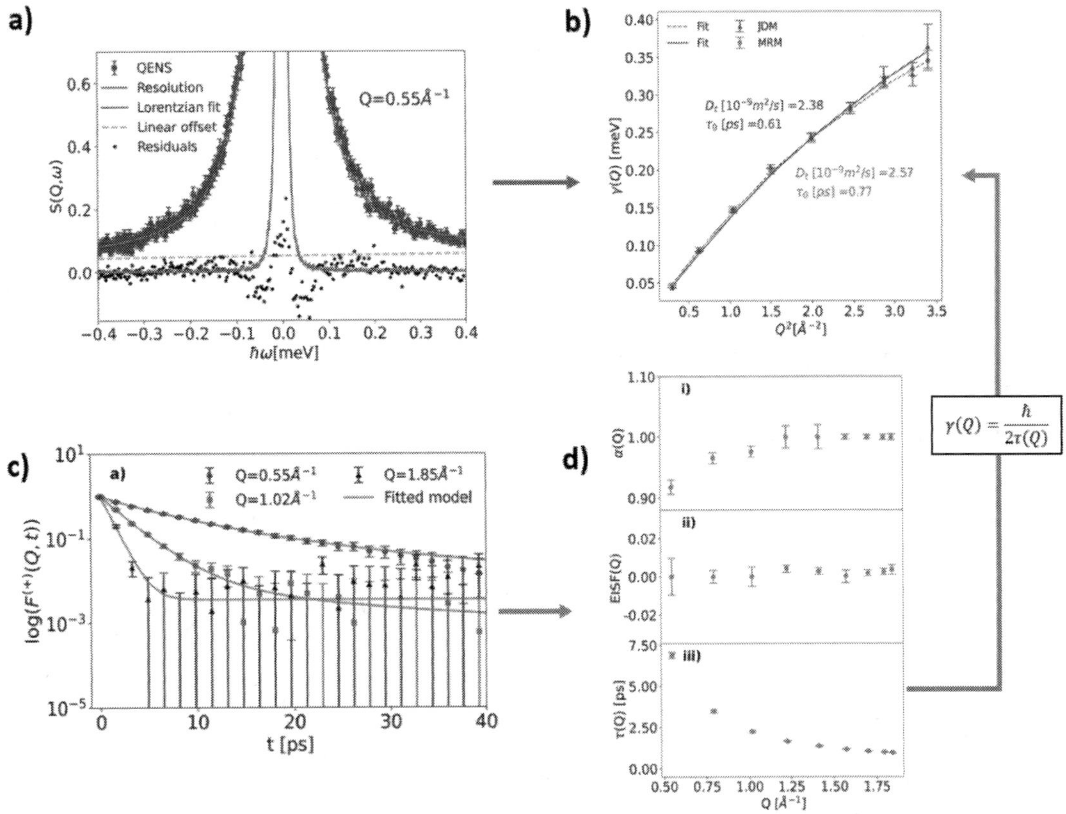

Figure 2. Overview of the data analysis performed on the QENS spectra for bulk-water at $300K$. a) The jump diffusion model (JDM) fit of the dynamic structure factor, $S(Q, \omega)$, according to Eq. 10. The residuals between the Lorentzian fit and the QENS spectrum are also plotted along with the resolution and the linear background. b) The half width at half maximum (HWHM) $\gamma(Q)_{JDM}$ values extracted from each JDM fit as a function of Q^2, compared to the HWHM $\gamma(Q)_{MRM}$ determined from the $\tau(Q)$ parameter of the multi-scale relaxation model (MRM). Both sets of HWHM data are fitted using, Eq. 12, where the fit results are presented above/below the fit in a corresponding color. c) MRM fit of the symmetrized, normalized, and deconvolved intermediate scattering function, $F^{(+)}(Q,t)$, according to Eq. 8. d) Water dynamics parameters of the QENS spectra as a function of Q obtained from the MRM model fit of $F^{(+)}(Q,t)$: i) $\alpha(Q)$, ii) the $EIS\,F(Q)$, and iii) $\tau(Q)$. The lack of Q-dependence and zero value for the $EIS\,F(Q)$ reflects that, as expected, the water molecules are not spatially constrained.

JDM describes the relaxation of the intermediate scattering function with a single exponential function, Eq. (13), whereas the MRM describes the relaxation with a stretched Mittag-Leffler function, Eq. (14). Trivially $\gamma(Q) = 1/\tau(Q)$ in the limit $\alpha \to 1$, where $\phi^{(+)}_{\mathrm{ML}}(t) = \phi^{(+)}_{\mathrm{JDM}}(t)$, but the interesting question is how $1/\tau(Q)$ compares to $\gamma(Q)$ for $\alpha < 1$. The answer is given in Fig. 2b, which shows that $\gamma(Q) \approx 1/\tau(Q)$ over the whole Q-range and that thus for both models (JDM and MRM) yield similar values are obtained for the parameters D_t and τ_0. This is an important result since it shows that the MRM can be considered as a refined form of the JDM, in the sense that the Chudely-Elliot model is a physical model for the median of the MRM,

$$\lambda_{1/2}(Q) \approx \gamma_{\mathrm{CE}}(Q). \tag{23}$$

This also further confirms that the multi-scale relaxation model is able to fully capture the complex and heterogeneous relaxation processes of the intermediate scattering function without prior assumptions and a minimal number of fit parameters.

5.2 Reproducing the experimental spectra by MD simulations

To verify that the MD simulation reproduces the QENS data, we compare the results graphically for three different Q-values in Fig. 4. Here the QENS data were transformed into $F^{(+)}(Q,t)$ following steps 1-2 described in Section 5.1, while the produced trajectories of the MD simulations were converted into the intermediate scattering function of the form $F(Q,t)$, using rewritten algorithms from nMoldyn [30]. Note that the coherent part of $F(Q,t)$ is omitted for the MD simulations, since we assume that only the incoherent contribution is probed in the experiment.

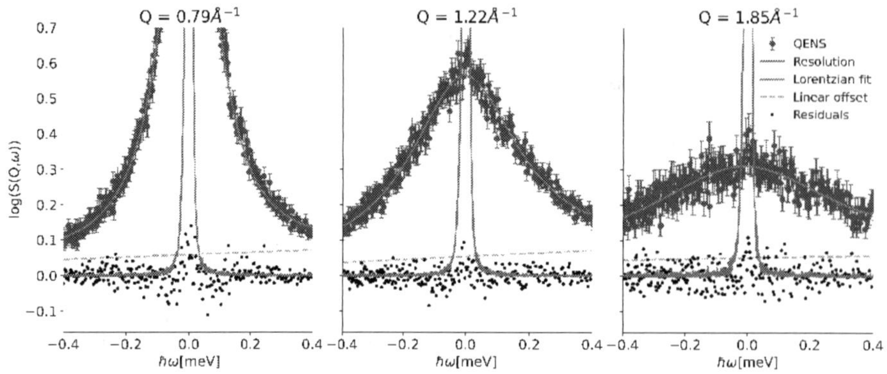

Figure 3. Zoom of the QENS experimental spectra for bulk-water at 300 K fitted using the jump diffusion model (JDM), according to Eq. 10 at three different Q-values. The residuals between the Lorentzian fit and the QENS spectrum are also plotted along with the resolution and the linear background.

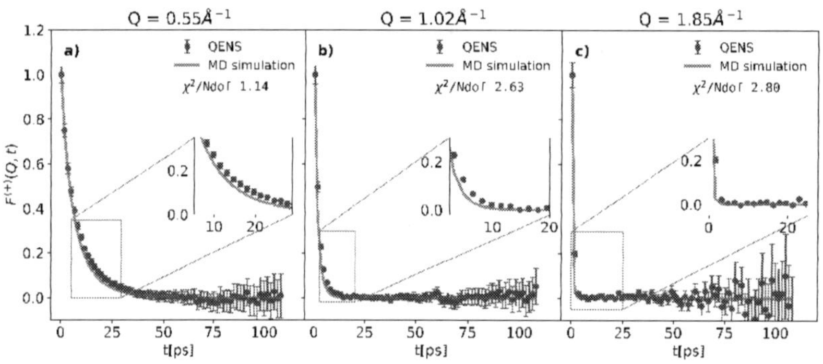

Figure 4. The symmetrized, normalised and deconvolved intermediate scattering function, $F^{(+)}(Q,t)$, for the QENS data and the MD simulation of bulk-water at $300K$, at three different Q-values.

Looking at Fig. 4, we observe that the MD simulation follows the QENS data at longer times. The large variation of $F^{(+)}(Q,t)$ with increasing computation time was previously noted in experimental QENS data [33] and is most likely related to the numerical Fourier transformation. Furthermore, considering that the dynamics of the system is hidden in the decay of $F^{(+)}(Q,t)$, we quantify the difference between the QENS data and MD simulation for all Q-values by the printed reduced χ^2-value. Here we observe that for higher Q-values the simulations do not reproduce the data equally well.

5.3 Reproducing the modelled spectra by MD simulations

Knowing that the MD simulation is comparable with the QENS data up to a certain point in time and space, we now evaluate the MD simulation output against the modelling approaches used to analyse the measured QENS data.

Using the MRM to inspect the MD results is straightforwardly done by fitting the simulated $F(Q,t)$ to Eq. 8. To consider the JDM, a free single exponential function is used to describe the diffusive motion. The results are depicted in Fig. 5a, where the two fitted models are plotted against the experimental data for two Q-values. We observe that the fitted curves follow the data in $F^{(+)}(Q,t)$ well. However, by analysing Fig. 5b, where the fitted models $F(Q,t)_{fit}$ are transformed into $S(Q,\omega)_{fit}$ and compared with the measured spectra, both models capture the measured QENS spectra particularly well for low Q-values only. This result can be better understood by analysing the evolution of $\gamma(Q)$, as a function of Q^2 obtained from the two models fits and depicted in Fig. 5d. We observe that for $Q > 1\text{Å}^{-1}$ the QENS data show asymptotic behavior, characteristic of jump diffusion of the atoms, while the MD simulations shows a linear growth indicating free diffusion with $D_t^{MD} \approx 3 \cdot 10^{-9}\frac{m^2}{s}$. As previously reported [6, 34], classic MD simulations have difficulties capturing the quantum effects of jump diffusion, making it plausible that the MD simulations continue probing free diffusion. This is somehow expected since the semi-classical approx-

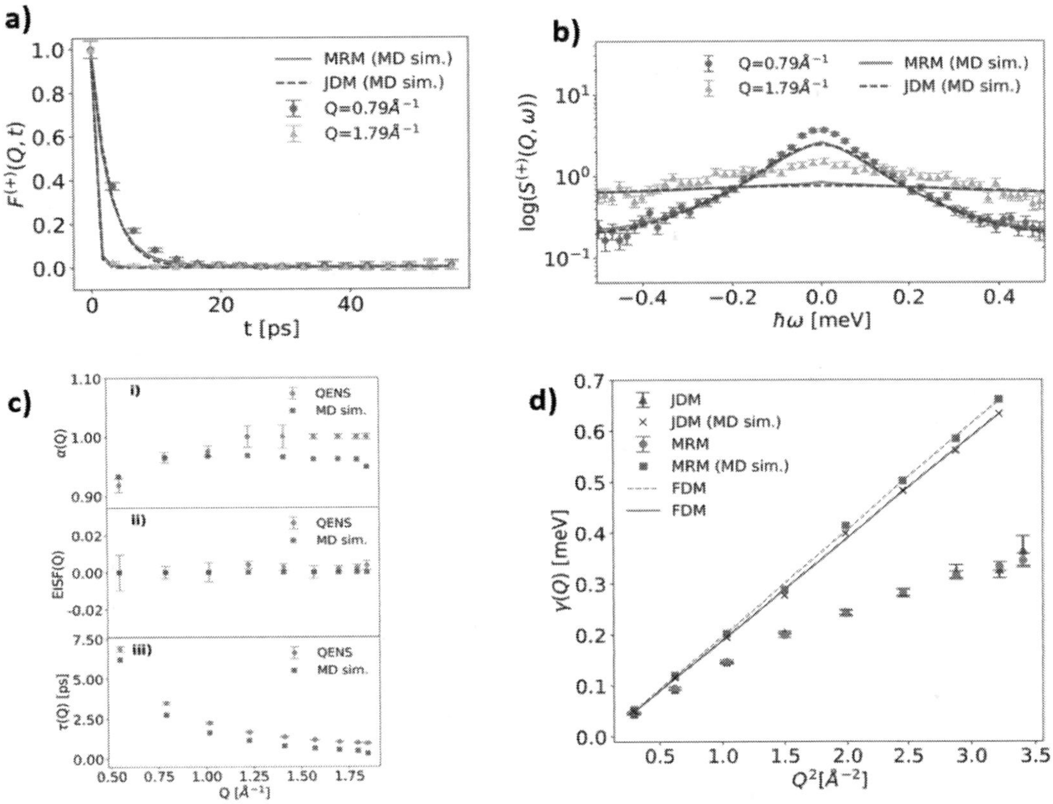

Figure 5. Overview of the data analysis performed on the MD simulation of bulk-water at 300K. a) The symmetrized, normalized, and deconvolved intermediate scattering function, $F^{(+)}(Q,t)$, of the QENS data for bulk-water at 300K plotted together with the jump diffusion model (JDM) fit and multi-scale relaxation model (MRM) fit of the MD simulation, according to a single exponential function and Eq. 8. b) The fitted models transformed into the symmetrized, normalized and convolved dynamic structure factor, $S^{(+)}(Q,\omega)$, plotted alongside the QENS spectra of bulk-water at 300K for comparison. c) Water dynamics parameters for the QENS data and MD simulation as a function of Q obtained from the multi-scale relaxation model fit of $F^{(+)}(Q,t)$: i) $\alpha(Q)$, ii) the $EIS\,F(Q)$, and iii) $\tau(Q)$. The lack of Q-dependence and zero value for the $EIS\,F(Q)$ reflects that, as expected, the water molecules are not spatially constrained. d) Extracted half width at half maximum (HWHM) $\gamma(Q)_{JDM(MD)}$ for each JDM fit of the QENS spectra and MD simulation as a function of Q^2, compared to the HWHM $\gamma(Q)_{MRM(MD)}$ determined from the $\tau(Q)$ parameter of the MRM fit of the QENS spectra and MD simulation. Both HWHM for the MD simulation are fitted using the free diffusion model (FDM), $\gamma(Q) = D_t Q^2 + const.$ resulting in $D_t^{JDM(MD)} = 3.04 \cdot 10^{-9} \frac{m^2}{s}$ and $D_t^{MRM(MD)} = 3.17 \cdot 10^{-9} \frac{m^2}{s}$. Both show minimal offset. Note that the HWHM for the last Q-value is omitted for the MD simulation, due to its larger size.

imation of the detailed balance relation, used to make the classical time correlation function equal to the quantum time correlation function, is only valid to the first order of $\hbar$ meaning that for high Q-values the approximation breaks down.

Now we turn to the evolution of the parameters of the MRM found by the fit of the MD simulation as a function of Q presented in Fig. 5c along with the fit parameters found for the QENS measured data. The following can be observed. (i) The value for the $\alpha(Q)$ parameter obtained by the two approaches agrees until $Q \approx 1.2\text{Å}^{-1}$, where it starts to diverge. This indicates that a slightly different diffusive motion of the hydrogen atoms is probed, where the MD simulation shows diffusion within a rougher harmonic potential, as the plateau value of $\alpha \approx 0.95$ suggests. (ii) $\tau(Q)$ shows the same Q-dependence with a 10% difference in magnitude, confirming that the MD simulation probes a faster diffusion.

6 Conclusions

Here, molecular dynamics (MD) simulations at 300K have been computed for bulk-water and compared to the measured quasi-elastic neutron scattering (QENS) spectra of the same system. The intermediate scattering function $F(Q,t)$ for the MD simulations, computed using updated nMoldyn algorithms, could be related to the measured QENS spectra either in terms of $F(Q,t)$ or its Fourier transform; known as the dynamic structure factor $S(Q,\omega)$. The comparison was done graphically, with supporting χ^2-values, along with analysis using both the jump diffusion model (JDM), consisting of fitting the measured spectra with a sum of Lorentzians, and a newly introduced "minimalistic" multi-scale relaxation model (MRM), which describes the the multi-scale relaxation of the hydrogen atoms of the measured spectra. The juxtaposition of the QENS data and the MD simulation for bulk-water showed that the MD simulation describes well the translational dynamics of the water relaxation. However, further investigation revealed that in reality the MD simulation cannot completely capture the discrete jumps occurring at more localized length scales where quantum effects becomes significant.

More importantly, by revisiting the modeling of QENS from bulk water, we show that the MRM, a new "minimalistic" energy landscape-based method for QENS analysis, can be used to accurately describe the data at all length scales. This is confirmed by comparing the Chudley-Elliot relaxation time and the relaxation time scale $\tau(Q)$ obtained from MRM, defining the *median relaxation rate* through $\lambda_{1/2}(Q) = 1/\tau(Q)$, and showing that these values follow on the same master curve.

References

[1] R.J. Elliott, C.T. Chudley, Proceedings of the Phys Society **77**, 353 (1961)
[2] P. Schofield, Phys Rev Letters **4**, 239 (1960)
[3] L. Van Hove, Phys Rev **95**, 249 (1954)
[4] J. Teixeira, Phys Rev. A, General Phys **31**, 913 (1985)
[5] H. Frauenfelder, P.W. Fenimore, R.D. Young, Proceedings of the National Academy of Sciences **111**, 12764 (2014)
[6] J. Qvist, et al., J Chem Phys **134**, 144508 (2011)
[7] H.N. Bordallo, et al., J Phys Chem C **112**, 13982 (2008)
[8] M.C. Berg, et al., ACS Applied Materials & Interfaces **10**, 9904 (2018)
[9] M.R. Harpham, et al., J Chem Phys **121**, 7855 (2004)
[10] R. Ignazzi, et al., J Phys Chem C **121**, 23582 (2017)
[11] T. Yamada, H. Seto, Front Chem. **8**, 8 (2020)
[12] G. Kneller, Proceedings of the National Academy of Sciences **115**, 9450 (2018)
[13] M. Saouessi, J. Peters, G.R. Kneller, J Chem Phys **150**, 161104 (2019)
[14] M. Saouessi, J. Peters, G.R. Kneller, J Chem Phys **151**, 125103 (2019)
[15] M.H. Petersen, et al., J Phys Chem C **125**, 15085 (2021)
[16] M. Childers, V. Daggett, J Phys Chem B **122**, 6673 (2018)
[17] G. Kneller, V. Keiner, M. Kneller, M. Schiller, Comp Phys Com **91**, 191 (1995)
[18] T. Róg, K. Murzyn, K. Hinsen, G. Kneller, J Comp Chem **24**, 657 (2003)
[19] V. Calandrini, E. Pellegrini, K. Hinsen, G. Kneller, École thématique de la Société Française de la Neutronique **12**, 201 (2011)
[20] K. Hinsen, E. Pellegrini, S. Stachura, G. Kneller, J Comp Chem **33**, 2043 (2012)
[21] R. Gorenflo, A.A. Kilbas, F. Mainardi, S.V. Rogosin, eds., *Mittag-Leffler Functions, Related Topics and Applications*, Springer Monographs in Mathematics (Springer, 2014)
[22] F.W.J. Olver, D.W. Lozier, R.F. Boisvert, *NIST handbook of mathematical functions hardback and CD-ROM* (Cambridge Univ Pr, 2010)
[23] A.N. Hassani, L. Haris, M. Appel, T. Seydel, A.M. Stadler, G.R. Kneller, J Chem Phys **156**, 025102 (2022)
[24] G.R. Kneller, Phys Chem Chem Phys **7**, 2641 (2005)
[25] O. Arnold, et al., Nuclear Instruments and Methods in Phys Research Section A: Accelerators, Spectrometers, Detectors and Associated Equipment **764**, 156 (2014)
[26] H.J.C. Berendsen, et al., J Phys Chem **91**, 6269 (1987)
[27] M.P. Allen, D.J. Tildesley, *Computer Simulation of Liquids* (Oxford, 2017)
[28] H.J.C. Berendsen, J.P.M. Postma, W.F. van Gunsteren, A. DiNola, J.R. Haak, J Chem Phys **81**, 3684 (1984)
[29] J.P. Ryckaert, G. Ciccotti, H.J. Berendsen, J Comp Phys **23**, 327 (1977)
[30] M.H. Petersen, *Overview of the used code*, Published at https://github.com/ in a private repository and access can be obtained by writing to marti nhp27@gmail.com (2022)
[31] *Computerome: The danish national supercomputer for life sciences*, https://www.computerome.dk/display/C2W/Computerome+2.0+Wiki
[32] H. Dembinski, P. Ongmongkolkul, C. Deil et al., Zenodo, doi **10** (2020)
[33] M.T.F. Telling, et al., Soft Matter **7**, 6934 (2011)
[34] T.O. Farmer, et al., J Phys Chem Letters **11**, 7469 (2020)

ICE-MAN the Integrated Computational Environment for Modeling and Analysis for Neutrons at ORNL

Anibal Ramirez-Cuesta[1,*], *Robert* Smith[2,**], *Eugene* Mamontov[3,***], and *Yongqiang* Cheng[1,****]

[1]Neutron Sciences Directorate, Neutron Scattering Division, Oak Ridge National Laboratory, Oak Ridge Tennessee 37831, USA
[2]Computing and Computational Sciences Directorate, Computer Science and Mathematics Division, Oak Ridge National Laboratory, Oak Ridge Tennessee 37831, USA
[3]Second Target Station, Instrument Systems, Oak Ridge National Laboratory, Oak Ridge Tennessee 37831, USA

Abstract. ICE-MAN is a modeling and analysis workbench for multi-modal studies, designed with neutron science in mind. It streamlines the workflow between different experimental techniques, computer modeling, and databases and reduces the time and learning curve needed to access them thus making a holistic approach to data interpretation more amenable and efficient.

1 Introduction

One of the most significant strengths of neutron scattering is the direct correlation between theoretical models and experiment; it is possible to calculate the neutron scattering cross-section rigorously from models [1]. However, most tools used to analyze data from atomistic modeling are qualitative rather than quantitative. Many standard fitting tools do not necessarily include atomistic detail, particularly those extracted from electronic structure calculations. This paper describes a platform for analyzing neutron scattering data that can be broadly applied across instrument suites at SNS and HFIR, building a suite of individual data analysis/processing engines that streamline workflows in data analysis for neutron scattering, the Integrated Computational Environment for Modeling and Analysis for Neutrons (ICEMAN).

Currently, there are two main components, OCLIMAX [2–7] and QClimax [8]. OCLIMAX is principally used on VISION (BL-16B) but can also be used to analyze phonon data from ARCS [9–12], CNCS, HySPEC [13], SEQUOIA [14–17], CORELLI, triple axes spectrometers (ORNL), LET, MAPS [18], MARI [19], MERLIN [20], TOSCA [18, 21–23], (ISIS), DCS (NIST), AMATERAS [19] (J-PARC), NERA [24] (Dubna) etc. OCLIMAX can calculate thermal neutron scattering cross-section of solids [4, 16, 17, 25]. QClimax, on the other hand, provides users with a flexible, straightforward environment to enable fitting to complex user-defined functions. It allows fits to be performed to data from multiple Q values simultaneously and provides global fitting parameters. QClimax can be used to analyze QENS data from any quasielastic spectrometer, including BASIS [26–33], CNCS (ORNL), DCS [34], HFBS [27, 29–31, 34, 35] (NIST), IRIS [36], OSIRIS (ISIS), TOFTOF [26, 37], SPHERES [37] (FRM II), etc. In addition. there are several small modules to generate input files, read output files and convert formats [6, 7].

2 The OCLIMAX module

The OCLIMAX software calculates the neutron scattering synthetic spectra, $S(\mathbf{Q}, \omega)$, for solids. When doing lattice dynamics calculations, and the harmonic approximation the results are rigurous, and it is necessary to provide the eigenvectors (displacement vectors) and eigenvalues (frequencies) for the calculation. OCLIMAX continues the naming tradition first established by Kearley and Tomkinson [38] and later by A.J. Ramirez-Cuesta [1, 39]. The initial version used force constants to fit INS spectra of simple molecules, the second was able to produce INS spectrum from the output of DFT modeling, for details see [2]. The conventional approach using density functional theory and lattice dynamics often falls short when the materials of interest are complex (e.g., defective, disordered, heterogeneous, amorphous, large-scale), for which molecular dynamics driven by interatomic force fields are a more common approach. OCLIMAX can directly convert molecular dynamics trajectories into simulated INS spectra, including not only fundamental but also higher order excitations [3].

2.1 Main features

OCLIMAX can calculate the spectra for extended solids, molecules in solids (using the isolated molecule approach). It can introduce solid state effects in the isolated molecule approach by using the shape of the phonon wings to approximately calculate overtones and combinations. For the calculation of the INS spectra of molecules

*e-mail: ramirezcueaj@ornl.gov
**e-mail: smithrw@ornl.gov
***e-mail: mamontove@ornl.gov
****e-mail: chengy@ornl.gov

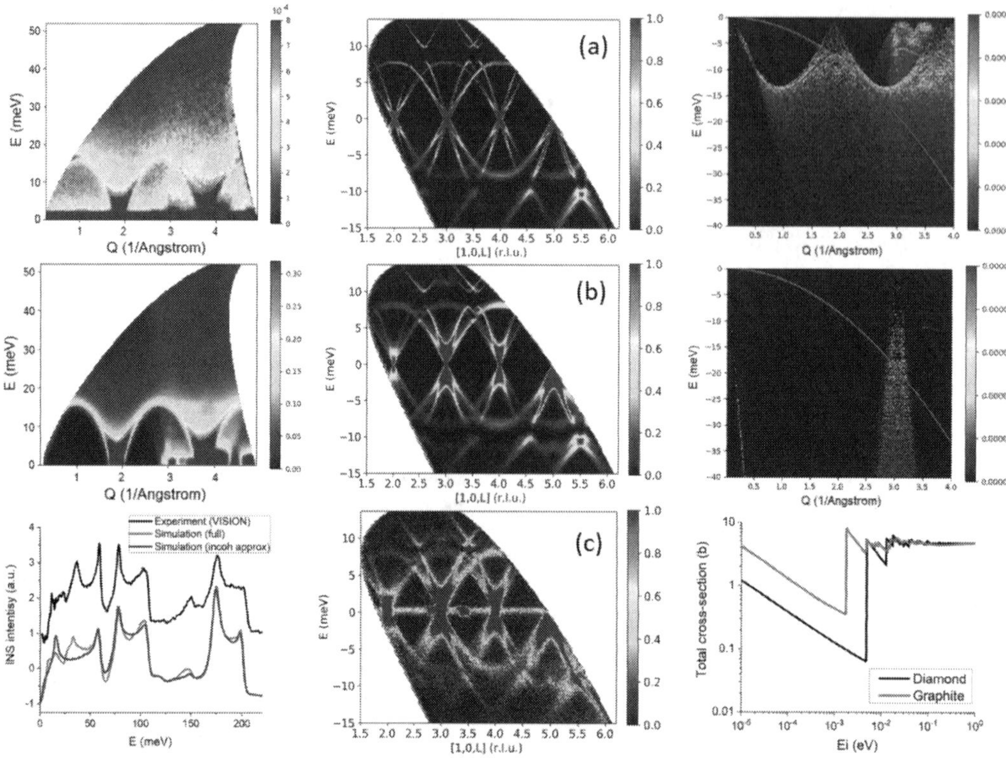

Figure 1. Examples of OCLIMAX applications. left column top and center, the S($\mathbf{Q}, \omega$) map for graphite powder, bottom the INS spectra calculated for graphite along the VISION trajectory. Middle column, phonon dispersion along the [1, 0, L] transverse cut using the HYSPEC spectrometer [13]. DFT simulations (a) without integration and (b) with same integration and range as measured data in (c). (c) Measured INS data integrated at $-0.12 < \zeta < 0.12$ in $(1 + \zeta, 0, 0)$ and $-0.12 < \eta < 0.12$ in $(1, \eta, 0)$. Right column, OCLIMAX calculation of the neutron energy gain part of the S($\mathbf{Q}, \omega$) map for (top) graphite and (center) diamond. The red line indicates the trajectory where nearby phonons have major contributions to the scattering cross-section for low-energy ($< 1meV$) neutrons. Note that to see the weak phonon excitations in diamond, the scale of the color bar for the right panel is only 1/10 of the left panel.

it uses the incoherent approximation, that essentially consists of using the total cross section of each atom.

For extended solids with periodic boundary conditions it is possible to calculate the incoherent and coherent contributions to S($\mathbf{Q}, \omega$). The contribution from elastic scattering can also added. The sum of all contributions is the total scattering function. The program can calculate single crystal and powder averaged spectra of S($\mathbf{Q}, \omega$), the whole map and along a given trajectory, like VISION. It can calculate direct and indirect geometry instrumentation. Instrument resolution can be entered as an empirical polynomial, or as in the case of direct geometry spectrometers, using instrumental parameters. It is also possible to calculate generic S($\mathbf{Q}, \omega$) maps covering given areas of $\mathbf{Q}$ and ω with arbitrary resolution.

2.2 Examples

A series of examples are shown in figure 1. On the left column is the experimental and calculated S($\mathbf{Q}, \omega$) for graphite as seen is SEQUOIA. At the bottom of the left column is the spectra of graphite as measured in VISION, with the calculated spectra using the incoherent approximation (blue trace) and the full coherent calculation (red trace). It can be easily seen that the coherent effects have to be taken into consideration to fully reproduce the experimental data.

In the mid column are the S($\mathbf{Q}, \omega$) from DFT calculation top and middle, at the bottom is the experimental data of a single crystal of $\alpha - TeO_2$ measured at HySPEC. Upon doing the same integration and range as measured data, i.e. including the correct momentum resolution, the calculation effectively reproduces the experimental data.

On the right column, the top two panels show the neutron energy gain for graphite and diamond at 300 K. The bottom panel shows the calculated total neutron cross section for both materials. The origin of the different transmission at low incident energy is clearly illustrated. Graphite has a low-lying phonon branch below 15 meV, which is the LA/LO mode and is highly populated at room temperature. The $\omega, \mathbf{Q}$ trajectory at low incident energy cuts through the branch in between 5 and 15 meV, contributing to its inelastic scattering cross-section. In contrast, diamond does not have a major phonon branching lying in the same $\omega, \mathbf{Q}$ trajectory. It only crosses a weak band between 15 and 20 meV at higher Q, resulting in the lower cross-section. It is worth noting that such insight can only be obtained through a full calculation of this kind including the coherent effects; the conventional approach using the phonon density of states will be far less revealing [4].

2.3 Workflows with OCLIMAX

The OCLIMAX program and its manual (including installation and user guide) can be found on the Web [5]. OCLIMAX requires an *.aclimax* (incoherent approximation simulation only) or *.oclimax* (required by coherent simulation) file, as well as an optional *.params* file to run. [2] To convert MD trajectories to INS spectra, OCLIMAX can be used in essentially the same way as reported previously, with the same format for the parameter file. Here instead of having the *.aclimax* or *.oclimax* file for phonon information, we need a *.tclimax* file for the trajectory. [3]

The *.aclimax*/*.oclimax* file contains information on the atomic structure, isotope, neutron scattering cross sections, and phonons, and the .params file contains parameters defining the simulation (if different from the default). The *.aclimax*/*.oclimax* file can be obtained from the output file(s) of a phonon calculation (e.g., using Phonopy), by running the conversion script provided in the program. Currently the script supports output files from CASTEP, VASP (either directly from VASP or through Phonopy), CP2K, Quantum Espresso, Phonopy, ORCA, DMol3, NWChem, etc.

In the *.params* file, users can specify how the calculation should be done. The default setup is for VISION/TOSCA spectrum simulation. One can modify the *.params* file to include combinations and overtones (up to a desired order), temperature effect, coherent effect, isotope substitution, and phonon wing calculation. The users can select to simulate the full S($\mathbf{Q}, \omega$) map or certain sampling trajectories corresponding to a (real or virtual) direct or indirect geometry instrument. A detailed explanation of the input parameters can be found in the manual. The program is currently distributed as a Docker image [40], allowing it to run on various platforms with minimum environment dependence. For most calculations with optimized phonon sampling density and parameters, it should take no more than a few minutes to finish. Also distributed on the Web site is a script to generate quick plot for visualization of the calculated spectra (*pclimax.py*).

The execution of OCLIMAX is a the command line call: *$ OCLIMAX run yourfile.oclimax*. A conversion tool is included in the OCLIMAX Docker distributions. For help with these, run *$ OCLIMAX convert* on the command line, the output will describe the available options.

The standard output data is in *.csv* format. The users can also select to generate a restart file (in binary format), an SPE file (for DAVE [41]), or (neutron weighted) phonon density of states. Overall it is much more powerful than its previous version aCLIMAX [39], as well as the recently developed ABINS algorithm [42] in Mantid [43].

3 The Qclimax module

Qclimax provides users with a flexible, straightforward environment to enable fitting to complex, user-defined functions. The power of QENS arises from the ability to access both momentum and energy transfer domains at the same time, providing information about the characteristic diffusion processes. It simultaneously allows fits to be performed to data from multiple $\mathbf{Q}$ values and provides global fitting parameters.

Typically, the measured intensity of the QENS experiments, $I(\mathbf{Q}, \omega)$, can be described as follows [44]:

$$I(\mathbf{Q}, \omega) = \{x(\mathbf{Q}) \times \delta(\omega) + (1 - x(\mathbf{Q}))S(\mathbf{Q}, \omega))\} \otimes R(\mathbf{Q}, \omega) + B(\mathbf{Q}, \omega) \quad (1)$$

where, $\mathbf{Q}$ is the momentum transfer and ω is the energy transfer, $x(\mathbf{Q})$ is the elastic fraction of the signal; $\delta(\omega)$ is a Dirac delta function accounting for the referred elastic scattering (zero energy transfer);$S(\mathbf{Q}, \omega)$ is the dynamic structure factor, i.e., the QENS model scattering function; $B(\mathbf{Q}, \omega)$ is a linear background; and $R(\mathbf{Q}, \omega)$ corresponds to the instrument resolution function, the symbol $\otimes$ represents the convolution operation with the resolution function . In general, $B(\mathbf{Q}, \omega)$ accounts for dynamical processes that are too fast for the dynamic range of the instrument (that is, with timescales shorter than a few picoseconds) and, possibly, sample- and temperature-dependent backgrounds. As an example, $S(\mathbf{Q}, \omega)$ can be described by a Lorentzian function:

$$S(\mathbf{Q}, \omega) = \frac{1}{\pi} \frac{\Gamma(\mathbf{Q})}{\Gamma(\mathbf{Q})^2 + \omega^2} \quad (2)$$

where $\Gamma(Q)$ is the half width at the half maximum of the curve.

In the simple case of a single Lorentzian process with a linear background, the equation to fit the experimental data is:

$$I(\mathbf{Q}, \omega) = \left\{x(\mathbf{Q}) \times \delta(\omega) + (1 - x(\mathbf{Q}))\frac{1}{\pi}\frac{\Gamma(Q)}{\Gamma(Q)^2 + \omega^2})\right\} \otimes R(\mathbf{Q}, \omega) + A(\mathbf{Q}) \times \omega + B(\mathbf{Q}) \quad (3)$$

The sequential fitting procedure consists in binning the data in a number of values in $\mathbf{Q}$ and fitting Equation 3 to the data for each individual value of $\mathbf{Q}$. The parameters of the fits as a function of $\mathbf{Q}$, are, in this case: $x(\mathbf{Q})$, $\Gamma(\mathbf{Q})$, $A(\mathbf{Q})$, and $B(\mathbf{Q})$. The parameters are then fitted to a given functional form in $\mathbf{Q}$. In this case, for example, the generic equation for $\Gamma(\mathbf{Q})$ is:

$$\Gamma(\mathbf{Q}) = f(\mathbf{Q}, \alpha, \beta, \ldots) \quad (4)$$

where $\alpha, \beta, \ldots$ are the parameters of interest that contain the physical meaning of the phenomena.

QClimax allows to include the functional form in $\mathbf{Q}$ within the fitting process. The results of the fitting are the parameters that fit Equation 3 given the constraints from

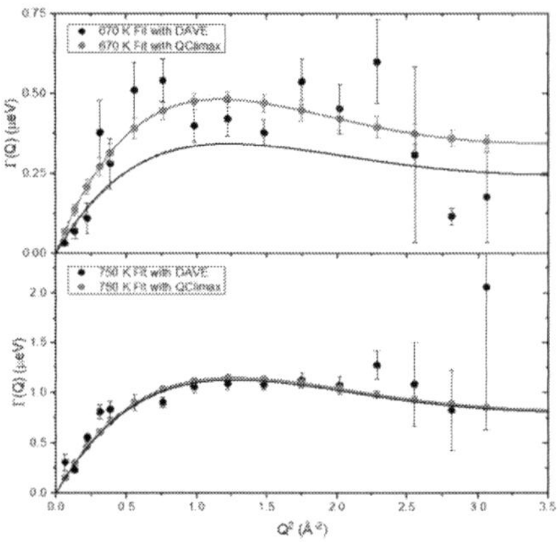

Figure 2. Example of fitting data using sequential and global fits in the case of BaH_2 [32], for T = 670K at the top and T = 750K at the bottom panel. The black points are the fitted values of Γ(**Q**) using sequential fitting (both DAVE and QClimax give the same values), the black trace is the fitting of the data to the Chudley–Elliott model, including all values. The red points are the fitted points using global fitting in QClimax, the red trace is the overall model.

Equation 4. This global fitting effectively directly gives the parameters $\alpha, \beta, \ldots$ from the fits.

Figure 2 shows how the use of global fitting. On the top panel, the sequential fit is not good on the last few points at large **Q**. The global fitting, on the other hand, introduces in the analysis a "prior knowledge" of the system, i.e., it behaves as a Chudley–Elliott model. This "prior knowledge" is an intrinsic Bayesian behavior that makes the fittings more robust. The bottom panel, T = 750K finds similar results using both methods, this is due to the better quality of the sequential fitting.

Qclimax has been written in python and uses the LMFIT python library [45]; it can be accessed through the ICEMAN homepage [46] (it is necessary to register for XCAMS access [47] to use it).

There are some predefined peak shapes built in QClimax. To add extra peaks shapes, please contact the authors. QClimax allows the user to create new constraint functions in Python. These functions once created are saved in the user's profile. All the data resulting from the fitting is saved in the profile. It also saves the initial data files, parameters, fitting functions, and constraint functions. The results can be downloaded as a *.zip* file and shared. All the information contained can be used to fully reproduce the fitting results.

The experimental files are in DAVE *.grp* format [41]. It is recommended to use μeV as units in energy. The initialization file *.ini* contains the required information for the fittings.

4 Conclusions

The ICEMAN project is still in development. The ICEMAN software with its two modules, OCLIMAX and QClimax, is available for use. OCLIMAX is available as a Docker container, QClimax can be accessed through the ICEMAN webpage. For QClimax there are Virtual Machine builds and soon it will be a Docker version available. For more information, suggestions, collaborations, etc., please contact any of the authors of this paper.

This research benefited from the use of the BASIS, HySPEC, SEQUOIA ,and VISION beam-lines at the Spallation Neutron Source a DOE Office of Science User Facility operated by the Oak Ridge National Laboratory. The computing and software resources were made available through the VirtuES and the ICEMAN projects, funded by the Laboratory Directed Researchand Development program (LDRDs 7739, 8237,10447) and Compute and Data Environment for Science (CADES) at the Oak Ridge National Laboratory, which is supported by the Office of Science of the DOE under Contract DE-AC05-00OR22725

References

[1] P.C. Mitchell, S.F. Parker, A.J. Ramirez-cuesta, J. Tomkinson, *Vibrational Spectroscopy With Neutrons-With Applications In Chemistry, Biology, Materials Science And Catalysis*, Vol. 3 (World Scientific, 2005)

[2] Y.Q. Cheng, L.L. Daemen, A.I. Kolesnikov, A.J. Ramirez-Cuesta, Journal of Chemical Theory and Computation **15**, 1974 (2019), `https://doi.org/10.1021/acs.jctc.8b01250`

[3] Y.Q. Cheng, A.I. Kolesnikov, A.J. Ramirez-Cuesta, Journal of Chemical Theory and Computation **16**, 7702 (2020), pMID: 33156633, `https://doi.org/10.1021/acs.jctc.0c00937`

[4] Y.Q. Cheng, A.J. Ramirez-Cuesta, Journal of Chemical Theory and Computation **16**, 5212 (2020), pMID: 32700910, `https://doi.org/10.1021/acs.jctc.0c00569`

[5] *Oclimax website*, https://sites.google.com/site/ornliceman/download, (Accessed on 07/18/2022)

[6] A.J. Ramirez-Cuesta, Y.Y. Cheng, hamon **30**, 154 (2020)

[7] A.J. Ramirez-Cuesta, Y.Q. Cheng, Neutron News **32**, 15 (2021), `https://doi.org/10.1080/10448632.2021.1875773`

[8] E. Mamontov, R. Smith, J. Billings, A. Ramirez-Cuesta, Physica B: Condensed Matter **566** (2019)

[9] R. Hanus, J. George, M. Wood, A. Bonkowski, Y. Cheng, D.L. Abernathy, M.E. Manley, G. Hautier, G.J. Snyder, R.P. Hermann, Materials Today Physics **18**, 100344 (2021)

[10] M.H. Karigerasi, K. Kang, G.E. Granroth, A. Banerjee, A. Schleife, D.P. Shoemaker, Phys. Rev. Materials **4**, 114416 (2020)

[11] D. Zhang, X. Hu, T. Chen, D.L. Abernathy, R. Kajimoto, M. Nakamura, M. Kofu, B.J. Foley, M. Yoon, J.J. Choi et al., Phys. Rev. B **102**, 224310 (2020)

[12] S. Mu, K.D. Dixit, X. Wang, D.L. Abernathy, H. Cao, S.E. Nagler, J. Yan, P. Lampen-Kelley,

D. Mandrus, C.A. Polanco et al., Phys. Rev. Research **4**, 013067 (2022)

[13] R. Juneja, S. Thébaud, T. Pandey, C. Polanco, D. Moseley, M. Manley, Y. Cheng, B. Winn, D. Abernathy, R. Hermann et al., Materials Today Physics **21**, 100548 (2021)

[14] S. Gao, A.F. May, M.H. Du, J.A.M. Paddison, H.S. Arachchige, G. Pokharel, C. dela Cruz, Q. Zhang, G. Ehlers, D.S. Parker et al., Phys. Rev. B **103**, 214418 (2021)

[15] X. Bai, F. Lechermann, Y. Liu, Y. Cheng, A.I. Kolesnikov, F. Ye, T.J. Williams, S. Chi, G.E. Granroth, A.F. May et al., *Antiferromagnetic fluctuations and orbital-selective mott transition in the van der waals ferromagnet fe3-xgete2* (2022), arXiv:2201.12375

[16] I.I. Al-Qasir, A.A. Campbell, G. Sala, J.Y. Lin, Y. Cheng, F.F. Islam, D.L. Abernathy, M.B. Stone, Carbon **168**, 42 (2020)

[17] I.I. Al-Qasir, Y. Cheng, J.Y. Lin, A.A. Campbell, G. Sala, K. Ramic, F.F. Islam, A. Qteish, B. Marsden, D.L. Abernathy et al., Annals of Nuclear Energy **161**, 108437 (2021)

[18] P.W. Albers, V. Leich, A.J. Ramirez-Cuesta, Y. Cheng, J. Hönig, S.F. Parker, Mater. Adv. **3**, 2810 (2021), publisher: RSC

[19] X. Li, P.F. Liu, E. Zhao, Z. Zhang, T. Guidi, M.D. Le, M. Avdeev, K. Ikeda, T. Otomo, M. Kofu et al., Nature Communications **11**, 942 (2020)

[20] A.S. Sukhanov, S.E. Nikitin, M.S. Pavlovskii, T.C. Sterling, N.D. Andryushin, A.S. Cameron, Y.V. Tymoshenko, H.C. Walker, I.V. Morozov, I.O. Chernyavskii et al., Phys. Rev. Research **2**, 043405 (2020)

[21] K. Drużbicki, R. Lavén, J. Armstrong, L. Malavasi, F. Fernandez-Alonso, M. Karlsson, The Journal of Physical Chemistry Letters **12**, 3503 (2021), pMID: 33792334, https://doi.org/10.1021/acs.jpclett.1c00616

[22] I. Vázquez-Fernández, K. Drużbicki, F. Fernandez-Alonso, S. Mukhopadhyay, P. Nockemann, S.F. Parker, S. Rudić, S.M. Stana, J. Tomkinson, D.J. Yeadon et al., The Journal of Physical Chemistry C **125**, 24463 (2021), https://doi.org/10.1021/acs.jpcc.1c05137

[23] I. Vázquez-Fernández, K. Drużbicki, F. Fernandez-Alonso, S. Mukhopadhyay, P. Nockemann, S.F. Parker, S. Rudić, S.M. Stana, J. Tomkinson, D.J. Yeadon et al., The Journal of Physical Chemistry C **125**, 24463 (2021), https://doi.org/10.1021/acs.jpcc.1c05137

[24] C. Cook, J.L. McKinley, G.J.O. Beran, The Journal of Chemical Physics **154**, 134109 (2021), https://doi.org/10.1063/5.0044385

[25] G. Romanelli, D. Onorati, P. Ulpiani, S. Cancelli, E. Perelli-Cippo, J.I.M. Damián, S.C. Capelli, G. Croci, A. Muraro, M. Tardocchi et al., Journal of Physics: Condensed Matter **33**, 285901 (2021)

[26] J.P. Melchior, N.H. Jalarvo, The Journal of Physical Chemistry C **123**, 14195 (2019), https://doi.org/10.1021/acs.jpcc.9b01873

[27] M.L. Martins, T.S. Mathis, X. Wang, L.L. Daemen, Y. Gogotsi, E. Mamontov, Phys. Rev. Materials **6**, 034001 (2022)

[28] M.L. Martins, R.L. Sacci, N.C. Sanders, J.L. Tyler, R. Matsumoto, I. Popov, W. Guo, S. Dai, P.T. Cummings, A.P. Sokolov et al., The Journal of Physical Chemistry C **124**, 22366 (2020), https://doi.org/10.1021/acs.jpcc.0c05311

[29] N.C. Osti, B.P. Thapaliya, S. Dai, M. Tyagi, E. Mamontov, The Journal of Physical Chemistry Letters **12**, 4038 (2021), pMID: 33881871, https://doi.org/10.1021/acs.jpclett.1c00461

[30] A.A. Koegel, E.M. Mozur, I.W.H. Oswald, N.H. Jalarvo, T.R. Prisk, M. Tyagi, J.R. Neilson, Journal of the American Chemical Society **144**, 1313 (2022), pMID: 35029372, https://doi.org/10.1021/jacs.1c11217

[31] M.L. Martins, R.L. Sacci, X. Lin, R. Matsumoto, I. Popov, J. Cui, T. Kobayashi, M. Tyagi, W. Guo, S. Dai et al., The Journal of Physical Chemistry C **126**, 2788 (2022), https://doi.org/10.1021/acs.jpcc.1c09352

[32] E. Novak, L. Daemen, A.J. Ramirez-Cuesta, Y. Cheng, R. Smith, T. Egami, N. Jalarvo, Scientific Reports **12**, 6194 (2022)

[33] H.W. Wang, T.R. Graham, E. Mamontov, K. Page, A.G. Stack, C.I. Pearce, The Journal of Physical Chemistry Letters **10**, 3318 (2019), https://doi.org/10.1021/acs.jpclett.9b01416

[34] N.C. Osti, B.P. Thapaliya, R.A. Matsumoto, A. Bansal, X. Lin, P.T. Cummings, M. Tyagi, S. Dai, E. Mamontov, The Journal of Physical Chemistry Letters **13**, 2845 (2022), https://doi.org/10.1021/acs.jpclett.2c00487

[35] B. Dyatkin, N.C. Osti, R.W. Smith, M. Tyagi, T. Butler, M. Laskoski, Journal of Polymer Science **58**, 3419 (2020), https://onlinelibrary.wiley.com/doi/pdf/10.1002/pol.20200626

[36] S. Pili, P. Rought, D.I. Kolokolov, L. Lin, I. da Silva, Y. Cheng, C. Marsh, I.P. Silverwood, V. García Sakai, M. Li et al., Chemistry of Materials **30**, 7593 (2018), https://doi.org/10.1021/acs.chemmater.8b02765

[37] J.P. Melchior, W. Lohstroh, M. Zamponi, N.H. Jalarvo, Journal of Membrane Science **586**, 240 (2019)

[38] G. Kearley, Nuclear Instruments and Methods in Physics Research Section A: Accelerators, Spectrometers, Detectors and Associated Equipment **354**, 53 (1995), proceedings of the Third Workshop on Neutron Scattering Data Analysis

[39] A.J. Ramirez-Cuesta, Computer Physics Communications **157**, 226 (2004)

[40] D. Merkel, Linux journal **2014**, 2 (2014)

[41] R.T. Azuah, L.R. Kneller, Y. Qiu, P.L. Tregenna-Piggott, C.M. Brown, J.R. Copley, R.M. Dimeo, Journal of research of the National Institute of Standards and Technology **114**, 341 (2009)

[42] K. Dymkowski, S.F. Parker, F. Fernandez-Alonso, S. Mukhopadhyay, Physica B: Condensed Matter **551**, 443 (2018)

[43] O. Arnold, J.C. Bilheux, J. Borreguero, A. Buts, S.I. Campbell, L. Chapon, M. Doucet, N. Draper, R.F. Leal, M. Gigg et al., Nuclear instruments and methods in physics research section a: accelerators, spectrometers, detectors and associated equipment **764**, 156 (2014)

[44] M.L. Martins, T.S. Mathis, X. Wang, L.L. Daemen, Y. Gogotsi, E. Mamontov, Phys. Rev. Materials **6**, 034001 (2022)

[45] M. Newville, T. Stensitzki, D.B. Allen, A. Ingargiola, *LMFIT: Non-Linear Least-Square Minimization and Curve-Fitting for Python* (2014), `https://doi.org/10.5281/zenodo.11813`

[46] *https://iceman.ornl.gov/ui/iceman/ui*, `https://iceman.ornl.gov/ui/ICEMAN/ui`, (Accessed on 07/18/2022)

[47] *Xcams faq*, `https://web.ornl.gov/xcams/xcamsfaq.htm`, (Accessed on 07/18/2022)

EPJ Web of Conferences 272, 01014 (2022) https://doi.org/10.1051/epjconf/202227201014
QENS/WINS 2022

Determining the relaxation time from a temperature-dependent scan of the neutron spin-echo signal amplitude

Eugene Mamontov[1*] and *Piotr* Zolnierczuk[1]

[1]Neutron Scattering Division, 37831 Oak Ridge, Tennessee, USA

Abstract. Temperature-dependent scans of the neutron scattering intensity are commonly employed in high energy-resolution quasielastic measurements. Besides serving as a useful diagnostic tool for identifying the temperature range that could give rise to a measurable relaxation signal, such scans of the "elastic" (resolution-defined) intensity could be employed for determining the temperature at which the relaxation time in the system becomes equal to the resolution-defined characteristic time of the spectrometer measurement. This is a model-independent alternative to the "traditional" approach, when, at a given measurement temperature, the relaxation time in the system is obtained from fitting the full dynamic spectra with a model scattering function. Here we introduce the temperature-dependent scan of the neutron spin-echo signal amplitude. Using a well-characterized system with a complex relaxation pattern, we demonstrate that the relaxation time obtained from the approach proposed herein maps well on the previous "traditionally" measured relaxation times. Thus, despite monitoring a different variable (neutron spin-echo signal amplitude vs. neutron scattering intensity), the benefits of the model-free temperature-dependent scan approach, traditionally utilized in neutron time-of-flight and backscattering experiments, can be extended to measurements of the very slow relaxations assessable only by high-resolution neutron spin-echo.

1 Introduction

High energy-resolution neutron scattering is a powerful technique for studying relaxational dynamics. Despite its unique advantages, such as weak interaction with the sample, H/D selectivity, and the spatial information on the motion accessible via the momentum transfer (Q) signal dependence, neutron scattering, especially when utilized in the high energy-resolution mode, is intensity limited. In quasielastic neutron scattering (QENS) [1], particularly, backscattering, which has very high energy resolution, second only to neutron spin echo (NSE) [2, 3], the severe intensity limitations in measurements of the small samples have prompted numerous attempts to circumvent the problem by measuring only the resolution-defined [stronger] "elastic" line instead of the whole dynamic spectrum with the [weaker] "wings".

It was proposed 20 years ago [4, 5] that measurements of small samples may benefit from "elastic resolution spectroscopy", when the relaxation function (traditionally obtained from the full range dynamic measurement) is reconstructed from the "elastic" incoherent scattering intensity. As originally proposed, this approach relied on the variable energy resolution easily possible with QENS-capable time-of-flight (TOF) neutron spectrometers. More recently this approach was adopted to the data collection from various backscattering spectrometers [6]. The main assumption of the "elastic resolution spectroscopy" approach is that the relaxation mechanism remains continuous when the temperature is changed [6].

An alternative approach, "resolution elastic neutron scattering", proposed [7, 8] that the relaxation time becomes equal to the resolution time of the spectrometer when an inflection point occurs, either in the temperature dependence of the "elastic" incoherent intensity (at a given resolution), or the resolution dependence of the "elastic" incoherent intensity (at a given temperature). The assumption of the "resolution elastic neutron scattering" approach is that the resolution function of the spectrometer is a Gaussian, and the relaxation function to be measured has a simple functional form (e.g., of Lorentz-type).

A newer method to assess the system dynamics by measuring the resolution-dependent "elastic" scattered intensity was proposed and extensively discussed more recently [9-11]. Nevertheless, the approach using an inflection point in the temperature dependence of the "elastic" incoherent scattering intensity at a few resolution settings of a neutron backscattering spectrometer [12] has been validated experimentally [13]. The relaxation times extracted from such analysis of the temperature dependence of the "elastic" neutron scattering intensity were successfully mapped onto the relaxation times previously obtained via the traditional QENS analysis of the same system [14]. The relaxation time mapping was successful even though the relaxation function for the measured system was non-Lorentzian, with a temperature-dependent stretch parameter [14, 15], and the spectrometer's resolution function was not a Gaussian, thus suggesting general applicability of this method of the temperature-dependent "elastic" scattering intensity scans.

In the current work, we would like to investigate possible extension of this approach to the longer

* Corresponding author: mamontove@ornl.gov

relaxation times, beyond those measurable with neutron backscattering spectrometers using QENS. NSE spectroscopy is a neutron scattering technique where the very high energy resolution is attained by harnessing neutron Larmor precession in a magnetic field [2, 3]. Since a detailed description of the NSE principles and practice are beyond the scope of this short paper, the reader is kindly referred to the available literature, e.g. [2, 3, 16].

Here we explore how the measurement time per a temperature point (in the course of a temperature-dependent scan) can be dramatically shortened to produce a variable whose temperature dependence can be used to identify the temperature at which the relaxation time in the sample becomes equal to the resolution time of the NSE spectrometer.

2 Methods

The preparation of the sample, an aqueous solution, $(H_2O)_{0.88}(LiCl)_{0.12}$, followed the previously described procedure [14] using deionized distilled water and anhydrous, ultra-dry lithium chloride powder (99.995 % purity) purchased from Alfa Aesar. The sample was loaded into a flat plate aluminum sample holder, 50 mm tall, 30 mm wide, and 0.25 mm thick, which was placed in a Janis SHI-950 (Janis, Woburn, MA, USA) closed cycle refrigerator connected to a LakeShore 336 (LakeShore, Westerville, OH, USA) temperature controller.

The NSE measurements (of the predominantly incoherent scattering signal from the protiated sample) were carried out at the SNS-NSE spectrometer [17] at the Oak Ridge National Laboratory. We chose neutron wavelength band $\Delta\lambda$ = 3-6 Å, the NSE field integral, $J = \int B\, dl$, was set to 0.01971 Tm (corresponding to τ = 0.45 ns at λ = 5 Å), and the scattering angle was 2θ=39.1°. That allowed us to cover the scattering vector Q from approximately 0.7 to 1.4 $Å^{-1}$.

The measurement outline was as follows:

1. Cooled the sample down to 4 K and measured a "full" echo establishing the symmetry phase (see Fig. 1).
2. Warmed up the sample to 300 K, again measured the full echo (confirmed that the echo "disappeared").
3. Gradually cooling the sample at the average rate of 0.1 K/min, while keeping all the precession currents constant and turning on/off only the flipper currents, we measured three neutron count rates:
    a. U - with all flippers turned off (I~1/3 incoherent, no spin-flip)
    b. D - only π −flipper turned on (I~2/3 incoherent, spin flip)
    c. E - all flippers turned on ("echo" condition).

The data reduction was carried out as follows:

1. Selected pixels with 0.3 ns ≤ τ ≤ 0.6 ns and 0.8 $Å^{-1}$ ≤ Q ≤ 1.0 $Å^{-1}$
    a. $\langle\tau\rangle$ = (0.45 ± 0.09) ns
    b. $\langle Q\rangle$ = (0.88 ± 0.05) $Å^{-1}$
2. For the pixels with these values of $\langle\tau\rangle$ and $\langle Q\rangle$:
    a. Calculated the scattering function:

$$F(Q,\tau) = \frac{2(E-\langle A\rangle)}{U-D}$$

    b. Here $\langle A\rangle = \frac{U+D}{2}$
3. Plotted F as a function of temperature, T (see Fig. 2).

Notes:

- Due to the time-of-flight and multipixel nature of the SNS-NSE, we cover a range of scattering angles, Q's, and τ.
- We take advantage of the mu-metal shielding of the spectrometer that allows excellent magnetic field (and, hence, echo phase) stability; see Fig. 3.
- Since this was a non-standard experiment, the data have been reduced and analysed using custom software created by one of the authors (P. Z.).

3 Results and discussion

Fig. 1 illustrates the standard NSE measurement and reduction [16]:

- Measure "full" echo, polarization (U,D)
- Determine the echo (symmetry) phase and echo amplitude A_{ECHO} by fitting a "convoluted cosine"
- Compute $F = \frac{2A_{ECHO}}{U-D}$

While the "traditional" approach relies on the measurement of the full echo at each chosen Fourier time, here instead we determined the echo amplitude at the symmetry phase. Importantly, this approach, bypassing the measurement of the full echo signal, was made possible by never changing the electric currents of the precession coils. Only the flipper currents were turned on and off.

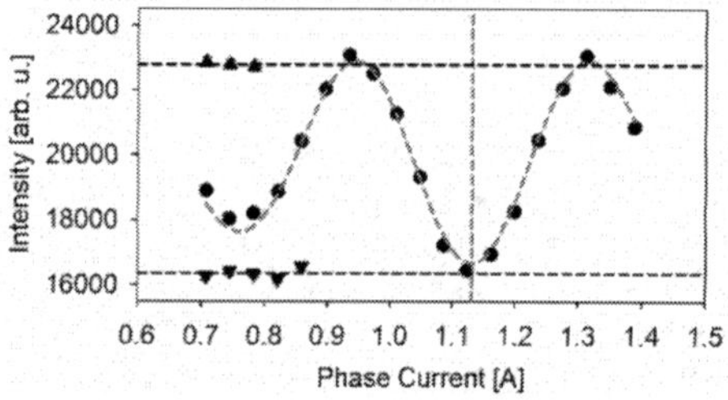

Fig. 1. Illustration of the standard measurement of the neutron spin-echo amplitude at a given temperature.

Fig. 2 shows the temperature dependence of the scattering function. The data (symbols) were fitted to the following expression (analogous to the approach adopted in [13]):

https://doi.org/10.1051/epjconf/202227201014

$$F(T) = \left[1 - \frac{1}{1+\exp(-G(T-T_0))}\right](a - bT) \quad (1)$$

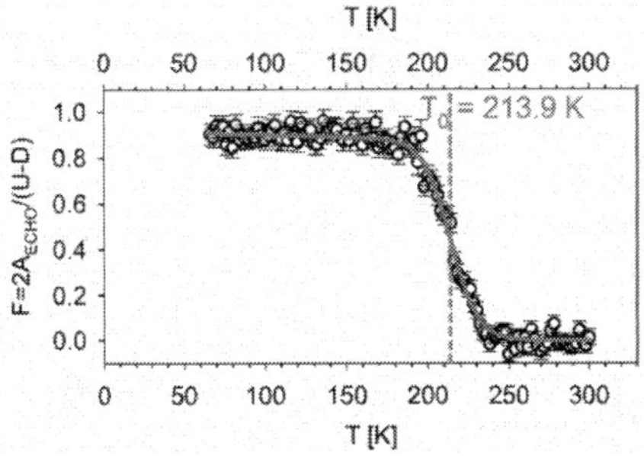

Fig. 2. Illustration of how the inflection temperature point was obtained.

That is, at T_0 (the inflection point), the resolution time measured in the sample is expected to become equal to $\langle\tau\rangle = (0.45 \pm 0.09)$ ns (for $\langle Q\rangle = (0.88 \pm 0.05)$ Å^{-1}). The parameter obtained from the fit with Equation 1 are presented in Table 1.

Table 1. Parameters of the fit shown as the red line in Fig. 2.

Parameter	Value
T_0	(213.9 ± 0.5) K
G	(0.120 ± 0.006) K^{-1}
a	(0.92 ± 0.02)
b	(19 ± 14) 10^{-5} K^{-1}
χ^2	0.69

As mentioned above, the excellent stability of the magnetic field at the sample position, as illustrated by the data presented in Fig. 3, is important for the measurement approach described herein.

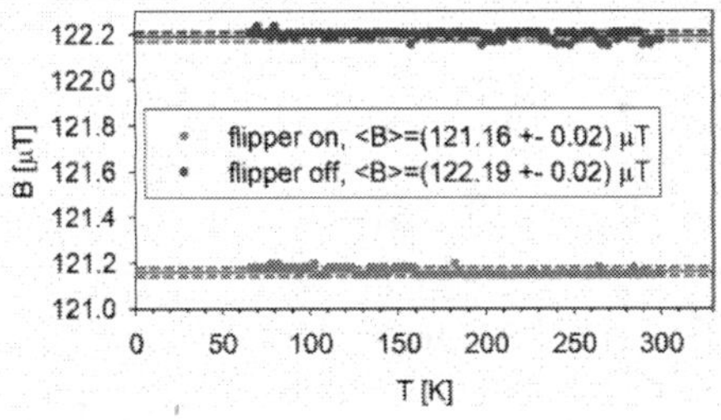

Fig. 3. Magnetic field at the sample position as a function of sample temperature. The field was stable to better than $2\cdot10^{-2}$ μT (or 0.2 mG).

Finally, this newly determined at T_0 = (213.9 ± 0.5) K ($\langle\tau\rangle$ = (0.45 ± 0.09) ns, $\langle Q\rangle$ = (0.88 ± 0.05) Å^{-1}) point in the (T, τ, Q) space has been added to the previously measured relaxation time map for this sample, as presented in Fig. 4. The data previously measured by neutron backscattering in the "traditional" way [14] show the relaxation times at preset temperatures. Conversely, the data previously measured by neutron backscattering from the elastic intensity temperature scans [13] show the temperatures at which the preset relaxation times are observed in the sample. Likewise, the data point obtained in the current NSE experiment shows the temperature (213.9 K) at which the preset relaxation time (0.45 ns) is observed in the sample. As one can see, the newly measured relaxation time maps well onto the previously measured relaxation times.

At a cooling rate of 0.1 K/min, it took about 2300 minutes, or 38 hours, to obtain the the (T, τ, Q) point of interest. It should be noted that the sample studied herein scatters predominantly incoherently, which necessitates long measurement times with "traditional" NSE measurements. Further, as one can infer from Fig. 2, the number of measured temperature points could have been decreased dramatically, because the 4-parameter fit with Equation 1 of the sigmoidal data set is robust and does not need many data points, especially at the extremes of the temperature range. Therefore, even though we conservatively allowed long measurements in the current proof-of-principle experiment, the measurement time necessary to obtain a (T, τ, Q) point using the approach described herein could be shortened dramatically. We expect that the biggest time saving using the proposed approach compared to the traditional NSE approach could be realized in measurements with the longest preset relaxation times, where not having to collect the full spin-echo spectra could be highly benefitial for shortening the experiment.

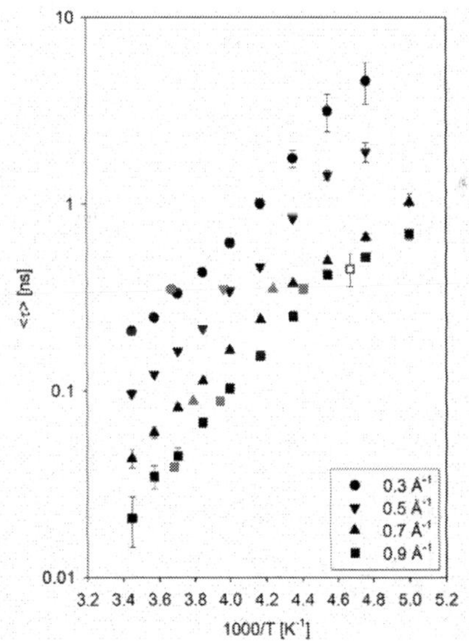

Fig. 4. Filled black symbols from neutron backscattering spectrometer [12] measurements: average relaxation times previously measured in the "traditional" way from QENS data [14] for $(H_2O)_{0.88}(LiCl)_{0.12}$. Filled red symbols from neutron backscattering spectrometer [12] measurements: average relaxation times obtained from the analysis of the temperature-dependent scans as described in [13]. Open blue symbol from SNS-NSE [17] measurements: average relaxation time s obtained in the current work. For the red and blue symbols, the uncertainties in the temperature points are shown with the horizontal bars (when they exceed the size of the symbols).

4 Conclusion

The traditional method of obtaining the relaxation times from NSE data involves fitting the spin-echo spectra measured at variable Fourier times and subsequent fitting with a model relaxation function of the $I(t)$ points resulting from the fitted amplitudes of the spin-echo spectra. In this work we demonstrate an alternative approach that relies of measuring the temperature dependence of the spin echo amplitude. Somewhat analogous to the temperature-dependent "elastic" intensity scans traditionally utilized in neutron time-of-flight and backscattering experiments, the proposed approach provides the benefit of the model-free analysis while greatly extending the measurable relaxation time range, to the energy/time domain inaccessible by any neutron scattering technique other than NSE. Remarkably, this approach is verified for a sample with a non-Lorentzian relaxation function, whose stretch parameter itself is a function of temperature. The method demonstrated in the present work thus holds a promise for time-efficient measurements of long relaxation times in the samples that may not be amenable to traditional full analysis of the NSE signal.

This research used resources at the Spallation Neutron Source, a DOE Office of Science User Facility operated by the Oak Ridge National Laboratory. Oak Ridge National Laboratory is managed by UTBattelle, LLC, for the U.S. DOE under Contract No. DE-AC05-00OR22725. This manuscript has been authored by UT-Battelle, LLC, under Contract No. DE-AC0500OR22725 with the U.S. Department of Energy. The United States Government retains and the publisher, by accepting the article for publication, acknowledges that the United States Government retains a non-exclusive, paid-up, irrevocable, world-wide license to publish or reproduce the published form of this manuscript, or allow others to do so, for the United States Government purposes.

References

1. M. Bee, Quasielastic Neutron Scattering, Principles and Applications in Solid State Chemistry, Biology and Materials Science, CRC Press (1988)
2. F. Mezei (ed.), Neutron Spin Echo, Lecture Notes in Physics, Vol. 128, Springer Verlag (1980)
3. F. Mezei, C. Pappas, T. Gutberlet (eds.), Neutron Spin Echo Spectroscopy - Basics, Trends and Applications, Lecture Notes in Physics, Vol. 601, Springer Verlag (2003)
4. W. Doster, M. Diehl, W. Petry, M. Ferrand, Physica B **301**, 65-68 (2001)
5. W. Doster, M. Diehl, R. Gebhardt, R.E. Lechner, J. Pieper, Chem. Phys. **292**, 487-494 (2003)
6. W. Doster, H. Nakagawa, M.S. Appavou, J. Chem. Phys. **139**, 045105 (2013)
7. S. Magazù, F. Migliardo, A. Benedetto, Rev. Sci. Inst. **82**, 105115 (2011)
8. S. Magazù, F. Migliardo, M.T. Caccamo, Adv. Mater. Sci. Eng. **2**, 695405-695411 (2013)
9. A. Benedetto, G. J. Kearley, Sci. Rep. **9**, 11284 (2019)
10. A. Benedetto, G. J. Kearley, Sci. Rep. **10**, 6350 (2020)
11. A. Benedetto, G. J. Kearley, Sci. Rep. **11**, 14093 (2021)
12. E. Mamontov, K.W. Herwig, Rev. Sci. Inst. **82**, 085109 (2011)
13. E. Mamontov, Nucl. Inst. Methods Phys. Res. A **949**, 162534 (2020)
14. E. Mamontov, J. Phys. Chem. B **113**, 14073-14078 (2009)
15. M. Nakanishi, P. Griffin, E. Mamontov, A. P. Sokolov, J. Chem. Phys. **136**, 124512 (2012)
16. P.A. Zolnierczuk, O. Holderer, S. Pasini, T. Kozielewski, L.R. Stingaciu, M. Monkenbusch, J. Appl. Cryst., **52**, 1022-1034 (2019)
17. M. Ohl, M. Monkenbusch, N. Arend, T. Kozielewski, G. Vehres, C. Tiemann, M. Butzek, H. Soltner, U. Giesen, R. Achten, H. Stelzer, B. Lindenau, A. Budwig, H. Kleines, M. Drochner, P. Kaemmerling, M. Wagener, R. Moller, E.B. Iverson, M. Sharp, D. Richter, Nucl. Inst. Methods Phys. Res. A **696**, 85-99 (2012)

Comparison of molecular dynamics simulations of water with neutron and X-ray scattering experiments

Veronika Reich[1,*], *Arnab* Majumdar[1,**], *Martin* Müller[1,2,3,***], and *Sebastian* Busch[1,****]

[1]German Engineering Materials Science Centre (GEMS) at Heinz Maier-Leibnitz Zentrum (MLZ), Helmholtz-Zentrum hereon GmbH, Lichtenbergstr. 1, 85748 Garching, Germany
[2]Institute of Materials Physics, Helmholtz-Zentrum hereon GmbH, Max-Planck-Str. 1, 21502 Geesthacht, Germany
[3]Institut für Experimentelle und Angewandte Physik (IEAP), Christian-Albrechts-Universität zu Kiel, Leibnizstr. 19, 24098 Kiel

Abstract. The atomistic structure and dynamics obtained from molecular dynamics (MD) simulations with the example of TIP3P (rigid and flexible) and TIP4P/2005 (rigid) water is compared to neutron and X-ray scattering data at ambient conditions. Neutron and X-ray diffractograms are calculated from the simulations for four isotopic substitutions as well as the incoherent intermediate scattering function for neutrons. The resulting curves are compared to each other and to published experimental data. Differences between simulated and measured intermediate scattering functions are quantified by fitting an analytic model to the computed values. The sensitivity of the scattering curves to the parameters of the MD simulations is demonstrated on the example of two parameters, bond length and angle.

1 Introduction

The structure and dynamics of matter can be examined on an atomic level with various methods. Neutron and X-ray scattering data [1] are generally interpreted by building analytical models that are then refined against the experimental data. With molecular dynamics (MD), simulations reflect the structure and dynamics on the same time and length scales. The simulations depend on a series of input parameters for the interaction potentials, which we propose optimizing with the measured scattering data. In the present contribution, we take the first step toward this goal by using liquid water as an experiment subject.

Water is an ideal candidate because it has been investigated in all possible states; from gaseous over liquid, supercooled, and glassy to the many different phases of ice. Its structure and dynamics can be – and often are – measured on the molecular level with neutron and X-ray scattering [2]. Water, in turn, is also often used to calibrate these instruments [3, 4]. Diffraction measurements do not analyze the energy transfer between probe and sample; instead, the scattered intensity as a function of momentum transfer contains information about the sample structure. Samples can be probed by neutrons, which scatter at the nuclei, or X-rays, which scatter at the electron shells [5, 6]. Spectroscopic measurements analyze not only the momentum but also the energy transfer between probe and sample, which makes it possible to observe the dynamics. For decades, quasielastic neutron scattering (QENS) has been the method of choice for studying the molecular motions of water [7–10]. Over the years, MD simulations have also been frequently used to study water on the molecular level [11–19], and the rapid advances in computing power have unlocked approaches that were hitherto not feasible.

Although MD simulations have been successfully used to evaluate the nanoscopic (picoseconds and Ångstroms) structure and dynamics probed by scattering data [8, 20–24], the interaction potentials used by the simulations are generally optimized to reproduce macroscopic, thermodynamic properties and densities [12, 13], sometimes with the aim of being able to cover a large portion of the phase diagram [13]. A model that is directly optimized to reproduce scattering experiments will likely yield a better description of the nanoscopic properties, albeit probably at the cost of generality. This principle is already employed for the evaluation of diffraction [25–27] and QENS [28, 29] experiments. So far, however, simulations have been refined either against diffraction or QENS measurements; we propose a simultaneous fit of both.

The current work uses two prominent water models: TIP3P [12, 30] and TIP4P/2005 [13]. Both are point models of a water molecule and simplify reality considerably but still depict the behavior of water molecules fairly accurately [15, 31, 32]. Both models can be rigid or flexible in consideration of bond length and angle. We will use TIP3P rigid as a base and compare it to TIP3P flexible as well as the TIP4P/2005 rigid model. All models will be explained in more detail in section 2.2. The experimental data, simulations, and analysis tools are introduced in section 2 with a particular focus on the calculation of scattering patterns described in section 2.5. Different varieties of the simulations are then compared to each other in sec-

*e-mail: veronika.reich@hereon.de
**e-mail: arnab.majumdar@hereon.de
***e-mail: martin.mueller@hereon.de
****e-mail: sebastian.busch@hereon.de

tion 3 as well as to experimental data. Since no perfect agreement between experimental data and simulation was achieved, a parameter optimization scheme is suggested in section 3.4. It demonstrates the sensibility of the scattering curves to MD parameter changes, thereby laying the groundwork for a parameter optimization scheme against scattering data to be presented in future work.

2 Materials and Methods

In this work we compare simulations and experiments as directly as possible: We use diffraction data in reciprocal space and spectroscopic data as intermediate scattering functions rather than derived quantities like the oxygen-oxygen pair correlation function or diffusion coefficient. This allows the scattering patterns to be calculated far more easily from the simulations as opposed to using a Fourier transform to convert the measured data into real space [33, 34].

The complete data analysis workflow has been made available in an accompanying data publication [35].

2.1 Experimental data

Neutron and X-ray diffraction data were taken from a meta analysis by Soper [36] at 298 K. For neutron scattering, diffractograms of H_2O, "null water" (an H_2O/D_2O mixture with an average coherent hydrogen scattering length of zero), HDO, and D_2O are provided. By averaging the coherent hydrogen scattering lengths to zero in null water, only oxygen-oxygen correlations are visible in the resulting diffraction data. For X-ray scattering, "normalization II" as defined by Soper [36] was employed:

$$D(Q) = \frac{I_x(Q)}{\sum_\alpha c_\alpha f_\alpha^2(Q)} - 1.0 \tag{1}$$

Quasielastic incoherent neutron scattering curves were taken from Teixeira et al. [37] at 293 K. The data were measured in energy space up to a maximal energy transfer of $\hbar\omega = 3\,\text{meV}$. This corresponds very roughly to a shortest probed time of $t_{\min} \approx 1\,\text{ps}$. All processes that occurred in the MD simulation at shorter correlation times cannot be expected to be visible in this dataset. Rather than taking the data directly, the fit function used to evaluate the data was also utilized to reconstruct the intermediate scattering function

$$I(Q,t) = DWF(Q) \cdot R(Q,t) \cdot T(Q,t) \tag{2}$$

The interpretation of this formula as a decoupling of fast vibrations, rotational motions, and translational motions is, for most of the present contribution, irrelevant – any fit function that describes the measured data with the same accuracy would suffice for the comparison with the simulation. The individual parts of the formula are the Debye–Waller factor due to fast vibrations

$$DWF(Q) = \exp\left[\frac{-Q^2\langle u^2\rangle}{3}\right] \quad , \tag{3}$$

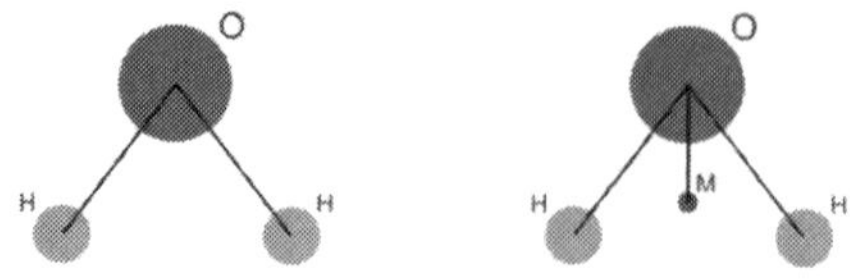

Figure 1. Sketches of TIP3P (left) and TIP4P/2005 (right).

the rotational diffusion

$$R(Q,t) = j_0^2(Qa) + 3j_1^2(Qa)\exp\left[\frac{-t}{3\tau_1}\right] + 5j_2^2(Qa)\exp\left[\frac{-t}{\tau_1}\right] \tag{4}$$

with

$$\tau_1 = \tau_1^0 \cdot \exp\left[\frac{E_A}{k_B T}\right] \tag{5}$$

where E_A is 1.85 kcal/mol, and the translational diffusion

$$T(Q,t) = \exp\left[-\Gamma(Q)t\right] \tag{6}$$

with

$$\Gamma(Q) = \frac{DQ^2}{1 + DQ^2\tau_0} \tag{7}$$

where

$$D = \frac{L^2}{6\tau_0} \quad . \tag{8}$$

Numerical values for these quantities are given in Table 2.

2.2 Water models

Most water models reduce the complexity of the nearly spherical and polarizable electron cloud of a water molecule [5] to the summation of a small number of interaction sites. How many of these sites are included and which interaction functions are employed varies between models. In this work, two widely used water models, the three-site TIP3P and the four-site TIP4P/2005, were employed. Their most important parameter values are summarized in Table 1 [38].

In the TIP3P model [12], the interaction sites are collocated with the positions of the nuclei, (see Figure 1). The O–H bond length and the H–O–H bond angle can be either rigid or flexible. While TIP3P reproduces some quantities quite well and is frequently used especially in biomolecular simulations, it is known to have deficiencies [30, 31, 39].

The TIP4P/2005 water model [13] is thought to be superior in many aspects [13, 14, 40–43]. Its four interaction sites are located at the three nuclear positions as well as a massless site M, which is located along the bisector of the H–O–H bond angle at 0.1546 Å from the oxygen atom. The negative partial charge is positioned at M rather than the oxygen nucleus, from where the van der Waals interaction is calculated. The TIP4P/2005 water model can be either rigid [13] or flexible [14]. In this contribution, only the rigid model was used.

Table 1. Table of the force field parameters used in the MD simulations. Hydrogen atoms did not participate in Lennard-Jones interactions. The flexible model used harmonic intramolecular potentials.

Parameter	TIP3P rigid	TIP3P flexible	TIP4P/2005 rigid
r(OH) / Å	0.9572	0.9572	0.9572
K r(OH) / (kcal/mol/Å^2)	—	450	—
∠ HOH / deg	104.52	104.52	104.52
K ∠ HOH / (kcal/mol/rad^2)	—	55	—
q(H) / e	0.415	0.415	0.5564
σ_O / Å	3.188	3.188	3.1589
ϵ_O / (kcal/mol)	0.102	0.102	0.1852

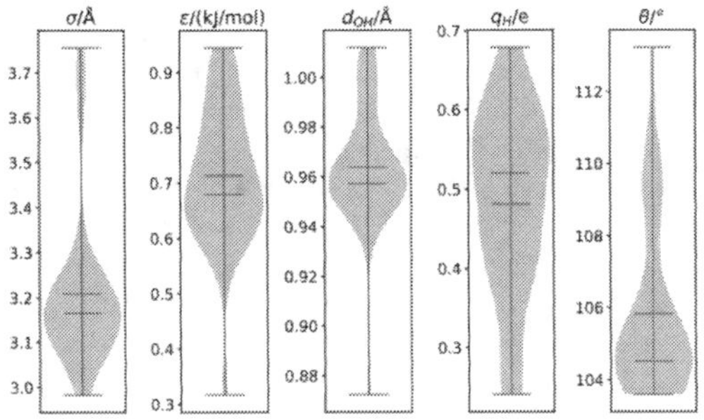

Figure 2. Violin plot of the distribution of water model parameters in the literature for a subset of parameters. The shaded area presents an approximated probability distribution to find a model with such a parameter value while the horizontal lines give minimum/maximum, mean, and median of the values used in this analysis.

Water models can either neglect or consider intramolecular vibrations, resulting in either a flexible or rigid model. Which version to use depends on computational cost (with flexible models being more costly) [14, 44], accuracy and artifacts [44], dielectric properties [45], or quantum treatment [46]. As will be illustrated in section 3.2.1, this choice changes the structure and dynamics of the models.

While water models may differ in the number and mathematical function of the interactions, some models with the same principal interactions may differ only in the functions' employed parameters. Figure 2 shows a distribution of a subset of parameter values for many models existing in the literature [11–14, 47–70]. The models used for this overview include differences in parameter values and functional forms. This ambiguity of parameter choice is due to the fact that water models can only reflect a subset of real water properties – and different models reproduce different properties [13].

2.3 MD simulations

The MD simulations were carried out with the program LAMMPS (Large-scale Atomic/Molecular Massively Parallel Simulator) [71]. The initial input files were taken from the literature [12, 30, 38, 40, 72], including the corresponding cutoff distances.

The cubic simulation box contained 8,000 water molecules (24,000 atoms). These molecules were placed in a large starting box that was subsequently compressed to an edge length of 62.086 Å, resulting in a density of 1.0000 g/cm^3 or 0.1 atoms/Å^3. A series of equilibration steps followed to ensure thermodynamical equilibrium of the simulation. The total setup runs cover a time span of 110 ps.

Once the simulation was stable, a production run with a length of 2 ns was performed; the coordinates were saved every picosecond. Subsequently, a further production run was performed covering 20 ps, in which the coordinates were saved every femtosecond. Both production runs were performed in the *NVT* ensemble; pressure coupling was consciously avoided to prevent seeing atom movements caused by the box resizing algorithm in the intermediate scattering function. All MD simulations in this work were run at 296 K. The pressure in the simulation box was 133 ± 83 bar (TIP3P rigid), -714 ± 170 bar (TIP3P flexible), and 112 ± 91 bar (TIP4P/2005 rigid).

The simulations were run with periodic boundary conditions. Two different versions of the coordinates were saved: 1) A *wrapped* set, where all coordinates were wrapped into the central image of the box. This box always had the correct density and was used for structural calculations. 2) An *unwrapped* set was saved where atoms are not kept in the central image of the periodic box but can travel to neighboring images. Since this version avoided unphysical movements of the atoms across the box length, it was used for the determination of all dynamic quantities.

2.4 Real space analysis

Radial distribution functions (RDFs, $g(r)$) and root mean square displacements (RMSDs) were calculated with the program Visual Molecular Dynamics (VMD) [73]. The RDFs were calculated taking the periodic boundary conditions of the wrapped simulation box into account; the RMSDs were averaged over all atoms and calculated from the unwrapped trajectory. Rotations and translations of the whole simulation box were removed by the algorithm before computing the RMSD, although we did not expect these to occur due to the careful setup of the MD simulations.

2.5 Calculation of scattering patterns

There are several open-source tools available to calculate scattering curves from MD simulations, such as nMOLDYN [74], LiquidLib [75], or Sassena [76, 77]. We chose Sassena for this work because of its flexibility and scalability.

Sassena can calculate neutron and X-ray single crystal Bragg diffraction, orientationally averaged powder diffraction, coherent and incoherent intermediate scattering functions, or elastic coherent and incoherent structure factors from MD simulations. Parallelization was used as a guiding design principle and was implemented in the used version via MPI and thread parallelism.

In order to increase the calculation speed, we have introduced vectorization to make full use of modern CPUs' wide registers as well as OpenMP shared memory parallelization. Vectorization improves the program speed of each core, thereby benefiting on local computers as well as high performance systems. Hybrid parallelization combining MPI and OpenMP retained the scalability of the code on complex multi-node systems in large clusters.

2.5.1 X-ray and neutron diffraction

X-ray and neutron diffraction data can be reduced to one term due to interference scattering (i. e., without single atom scattering)

$$D(Q) \sim \left\langle \sum_{j}^{N} \sum_{k>j}^{N} b_j^c b_k^c \exp\left[\mathrm{i}\vec{Q} \cdot \left(\vec{R}_j(t_0) - \vec{R}_k(t_0)\right)\right] \right\rangle_{t_0,\Omega} \quad (9)$$

where Q is the modulus of the scattering vector, N is the number of atoms in the sample, b_l^c is the (Q-independent) coherent neutron scattering length [78] or (Q-dependent) atomic X-ray scattering factor [79] of scatterer l, i is the imaginary number, $\vec{R}_l(t)$ is the position of scatterer l at time t, and t_0 is a starting time. $\langle \cdots \rangle_{t_0,\Omega}$ denotes an average over all possible starting times t_0 and selected orientations Ω of the Q-vectors.

This quantity is obtained from the Sassena result $F(Q)$ in the neutron case via

$$D(Q) = \left[\frac{F(Q)}{N} - \sum_i c_i b_i^2\right] / 100 \quad (10)$$

since the single atom scattering is included in the Sassena result and has to be subtracted; c_i are the concentrations of nuclei with coherent neutron scattering length b_i. The factor 100 is due to a unit conversion from fm^2 to barn. In the case of X-ray scattering, the single atom scattering is also subtracted and the interference scattering is divided by the square of the mean atomic scattering factor [36]:

$$D(Q) = \frac{F(Q)}{N \cdot \sum_i c_i f_i(Q)^2} - 1 \quad (11)$$

where in the present case the atomic scattering factors were calculated from the analytical approximation as a sum of Gaussians [79].

Sassena was used to compute the diffraction patterns from the MD simulations up to $Q = 30\,\text{Å}^{-1}$ with a spacing between points of $0.05\,\text{Å}^{-1}$. A Monte Carlo scheme was used to average over the different orientations of Q-vectors; the X-ray diffractograms were averaged over 1,000 random orientations and the neutron diffractograms over 100. We observe that the results were not particularly sensitive to the number of Q-orientations in these disordered systems, in contrast to the situation in crystalline structures.

Different isotopical substitutions were evaluated. $b_O = 5.803$ fm was kept the same through all evaluations but the original value of $b_H = -3.7390$ fm was changed to $b_{null} = 0$ fm, $b_{HD} = 1.466$ fm, and $b_D = 6.671$ fm for the evaluation of each isotope as described in section 2.1.

Sassena offers the option to subtract the average scattering length from all atoms in the system. While this removes the small-angle scattering caused by the finite simulation box, it distorts the wide-angle diffractogram and was therefore not used for this contribution. Due to the relatively large size of the simulation box, the corresponding small-angle scattering signal was only present at very small Q.

2.5.2 Incoherent quasielastic neutron scattering

The incoherent intermediate scattering function can be calculated as

$$I(Q,t) \sim \left\langle \sum_{j}^{N} b_j^i b_j^i \exp\left[\mathrm{i}\vec{Q} \cdot \left(\vec{R}_j(t_0 + t) - \vec{R}_j(t_0)\right)\right] \right\rangle_{t_0,\Omega} \quad (12)$$

where Q is the magnitude of the scattering vector, t the time, and b_j^i the incoherent scattering length of scatterer j [78]. $\vec{R}_j(t)$ is the position of scatterer j at time t.

Sassena was used to compute the intermediate scattering function at Q-values from 0.1 to $5.0\,\text{Å}^{-1}$ with a step size of $0.1\,\text{Å}^{-1}$, each averaging over 10 orientations of the Q-vector. The real part of the result was used and normalized to the value at $I(Q, t = 0)$.

2.6 Analytical evaluation of the computed intermediate scattering functions

Albeit not the main point of this contribution, the fit formula commonly used for water [37] given in equation 2 was also used to evaluate the computed intermediate scattering functions. The values of $\sqrt{u^2}$, τ_1^0, τ_0, and L were fitted by the method of least squares, the bond length a was changed from the value of 0.98 Å used in the literature [37] to the bond length used in the simulations, 0.9572 Å. The fit was carried out in the time range from 1 ps to 2 ns and for Q-values from $0.1\,\text{Å}^{-1}$ to $5.0\,\text{Å}^{-1}$ in steps of $0.1\,\text{Å}^{-1}$.

The error bars of the fitted parameters were approximated by assuming the Hessian matrix H to be $J^T * J$, where J is the Jacobian matrix returned from the fit routine. H^{-1} was multiplied with the reduced χ^2 and diagonalized. The square roots of the diagonal elements are expressed as error bars in Table 2. It should be noted that

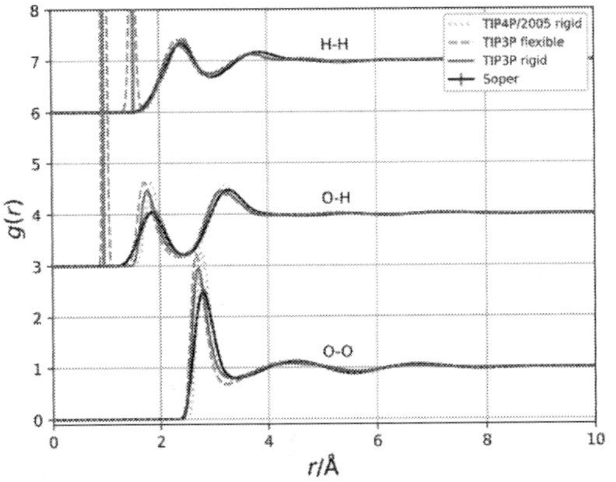

Figure 3. RDFs $g(r)$ of TIP3P rigid, TIP3P flexible, and TIP4P/2005 rigid for O–O, O–H, and H–H correlations. O–H was shifted by an offset of 3 on the y-axis and H–H was shifted by an offset of 6 on the y-axis. The computed RDFs are compared to literature values from [36]

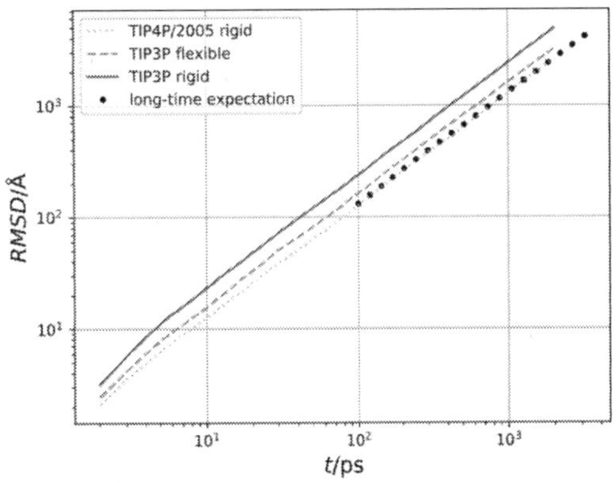

Figure 4. RMSD averaged over all atoms compared to the long-time expectation given by the macroscopic diffusion coefficient of water.

the input data – the evaluated fit function found in the literature [37] – do not have error bars.

3 Results

3.1 Real space analysis

The structure of liquid water in real space can be characterized by the RDFs $g_{OO}(r)$, $g_{OH}(r)$, and $g_{HH}(r)$. The functions computed from the simulations are compared to the RDFs obtained for water at 298 K by Soper [36] in Figure 3. It should be noted that these $g(r)$ should not be referred to as *experimental data* since they are already the result of a computer simulation that reproduced the actual experimental data, which are taken in reciprocal space.

The first sharp peak in the O–H and H–H RDFs is caused by intramolecular correlations and is therefore much sharper than the following intermolecular features. The rigid models exhibit very sharp peaks while the distribution is more smeared out in the flexible model. By and large, there is a good agreement of the intermolecular $g(r)$ with the literature values. The flexible TIP3P model produces slightly more pronounced features in the curves at minimally shorter distances than the rigid model. TIP4P/2005 does not reproduce the literature values substantially better than TIP3P.

The dynamics of a diffusing liquid was extracted from the simulations by the RMSD, averaged over all atoms. The resulting curves are shown in Figure 4 together with the experimental long-time behavior, given by the relation

$$\mathrm{RMSD}(t) = \sqrt{\left\langle \frac{1}{N} \sum_{j=1}^{N} \left(\vec{R}_j(t_0 + t) - \vec{R}_j(t_0) \right)^2 \right\rangle_{t_0}} \quad (13)$$

where N is the number of atoms and $\vec{r}_k(t)$ the position of atom k at time t. $\langle \cdots \rangle_{t_0}$ denotes an average over all starting times t_0. The macroscopic water diffusion coefficient for 296 K was taken to be $2.1745 \cdot 10^{-3}$ mm^2/s [80].

None of the considered models shows a slower long-range diffusive motion than expected from the macroscopic value. The TIP4P/2005 rigid model reproduces the experimentally determined macroscopic diffusion coefficient nearly perfectly. In the case of TIP3P, the diffusion slows when flexibility is introduced to the model – and is in either case further away from the macroscopic value than TIP4P/2005.

3.2 Reciprocal space analysis

Figure 5 shows the computed diffraction patterns and intermediate scattering functions of the three simulated water models together with published experimental neutron and X-ray diffraction data [36] and the published intermediate scattering functions [37].

It should be noted that all curves of the given models were computed from the same MD simulation of H_2O. For diffraction, it is possible in Sassena to vary the probe between X-rays and neutrons and even assign the hydrogen atoms different scattering lengths in the case of neutron scattering, reproducing a measurement series with isotopic substitution. Rather than assigning individual hydrogen atoms random scattering lengths of H and D, all hydrogen atoms were given the same averaged scattering length.

3.2.1 TIP3P rigid versus TIP3P flexible

The O–O correlations that dominate the "null water" neutron diffraction pattern in Figure 5 are hardly affected by the choice between rigid and flexible water molecules. The O–H correlations, particularly visible in the H_2O and D_2O neutron diffraction patterns, show small differences at low Q-values (up to ~6 Å^{-1}) and a slight shift to smaller Q-values of the features at higher Q when switching from rigid to flexible. The oscillation amplitude of the high-Q features decreases only very slightly when changing from

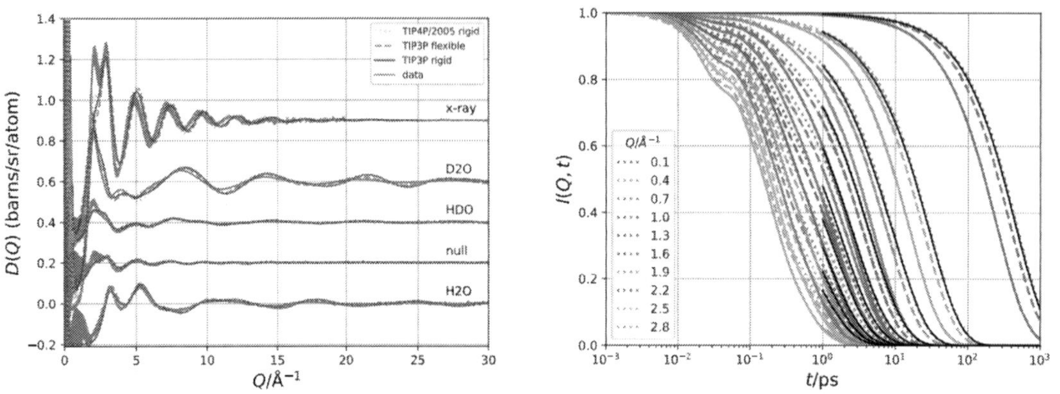

Figure 5. Left: diffractograms for X-rays and neutrons with four different isotopic substitutions for each of the three water models as well as diffraction data from [36] with different offsets for better visibility respectively. Right: intermediate scattering functions, black solid lines refer to intermediate scattering functions from [37], solid colored lines refer to TIP3P rigid, dashed lines to TIP3P flexible, and dotted lines to TIP4P/2005 water model.

a rigid to a flexible model and is in either case much more pronounced than in the data.

The long-range diffusion of the water hydrogens (and therefore the water molecules) is probed by the intermediate scattering functions with small Q and long times, while localized motions dominate the signal at higher Q and short times. The motions slow upon flexibilization – i. e., $I(Q, t)$ decays at longer times compared to the rigid model. This effect is visible across the whole time scale.

3.2.2 TIP3P rigid versus TIP4P/2005 rigid

The diffraction patterns in Figure 5 show a marked difference at the lowest Q values where TIP4P/2005 has a much more pronounced peak around $2\,\text{Å}^{-1}$ than TIP3P. While the neutron H_2O and D_2O diffractograms show very similar O–H correlations between the two models at high Q (i. e., predominantly short distances), the O–O correlations probed by X-ray diffractogram have their oscillations shifted to lower Q in TIP4P/2005.

The intermediate scattering functions of both rigid water models differ strongly. This variation was already indicated in the display of the RMSD curves shown in Figure 4.

3.2.3 Simulations versus scattering experiments

In Figure 5 one can clearly see a discrepancy in all diffractograms between the simulation and experiment at low Q. Particularly striking is the difference for X-ray data, where the first peak in the diffraction pattern is missing altogether for the simulation. This first peak is also not well reproduced for the neutron measurements. It is obvious that the computed diffractograms of TIP4P/2005 align the best with the low Q region of the data and at least the computed curves of the X-ray data have a small feature in this range. At higher Q the oscillations of all three models have roughly the correct amplitude for the X-ray and "null water" measurements, where O–O correlations dominate, with TIP4P/2005 having the best alignment. On the other hand all are shifted to larger Q than the experimental data, indicating packing that is too close. The H_2O and D_2O curves of both rigid models in particular exhibit oscillations that are too large at high Q, indicating an O–H distance that is too defined.

For small Q, the intermediate scattering function of the flexible TIP3P simulation shows better agreement with the experiment than the rigid one. The intermediate scattering function, dominated by long-range diffusion at low Q, decays faster than the literature values and the same holds for local motions probed at higher Q. TIP4P/2005 seems however to have managed to match the experimental data in the long-range regime better than TIP3P. Still, also TIP4P/2005 deviates significantly from the data at higher Q.

3.3 Interpretation of the simulated intermediate scattering function

The fit formula given in section 2.5.2 was used for a global fit of the computed intermediate scattering functions over the whole Q-range. As shown in Figure 6, the formula is able to reproduce the simulated intermediate scattering functions very well in this time and Q-range. Table 2 compares the original parameters from [37] to the current fit results.

The first component of the fit formula is the Debye-Waller factor where the simulations exhibited a vibrational amplitude $\sqrt{\langle u^2 \rangle}$ that was about 1.3 times as large as the literature value [37]. The rotational motion includes the radius of rotation as a parameter, which was not fit but set to 0.98 Å in the literature and 0.9572 Å here, in accordance with the water models' O–H bond length. The relaxation time of the rotational motion τ_1 was two to three times larger than in the literature, indicating that the rotational motions of the simulation were too slow. The residence

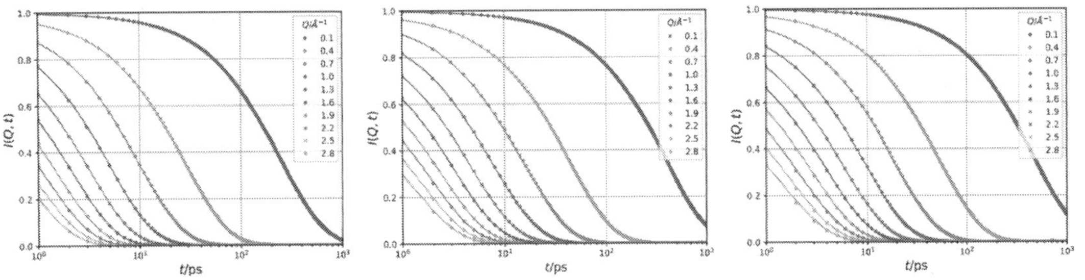

Figure 6. Fit of $I(Q,t)$ computed from the TIP3P rigid (left), TIP3P flexible (middle), and TIP4P/2005 rigid (right) simulations, shown as dots, with an analytic expression from the literature [37]. The best global fit of this formula is shown as drawn-out lines.

Table 2. Parameters from fits to experimental data published in the literature [37] compared to the values obtained from fits to the intermediate scattering patterns computed in this work for both TIP3P and the TIP4P/2005 rigid water models.

Parameter	lit. [37]	TIP3P rigid	TIP3P flexible	TIP4P/2005
$\sqrt{\langle u^2 \rangle}$ / Å	0.48	0.594 ± 0.001	0.617 ± 0.000	0.618 ± 0.001
τ_1^0 / ps	0.0485	0.124 ± 0.001	0.095 ± 0.000	0.150 ± 0.001
τ_0 / ps	1.25	0.277 ± 0.001	0.623 ± 0.001	0.581 ± 0.002
L / Å	1.29	0.820 ± 0.001	0.991 ± 0.001	0.864 ± 0.002

time of the jump diffusion was between one half and one fifth in the simulations compared to the literature, meaning that the water molecules are "stuck" for a much shorter time in the simulation than the analysis of the experiment suggests. If they moved, however, their "jump length" L was about 1.5 times smaller in the simulation than in the literature.

3.4 Parameter variation

As a last step, the input parameters of the TIP3P rigid water model were varied manually to investigate their impact on the structure and the dynamics. The result is shown in Figure 7 for two of these parameter changes, O–H bond length and H–O–H bond angle. The purpose of this procedure is to evaluate if and how sensitive the scattering curves react to parameter changes. It will be shown that the scattering curves react with different sensitivities and in different manners to these changes, which we will use in future works to create an optimization algorithm for the simulation input parameters. The success of a coupled change of these two parameters was recently shown in comparison to neutron spectroscopy [10, 81].

The first parameter variation concerns the O–H bond length. In the literature there are many different values for the bond length of water, which sometimes depend on the temperature or even the experimental technique used to measure it. We employed the bond length of 0.9572 Å originally used in TIP3P [12, 82] and 1.00 Å [83] (which is the same as in the SPC water model [47]).

The "null water" and X-ray diffractograms sensitive to O–O correlations were not very sensitive to this parameter variation (see Figure 7). A slight shift of the high-Q oscillations to lower Q as the bond length was increased shows that the molecules grew slightly. The change of the O–H bond length had a slightly larger effect on the H_2O and D_2O diffractograms, where a change of bond length resulted in a shift of the oscillations at large Q.

In contrast to the rather small changes of the diffraction patterns, the influence of the O–H bond length on the intermediate scattering function shown in Figure 7 is drastic. Prolonging the bond length led to a decrease of dynamics in a time range from a few tens of femtoseconds to nanoseconds. The final decay of the correlations at long times, probably corresponding to the long-range diffusion of the molecules, is more affected than the processes at sub-picosecond times. Comparing the whole intermediate scattering function to the data clearly yields much more information than looking only at the macroscopic diffusion coefficient.

The second example of a parameter change is the bond angle. The original TIP3P value was 104.52° [12, 82], and the comparison simulation was run with an angle of 109.50° (the angle of a tetrahedron), which is the same value used in the SPC water model [47]. The corresponding plots are also shown in Figure 7.

The diffractograms were very insensitive to this parameter change. Only the scattering patterns corresponding to the D_2O sample are slightly affected, since the H–H correlations weigh more than the other isotope combinations. However, it is questionable if this difference would be noticeable when comparing the computed curves to real data.

The influence on the intermediate scattering functions is much smaller than when varying the bond length. An increase of the bond angle leads to an increase in the dynamics. This acceleration primarily affects the long-time processes and leaves fast motions with correlation times below ~1 ps basically unaltered.

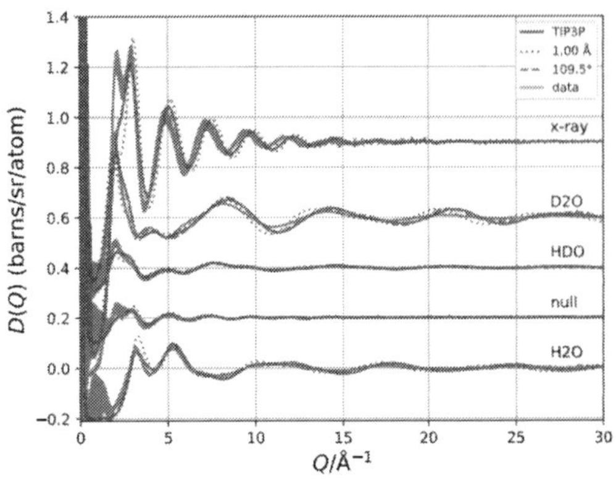

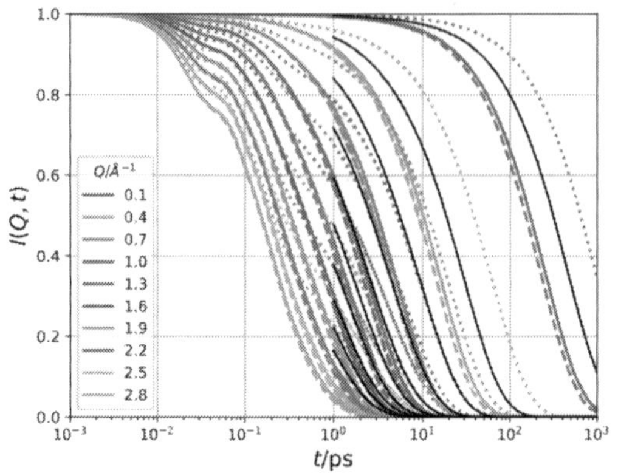

Figure 7. Parameter change in TIP3P rigid: 0.9572 Å O−H bond length versus 1.00 Å O−H bond length, and 104.25° H−O−H angle versus 109.50° H−O−H angle. Left: diffractograms for X-rays and neutrons with four different isotopic substitutions for each of the parameter changes as well as data [36] with different offsets for better visibility respectively. Right: intermediate scattering functions, black solid lines refer to literature data [37], solid colored lines refer to the standard TIP3P with a 104.25° angle and a 0.9572 Å bond length, dotted lines to a 1.00 Å bond length, and dashed lines refer to a 109.50° angle.

4 Discussion

Although no model is perfect, good models simplify complex situations while reproducing relevant observables. Which observables are relevant depends on the purpose of the model. This results in different possible parameter values for MD simulations, depending on if the model's creator placed more emphasis on reproducing for example the phase diagram or the nanoscopic structure and dynamics.

The experimental data used in this contribution were taken completely from the literature [36, 37]. The QENS experiments were performed at 293 K, the simulations were run at 296 K, and the diffraction data were taken at 298 K. Due to the slightly different temperatures, small variations between the simulation and experiment can be expected; however, because the density of the simulation was constrained and the atoms in the simulation moved faster than in the experiment, we do not expect the effects described in this contribution to be merely due to the temperature. The QENS data were used in the form of the intermediate scattering function reconstructed from the fit formula utilized in the literature [37]. This procedure circumvents problems associated with using the measured data of the scattering function $S(Q, \omega)$ directly, such as cutoff effects. However, we will attempt in the future to perform the comparison between simulation and experiment closer to the actually measured data.

Some parameters in simulations are mostly independent of the the water model chosen, such as cutoff distances or box sizes. For most of these parameters, we chose the consensus values published in the literature [12, 30, 38, 40, 72]. The size of the simulated box was considerably bigger than in many previous studies [8, 12, 13, 15, 31, 33], but simulations of even larger boxes have been published [19]. The use of a fairly large simulation box is important since too small of a box size is expected to influence both the structure and dynamics in the resulting scattering curves. While the structure in simulations is not expected to be influenced strongly by periodic boundaries, the calculation of $D(Q)$ is: peaks will broaden and small-angle scattering will become visible. For a disordered system like a liquid, the peak broadening effect is assumed to be negligible. The small-angle scattering, caused by the cubic simulation box effectively floating in vacuum, is confined to very small Q-values for a box of this size. For the incoherent intermediate scattering function, problems will arise in the simulation itself [84] while computation of the scattering function from the trajectories should not be hampered. A production run with an *NPT* ensemble is planned for future work to evaluate the impact of the pressure coupling on the intermediate scattering functions. Barostats regulate the pressure in the simulation box by resizing the box and scaling all atom coordinates accordingly. This change in coordinates is probably negligible but an *NVT* ensemble was used for this contribution to exclude any possible influences on the intermediate scattering function.

After minimizing influences due to the general setup of the simulations, we could concentrate on comparing different water models. All models do not reproduce the diffraction data over the whole Q-range; the TIP4P/2005 model is at low Q better than the TIP3P variants for all probes and is closer to the X-ray data also at higher Q-values. No significant change between the TIP3P rigid and flexible model was visible in the diffractograms. In particular, the O−H bond length appeared to be too well defined in either case, resulting in oscillations in the diffractograms at high Q with a too high amplitude. Since the bond stiffness was also constrained by other measurements such as Raman spectroscopy, this could be connected to nuclear quantum effects, which were not accounted for in the current simulation [46, 85, 86].

TIP4P/2005 captures the dynamics very well at low Q, which is consistent with the good reproduction of the long-range diffusion coefficient. The long-range mobility of TIP3P increases upon flexibilization. The local dynamics probed at higher Q is not well reproduced by any of the models shown here. Interestingly, the effect of introducing flexibility into TIP3P manifests itself mainly at the picosecond time scale, although the bond vibrations take place on a much shorter time scale.

A more detailed characterization of the simulated intermediate scattering function was done in section 2.6. The formula was able to reproduce the computed points in the covered time and Q-range. However, as seen in Table 2, the fitted values differ from the values in the literature by significant amounts. This disagreement is visible for both water models, highlighting that the simulated microscopic dynamics can deviate significantly even if the macroscopic diffusion coefficient is replicated correctly by the model. A further analysis of the individual values was not undertaken in the current contribution.

The rigid TIP3P model was modified by changing some exemplary parameters individually. Examples of the variation in the O−H bond length and H−O−H angle are shown in this work. We observed that a small variation in bond length can significantly impact the structure and dynamics, whereas a change in bond angle does not. The sensitivity of the computed scattering patterns to simulation parameter adaptations implies that an optimization of the agreement between computed and measured scattering curves could help refine the set of simulation parameter values. It should be noted that a water model consists of a set of parameters, so optimizing only one parameter is not expedient.

5 Summary and outlook

This work established a functional workflow of running MD simulations, calculating X-ray and neutron scattering data, and comparing these curves to already published experimental data. It was found that the scattering curves do not agree completely with the measured data; this disagreement was quantified in the case of dynamics. It was also observed that the scattering curves reacted with different sensitivity to different parameter changes.

Future work will compare the simulated curves with experimental data to use the differences between the two to optimize the set of simulation parameters. A multi-parameter fitting with Monte Carlo methods for this purpose is already in development, and possible machine learning techniques will also be evaluated. The aim is to obtain a procedure that can find a set of parameters for a system to reproduce the real nanoscopic structure and dynamics as probed by neutron and X-ray scattering.

References

[1] D.S. Sivia, *Elementary scattering theory* (Oxford University Press, London, England, 2011)

[2] K. Amann-Winkel, M.C. Bellissent-Funel, L.E. Bove, T. Loerting, A. Nilsson, A. Paciaroni, D. Schlesinger, L. Skinner, Chemical Reviews **116**, 7570 (2016)

[3] K. Nishikawa, N. Kitagawa, Bulletin of the Chemical Society of Japan **53**, 2804 (1980)

[4] P. Lindner, F. Leclercq, P. Damay, Physica B: Condensed Matter **291**, 152 (2000)

[5] A.H. Narten, H.A. Levy, The Journal of Chemical Physics **55**, 2263 (1971)

[6] A.H. Narten, M.D. Danford, H.A. Levy, Discuss. Faraday Soc. **43**, 97 (1967)

[7] F. Cavatorta, A. Deriu, D.D. Cola, H.D. Middendorf, Journal of Physics: Condensed Matter **6**, A113 (1994)

[8] D.D. Cola, A. Deriu, M. Sampoli, A. Torcini, The Journal of Chemical Physics **104**, 4223 (1996)

[9] B.J. Niebuur, W. Lohstroh, M.S. Appavou, A. Schulte, C.M. Papadakis, Macromolecules **52**, 1942 (2019)

[10] A. Arbe, G.J. Nilsen, J.R. Stewart, F. Alvarez, V.G. Sakai, J. Colmenero, Physical Review Research **2** (2020)

[11] F.H. Stillinger, A. Rahman, The Journal of Chemical Physics **60**, 1545 (1974)

[12] W.L. Jorgensen, J. Chandrasekhar, J.D. Madura, R.W. Impey, M.L. Klein, The Journal of Chemical Physics **79**, 926 (1983)

[13] J.L.F. Abascal, C. Vega, The Journal of Chemical Physics **123**, 234505 (2005)

[14] M.A. González, J.L.F. Abascal, The Journal of Chemical Physics **135**, 224516 (2011)

[15] J. Zielkiewicz, The Journal of Chemical Physics **123**, 104501 (2005)

[16] Y. Ding, A.A. Hassanali, M. Parrinello, Proceedings of the National Academy of Sciences **111**, 3310 (2014)

[17] T.A. Ho, A. Striolo, Molecular Simulation **40**, 1190 (2014)

[18] A. Henao, S. Busch, E. Guàrdia, J.L. Tamarit, L.C. Pardo, Physical Chemistry Chemical Physics **18**, 19420 (2016)

[19] F. Alvarez, A. Arbe, J. Colmenero, The Journal of Chemical Physics **155**, 244509 (2021)

[20] G. Smith, W. Paul, M. Monkenbusch, D. Richter, Chemical Physics **261**, 61 (2000)

[21] M. Brodeck, F. Alvarez, A. Arbe, F. Juranyi, T. Unruh, O. Holderer, J. Colmenero, D. Richter, The Journal of Chemical Physics **130**, 094908 (2009)

[22] J.C. Smith, M. Krishnan, L. Petridis, N. Smolin, in *Dynamics of Soft Matter* (Springer US, 2011), pp. 189–204, `https://doi.org/10.1007/978-1-4614-0727-0_7`

[23] A. Arbe, F. Alvarez, J. Colmenero, Polymers **12**, 3067 (2020)

[24] N. Zec, G. Mangiapia, A.C. Hendry, R. Barker, A. Koutsioubas, H. Frielinghaus, M. Campana, J.L. Ortega-Roldan, S. Busch, J.F. Moulin, Membranes

11, 507 (2021)
[25] R.L. McGreevy, L. Pusztai, Molecular Simulation **1**, 359 (1988)
[26] A.K. Soper, Molecular Simulation **38**, 1171 (2012)
[27] T. Youngs, Molecular Physics **117**, 3464 (2019)
[28] A.J. Markvardsen, R.L. McGreevy (2017)
[29] J. Armstrong, Journal of Physics Communications **6**, 102002 (2022)
[30] D.J. Price, C.L. Brooks, The Journal of Chemical Physics **121**, 10096 (2004)
[31] P. Mark, L. Nilsson, The Journal of Physical Chemistry A **105**, 9954 (2001)
[32] A. Pérez, A. Rubio, The Journal of Chemical Physics **135**, 244505 (2011)
[33] K.T. Wikfeldt, M. Leetmaa, M.P. Ljungberg, A. Nilsson, L.G.M. Pettersson, The Journal of Physical Chemistry B **113**, 6246 (2009)
[34] M. Rovira-Esteva, N.A. Murugan, L.C. Pardo, S. Busch, J.L. Tamarit, S. Pothoczki, G.J. Cuello, F.J. Bermejo, Physical Review B **84** (2011)
[35] V. Reich, A. Majumdar, M. Müller, S. Busch, *Data publication to comparison of molecular dynamics simulations of water with neutron and x-ray scattering experiments* (2022), `https://dx.doi.org/10.5281/zenodo.7225368`
[36] A.K. Soper, ISRN Physical Chemistry **2013**, 1 (2013)
[37] J. Teixeira, M.C. Bellissent-Funel, S.H. Chen, A.J. Dianoux, Physical Review A **31**, 1913 (1985)
[38] M. Orsi, Molecular Physics **112**, 1566 (2013)
[39] E.E. Ong, J.L. Liow, Fluid Phase Equilibria **481**, 55 (2019)
[40] C. Vega, J.L.F. Abascal, M.M. Conde, J.L. Aragones, Faraday Discuss. **141**, 251 (2009)
[41] L. Pusztai, O. Pizio, S. Sokolowski, The Journal of Chemical Physics **129**, 184103 (2008)
[42] R.B. Best, J. Mittal, The Journal of Physical Chemistry B **114**, 14916 (2010)
[43] G. Guevara-Carrion, J. Vrabec, H. Hasse, The Journal of Chemical Physics **134**, 074508 (2011)
[44] I.G. Tironi, R.M. Brunne, W.F. van Gunsteren, Chemical Physics Letters **250**, 19 (1996)
[45] D.E. Smith, A.D.J. Haymet, The Journal of Chemical Physics **96**, 8450 (1992)
[46] S. Habershon, T.E. Markland, D.E. Manolopoulos, The Journal of Chemical Physics **131**, 024501 (2009)
[47] H.J.C. Berendsen, J.P.M. Postma, W.F. van Gunsteren, J. Hermans, in *The Jerusalem Symposia on Quantum Chemistry and Biochemistry* (Springer Netherlands, 1981), pp. 331–342, `https://doi.org/10.1007/978-94-015-7658-1_21`
[48] H.J.C. Berendsen, J.R. Grigera, T.P. Straatsma, The Journal of Physical Chemistry **91**, 6269 (1987)
[49] M.W. Mahoney, W.L. Jorgensen, The Journal of Chemical Physics **112**, 8910 (2000)
[50] J.R. Grigera, The Journal of Chemical Physics **114**, 8064 (2001)
[51] S.W. Rick, The Journal of Chemical Physics **114**, 2276 (2001)
[52] H.A. Stern, F. Rittner, B.J. Berne, R.A. Friesner, The Journal of Chemical Physics **115**, 2237 (2001)
[53] H.W. Horn, W.C. Swope, J.W. Pitera, J.D. Madura, T.J. Dick, G.L. Hura, T. Head-Gordon, The Journal of Chemical Physics **120**, 9665 (2004)
[54] S.Y. Liem, P.L.A. Popelier, M. Leslie, International Journal of Quantum Chemistry **99**, 685 (2004)
[55] S.W. Rick, The Journal of Chemical Physics **120**, 6085 (2004)
[56] H. Yu, W.F. van Gunsteren, The Journal of Chemical Physics **121**, 9549 (2004)
[57] J.L.F. Abascal, E. Sanz, R.G. Fernández, C. Vega, The Journal of Chemical Physics **122**, 234511 (2005)
[58] P. Paricaud, M. Předota, A.A. Chialvo, P.T. Cummings, The Journal of Chemical Physics **122**, 244511 (2005)
[59] G.S. Fanourgakis, S.S. Xantheas, The Journal of Physical Chemistry A **110**, 4100 (2006)
[60] Y. Wu, H.L. Tepper, G.A. Voth, The Journal of Chemical Physics **124**, 024503 (2006)
[61] P.T. Kiss, A. Baranyai, The Journal of Chemical Physics **138**, 204507 (2013)
[62] L.P. Wang, T. Head-Gordon, J.W. Ponder, P. Ren, J.D. Chodera, P.K. Eastman, T.J. Martinez, V.S. Pande, The Journal of Physical Chemistry B **117**, 9956 (2013)
[63] W. Yu, P.E.M. Lopes, B. Roux, A.D. MacKerell, The Journal of Chemical Physics **138**, 034508 (2013)
[64] S.J. Bachmann, W.F. van Gunsteren, The Journal of Chemical Physics **141**, 22D515 (2014)
[65] S. Izadi, R. Anandakrishnan, A.V. Onufriev, The Journal of Physical Chemistry Letters **5**, 3863 (2014)
[66] R. Qi, L.P. Wang, Q. Wang, V.S. Pande, P. Ren, The Journal of Chemical Physics **143**, 014504 (2015)
[67] R. Fuentes-Azcatl, M.C. Barbosa, Physica A: Statistical Mechanics and its Applications **444**, 86 (2016)
[68] S. Izadi, A.V. Onufriev, The Journal of Chemical Physics **145**, 074501 (2016)
[69] Y. Khalak, B. Baumeier, M. Karttunen, The Journal of Chemical Physics **149**, 224507 (2018)
[70] C.L. Zhao, D.X. Zhao, C.C. Bei, X.N. Meng, S. Li, Z.Z. Yang, The Journal of Physical Chemistry B **123**, 4594 (2019)
[71] A.P. Thompson, H.M. Aktulga, R. Berger, D.S. Bolintineanu, W.M. Brown, P.S. Crozier, P.J. in 't Veld, A. Kohlmeyer, S.G. Moore, T.D. Nguyen et al., Comput. Phys. Commun. **271**, 108171 (2022)
[72] C. Vega, E. de Miguel, The Journal of Chemical Physics **126**, 154707 (2007)
[73] W. Humphrey, A. Dalke, K. Schulten, Journal of Molecular Graphics **14**, 33 (1996)
[74] T. Róg, K. Murzyn, K. Hinsen, G.R. Kneller, Journal of computational chemistry **24**, 657 (2003)
[75] N.P. Walter, A. Jaiswal, Z. Cai, Y. Zhang, Computer Physics Communications **228**, 209 (2018)

[76] B. Lindner, J.C. Smith, Computer Physics Communications **183**, 1491 (2012)

[77] B. Lindner, dissertation, University of Tennessee (2012), https://trace.tennessee.edu/utk_graddiss/1589/

[78] V.F. Sears, Neutron News **3**, 26 (1992)

[79] P.J. Brown, A.G. Fox, E.N. Maslen, M.A. O'Keefe, B.T.M. Willis, in *International Tables for Crystallography* (International Union of Crystallography, 2006), pp. 554–595, https://doi.org/10.1107/97809553602060000600

[80] O. Dietrich, *Diffusion coefficients of water*, https://dtrx.de/od/diff/

[81] L.P. Wang, T.J. Martinez, V.S. Pande, The Journal of Physical Chemistry Letters **5**, 1885 (2014)

[82] J.B. Hasted, in *The Physics and Physical Chemistry of Water* (Springer New York, 1972), pp. 255–309, https://doi.org/10.1007/978-1-4684-8334-5_7

[83] A. Zeidler, P.S. Salmon, H.E. Fischer, J.C. Neuefeind, J.M. Simonson, T.E. Markland, Journal of Physics: Condensed Matter **24**, 284126 (2012)

[84] I.C. Yeh, G. Hummer, The Journal of Physical Chemistry B **108**, 15873 (2004)

[85] I. Bakó, Á. Madarász, L. Pusztai, Journal of Molecular Liquids **325**, 115192 (2021)

[86] A. Eltareb, G.E. Lopez, N. Giovambattista, Physical Chemistry Chemical Physics **23**, 6914 (2021)

Complementary approaches to obtaining thermodynamic parameters from protein ligand systems-challenges and opportunities

Mona Sarter[1,*], *Doreen* Niether[2], *Simone* Wiegand[2,3], *Joerg* Fitter[4,5], and *Andreas M.* Stadler[6,7]

[1]ISIS Neutron and Muon Facility, STFC Rutherford Appleton Laboratory, Chilton, Didcot, OX11 0QX, UK
[2]Institute of Biological Information Processing (IBI-4: Biomacromolecular Systems and Processes) & JARA-SOFT, Forschungszentrum Jülich GmbH, D-52428 Jülich, Germany
[3]Department für Chemie - Physikalische Chemie, Universität zu Köln, 50939 Cologne, Germany
[4]I. Physikalisches Institut (IA), AG Biophysik, RWTH Aachen, Sommerfeldstrasse 14, Aachen D-52074, Germany
[5]Institute of Biological Information Processing (IBI-6: Molecular Biophysics), Forschungszentrum Jülich GmbH, D-52428 Jülich, Germany
[6]Jülich Centre for Neutron Science (JCNS-1) and Institute of Biological Information Processing (IBI-8: Neutron Scattering and Soft Matter), Forschungszentrum Jülich GmbH, D-52428 Jülich, Germany
[7]Institute of Physical Chemistry, RWTH Aachen University, Landoltweg 2, Aachen D-52056, Germany

Abstract. Protein ligand interactions play an important role in biology. Increasingly the aim is to understand and influence protein ligand binding. The binding process is heavily influenced by its thermodynamic parameters. In order to understand how the whole system thermodynamics work it is important to characterise the individual contribution of each of the systems components. While the change in conformational entropy of the protein can be determined using QENS complementary methods are necessary in order to characterise all components. This paper will describe the challenges that can occur when combining the different methods, as well as how they can be overcome.

1 Protein ligand interactions

Protein ligand interactions play an important role in biology [1]. A ligand is any particle that forms a complex with a biomolecule for a biological purpose. These particles can be vitamins, minerals, sugars, or other small particles [2]. Protein ligand interactions induce structure changes, facilitate ligand transport or ligand trapping, as well as signal transfer [3–5].

The strength of a protein ligand interaction is described by the equilibrium dissociation constant $K_d = K_a^{-1} = k_{on}/k_{off}$, here K_a is the association constant, k_{on}, k_{off} are the binding and release rates respectively. This dissociation constant describes the concentration at which the ligand free and ligand bound state exist in solution. If a complex is diluted below this threshold the ligand will be continuously released [6–8].

Protein ligand interactions are governed by thermodynamic parameters. Any protein ligand binding that happens spontaneously needs to be accompanied by an advantageous change in Gibb's free energy $\Delta G = \Delta H - T\Delta S < 0$, which means that the reaction is exergonic. A reaction can be favourable even if ΔH or ΔS are not favourable due to entropy enthalpy compensation [7, 9]. In addition to the whole system contributions to these parameters, each parameter contains the individual contributions of each component to the system. These components are not easily accessible and require a combination of different techniques to be measured. While it is tempting to assume that all individual contributions are either opposing or favourable for the reaction this is not the case, as has been described for the binding of biotin (B) to streptavidin (STV), where entropy-entropy compensation was observed [10, 11].

This paper will focus on discussing the different components of a protein ligand system and what their contributions are to the total change in entropy ΔS. For this purpose, neutron scattering, as well as other complementary methods will be discussed, since the different components cannot be separated from the whole change in entropy ΔS using only calorimetric techniques. Using different methods that were not originally designed to be used on the same systems poses several different challenges. However, some of these differences also provide unique opportunities and information that might not have been explored without these methods.

In order to illustrate this approach, the system of STV and B will be used as an example. STV is a homotetramer with a molecular weight of 13271.6 Da [10, 12]. Its ligand B, which has a molecular mass of 244.3 g mol^{-1}, binds to it with a stoichiometry of 4:1, see Fig. **1** [13]. This binding is among the strongest bindings known in biology with a

* Corresponding author: mona.sarter@stfc.ac.uk

dissociation constant of $K_d \approx 10^{-14}$M [14, 15]. The biological function in this case is the trapping of the ligand biotin, which creates a slightly antibacterial environment, and gives the bacterium Streptomyces avidinii an advantage over competing bacteria growing nearby[16].

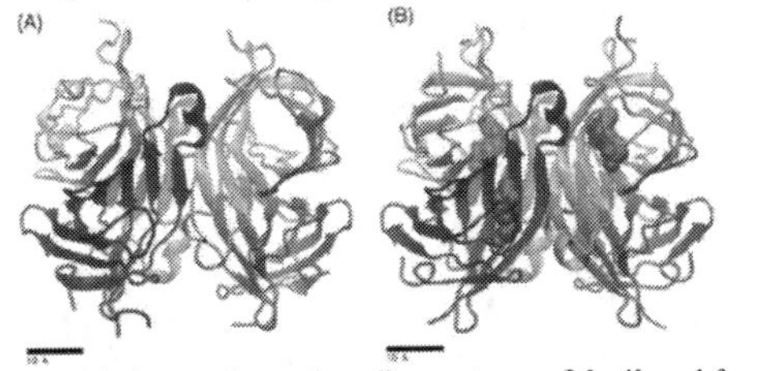

Fig. 1. This image shows the pdb structures of the ligand free STV tetramer (A), and of the complex STV+B (B), with one B per STV binding pocket [10]. Reprinted with permission from [10] M. Sarter et al., "Strong Adverse Contribution of Conformational Dynamics to Streptavidin-Biotin Binding," J. Phys. Chem. B, vol. 124, no. 2, pp. 324–335, 2020 American Chemical Society.

While this biological cause for the binding is not of particular relevance for current applications, the strong binding affinity of STV and B has led to a specific application. In order to fix a molecule to a surface the surface has biotin attached to it, and the molecule has biotin attached as well, while STV is added. STV then binds to the surface and the molecule so that the molecule is affixed [14].

The adverse change in entropy observed during the STV+B complex formation is of particular interest, as it is such a strong binding and contains an adverse component [17, 18]. Therefore, we would like to persue how the different components contribute.

2 Entropic components and corresponding experimental approach

Any spontaneously happening reaction requires an advantageous change in Gibb's free energy ΔG. To understand this change the change in enthalpy ΔH and in entropy ΔS need to be considered. To understand the change in enthalpy knowledge of the different bonds and during the protein ligand interaction, as well as the structures is sufficient to perform calculations [19]. The situation is more complex and involves more parameters for the change in entropy ΔS, any simulation will require full knowledge and understanding of the components involved. We have therefore chosen to focus on the components to the change in entropy ΔS here.

Before discussing how to experimentally obtain the different components that make up the total change in entropy during protein ligand interactions it is important to define them.

Protein ligand systems are analysed in solution, so the influence of the solvent needs to be considered, in addition to that of the protein, its ligand, and any conformational changes upon complex formation.

$$\Delta S = \Delta S_{protein} + \Delta S_{ligand} + \Delta S_{hydr} \quad (1)$$

$$\Delta S_{protein} = \Delta S_{conf,protein} + \Delta S_{trans,rot,protein} \quad (2)$$

$$\Delta S_{hydr} = \Delta S_{displaced} + \Delta S_{surrounding} \quad (3)$$

$$\Delta S_{ligand} = \Delta S_{conf,ligand} + \Delta S_{trans,rot,ligand} \quad (4)$$

Equation 1 shows that there are three main contributions to consider for a protein ligand system, each of which has its own parts. The main contributions are any entropy change experienced by the protein $\Delta S_{protein}$, and its ligand ΔS_{ligand}, as well as entropy changes to the solvent ΔS_{hydr}. Table 1 provides an overview of the methods which are appropriate for each component, which will be discussed in detail below.

In addition to the individual contributions the whole system also needs to be measured. A method that is able to obtain all the relevant experimental contributions is isothermal titration calorimetry (ITC). In addition to the change in entropy ΔS, it also provides the binding stoichiometry, binding affinity K_a, change in enthalpy ΔH, and change in Gibbs free energy ΔG[20]. If the aim is to only get K_a without the thermodynamic parameters a labelled competition experiment is a viable alternative to ITC [21].

The first part of ΔS is the protein's contribution $\Delta S_{protein}$, the components of which are given in Equation 2[20]. To analyse the change in conformational entropy $\Delta S_{conf,protein}$ quasi elastic neutron scattering (QENS) experiments in a D_2O based buffer are used [22]. If the energy resolution for the experiment is chosen appropriately, diffusive contributions for $\Delta S_{trans,rot,protein}$ can also be observed [23]. However, it might also be opportune to use dynamic light scattering (DLS) to determine the proteins' diffusive behaviour [24]. All QENS experiments proposed above are based on the assumption that the ligand is small compared to the protein, and will not provide a relevant scattering contribution. If the ligand should have a high number of H atoms compared to the number in the protein deuteration of the ligand becomes necessary.

With all experiments carried out in solution the hydration layer surrounding the proteins provides a relevant contribution to the thermodynamics of the whole system [25]. Its change in entropy ΔS_{hydr} and the contributing components are given in Equation 3. In order to determine the influence of $\Delta S_{displaced}$ it is at first necessary to know which water molecules are hydrogen bonded to the binding site, and how many of these will be displaced upon binding. This can be achieved by comparing the free protein and protein ligand bound crystal structures, as well as by MD simulations. The entropic component is calculated from this information. Another contributing component is $\Delta S_{surrounding}$, which gives the entropy change of the hydration layer surrounding the protein. It can be experimentally accessed using thermal diffusion forced Rayleigh scattering (TDFRS) [26, 27].

The last contributions to the total change in entropy are those provided by the ligand ΔS_{ligand}, as given in Equation 4. A change in the structure of the ligand $\Delta S_{conf,ligand}$ can often be ruled out due to the ligands size and structure, if it is a small compact ligand such as most vitamins or minerals. If this is not the case, QENS experiments on the free ligand, as well as on the protein ligand complex with the protein deuterated are necessary to determine $\Delta S_{conf,ligand}$. A reduction in

$\Delta S_{trans,rot,ligand}$ will invariably take place, since the ligand will go from free movement in solution to being bound to the protein. Depending on the ligands size and the amount required it might be negligible or not. If QENS experiments are carried out on the ligand they can be used as discussed for the protein. Otherwise DLS experiments on the free ligand can determine its diffusion coefficient, which can be used to calculate how much the entropy is reduced if that ligand diffusion is slowed down upon ligand binding.

Table 1 This table provides an overview about the methods which are appropriate for which entropy component.

	ΔS	$\Delta S_{protein}$	$\Delta S_{hydration}$	ΔS_{ligand}
ITC	Yes	No	No	No
QENS	No	Yes	No	Using deuteration
TDFRS	No		Yes	No

3 Challenges posed by the different methods

Before discussing the challenges of each method a brief description of how they are experimentally applied and how they work will be given.

ITC, used to determine the thermodynamic parameters of the whole system, is a calorimetric method. As such the protein solution is placed in one sample cell, while the buffer solution is placed in a reference cell. The cells are temperature controlled, with a highly concentrated solution of the ligand being titrated to the sample cell [11]. Due to the great sensitivity of the method it is important to determine the concentration of the protein solution and the ligand solution to a very high level of accuracy. Additionally, avoiding aggregation of the protein in solution is of utmost importance, so this has to be checked beforehand.

For data analysis it is first important to utilise the correct binding model during data analysis. While it can be 1:1, meaning that each ligand binds with the same affinity to the protein this is not necessarily the case for all systems [28]. A literature analysis and comparison of binding sites if a crystal structure is available are of great use.

Here the data analysis will be discussed for the simplest 1:1 model, which was used for the example data. ΔH is determined from the difference in energy required to keep the reference and sample cell at the same temperature. The binding stoichiometry can be determined from the molar ratio of ligand and protein at the point where the binding reaction stops. To determine the change in Gibbs free energy $\Delta G = -RTln(K_a)$ it is necessary to know the binding affinity K_a and the temperature at which the experiment is carried out [20]. Titrating the ligand to the protein solution provides a sigmoidal curve, with the change of energy plotted against the molar ratio of ligand to protein. By fitting the sigmoidal curve K_a can be determined as the turning point of this curve. With this information the total change in entropy $\Delta S = T^{-1}(\Delta G - \Delta H)$ can now be calculated [28].

Now that the method has been introduced it is time to talk about the associated challenges. Due to the high sensitivity of ITC it is important to keep the temperature in the lab where the ITC stands stable, otherwise a drift in the data will make the analysis more difficult. To ensure that no concentration effects exist for the observed system experiments should be performed at different concentrations and both data sets used simultaneously during data analysis. In order to obtain good results, it is helpful to repeat the experiments several times and perform a statistical analysis of the results, as the sensitivity of the method means that the slightest variation in cell filling or room temperature might affect the results. If a reaction is very strong, meaning it has a small K_d, or very weak, it is difficult analyse as the reaction will not be slow enough to allow for thorough analysis [29][30]. This holds for the later discussed example of STV+B. In these cases a solution is to obtain K_d from additional experiments, such as labelled ligand displacement experiments, and use it as a known quantity to calculate ΔG and thus ΔS, or to perform ligand displacement ITC [21, 30].

The second method mentioned as a mayor experimental approach is QENS [31]. These experiments require a high protein concentration of around 50 to 100 mg/ml in order to achieve sufficient signal intensity while at the same time avoiding multiple scattering contributions to the recorded spectra. In addition, QENS requires a sample volume of around 1 to 2 ml that is large when compared to several other biological methods. All samples are prepared in a D_2O based buffer, so that the protein's and ligand's hydrogen provide the main scattering contribution [23, 32]. The buffer is still measured separately from the sample and subtracted as background. Experimental data is typically fitted with one or more Lorentzians to describe the protein's internal dynamics, as well as the global diffusion if an additional Lorentzian is fitted. Which dynamics can be observed depends on the timescale analysed. In addition to the peak broadening of the Lorentzian to determine the dynamics the system's mean-square displacement (MSD) is also analysed. This value can be used to calculate

$$\Delta S_{conf,protein} = \text{Res} \cdot 3R \ln\left(\sqrt{\frac{MSD_{protein+ligand}}{MSD_{protein}}}\right)$$

, here Res is the number of residues of the protein[33]. DSC and QENS experiments yielded the same values for $\Delta S_{conf,protein}$ for the thermal unfolding of myoglobin, which are in quantitative agreement within the errors and hence, corroborate the QENS approach described above [34–36].

Having briefly summarised the experimental and data analysis approach for QENS the focus will now be on the challenges this method poses. As mentioned above, for larger ligands it can become necessary to deuterate both the protein, as well as the ligand. In this case access to a deuteration lab, and the knowledge of how to express the protein in a deuterated state will be necessary. The same will have to be achieved for the ligand. Another related issue is sample availability, due to the comparatively large amount of sample required (100 to 200 mg protein per sample) it can be difficult to produce the required amount of protein. Even with the high sample concentration and volumes the time necessary to complete one data set is still around 12 h for a QENS experiment[10, 26]. If a protein ligand system has a very high binding affinity this is not a problem, as the complex will not dissolve during the measurement. Therefore, the ligand will not need to be present at oversaturation levels to ensure complex formation. For lower binding affinities over saturation with the ligand might be required to ensure that complex formation occurs. In this case the ligand free buffer needs to be measured, as well as the buffer at oversaturation levels. The ligand free buffer spectrum is then subtracted from the protein spectrum, while the buffer with the ligand at oversaturation levels is subtracted from the protein ligand complex. In the case of protein ligand systems with even lower binding affinities the complex will not be the only population present in the sample, instead a certain percentage of free protein will always be a part of the solution. For these samples it is important to calculate the percentage of ligand free protein that is part of the sample. After that calculation is performed during data treatment the spectrum of the free protein and the buffer spectrum are subtracted each with the appropriate factor. To do this the buffer data subtracted protein dataset will be used for subtraction, with the correct percentage applied as intensity scaling. Another challenge is that the measurements are performed in a D_2O based buffer, while for many proteins this does not have any effects there have been reports about proteins being affected by the change in solvent from H_2O to D_2O. Therefore, it is necessary to check that the protein's, as well as the protein ligand complexes, properties are not affected by the change in solvent. For this purpose, small angle scattering (SAS) experiments to analyse structural changes, ITC experiments in both buffers, TDFRS experiments in both buffers, as well as circular dichroism (CD) to check for structural changes, or DLS to check on the diffusive behaviour can be used. An appropriate selection of these methods should be made depending on the system. To confirm that the high concentration required does not affect the results it is necessary to perform dilution experiments, such as using DLS to ensure that the concentration has no negative influence and causes no aggregation. In addition, SAS, ITC, and TDFRS experiments at different concentrations can be used.

In order to determine the behaviour of ΔS_{hyd} TDFRS can be used [27]. The method is based on thermophoresis, which is the transport of matter caused by a temperature gradient and shows a high sensitivity to the number and strength of hydrogen bonds formed between solute and water [37]. This transport is analysed in solution using the Soret coefficient $S_T = D_T/D$, with D_T as the thermal diffusion coefficient, and D the mutual diffusion coefficient. In order to induce a temperature gradient infrared TDFRS (IR-TDFRS) is applied. Interference of two infrared laser beams with a wavelength of $\lambda = 980\ nm$ is used to create a thermal grating, which triggers the thermophoresis [26, 38, 39]. This wavelength corresponds to an absorption band in H_2O, but not in D_2O. Experiments are carried out at different temperatures, to analyse the temperature sensitivity of S_T. S_T is determined from the amplitude of the concentration signal A as follows, $A = \left(\frac{\partial n}{\partial c}\right)_{p,T}\left(\frac{\partial n}{\partial T}\right)_{p,c}^{-1} S_T c(1-c)$. Here c is the concentration, n the refractive index increment with concentration, $\left(\frac{\partial n}{\partial c}\right)_{p,T}$ was obtained from a refractometer and the refractive index increment with temperature $\left(\frac{\partial n}{\partial T}\right)_{p,c}$ from interferometry. Further details can be found elsewhere [40].

In systematic experiments a strong correlation between thermodiffusion and hydrophilicity of the solute was found [37]. The more hydrophilic solutes have a stronger tendency to accumulate in the cold. This is illustrated in Figure 2 The free STV forms more hydrogen bonds with the surrounding water and shows a stronger tendency to accumulate in the cold, while biotin blocks some of the hydrogen bonding sites leading to a less hydrophilic complex and higher entropy ΔS_{hydr} of the water molecules in the hydration shell. The lower hydrophilicity is also reflected by a larger temperature sensitivity of S_T often described by ΔS_T the difference between S_T at two temperatures.

For the TDFRS experiments the protein needs to be measured in solution with and without the ligand, as well as the ligand in solution. If oversaturation is required for complex formation then the oversaturated buffer also needs precise analysis. Different temperatures and protein concentrations are used for the experiments. To ensure comparability with QENS the buffers need to be D_2O based, which leads straight to the challenges faced with this method [26]. Due to the problems measuring in D_2O mentioned above a pure D_2O based buffer is not possible, instead the D_2O concentration has to be steadily increased. An increase of up to 50% was found to be possible. If no D_2O concentration dependent change of T dependent S_T behaviour can be observed it is safe to assume that no solvent dependence exists. However, data acquisition gets more difficult with increasing D_2O concentration due to a lower refractive index contrast [38].

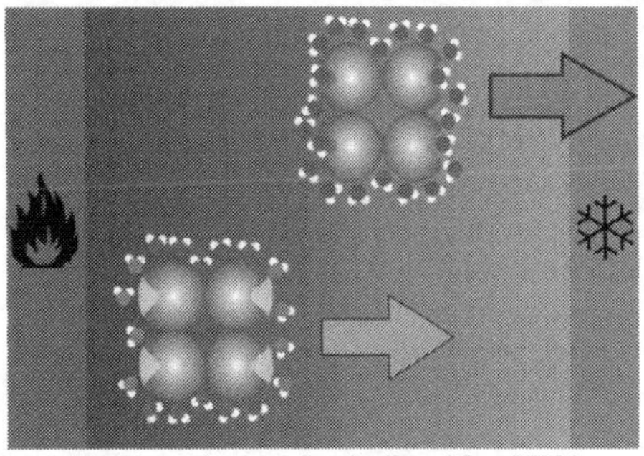

Fig. 2. Here it can be seen that the ligand free STV is more likely to accumulate in colder areas than the ligand bound STV+B complex. This is thought to be related to the higher hydrophilicity of free STV [26].

4 Experimental approaches discussed for the example of STV+B

STV+B will now be used as an example to explore the challenges faced for each method, the results obtained for this system will also be presented. The solutions chosen to face them, as well as reflections on which additional approaches might have been helpful in different circumstances will be presented. All experiments were performed in a 25 mM TRIS-HCl, 120 mM NaCl, 5 mM KCl, 3 mM MgCl, pH 7.4 buffer, either H_2O, or 99.9% atom D based, or buffer mixtures depending on the method used.

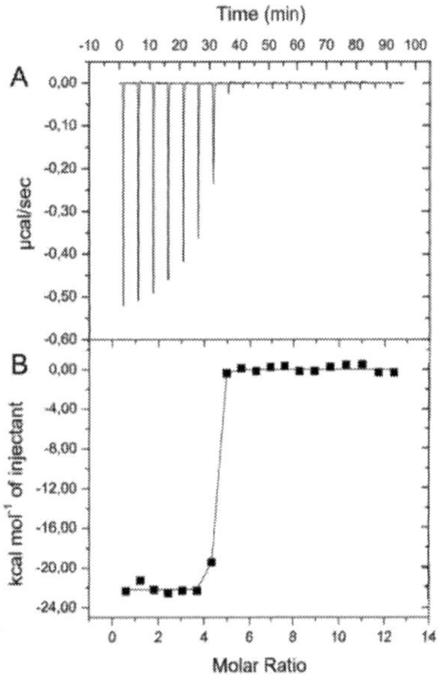

Fig. 3. (A) shows the raw data obtained from titrating biotin to STV plotted against time. The change in energy per second can be observed. It is also visible that between each titration sufficient time has passed that the reaction is finished and the solution equilibrated. (B) shows the energy change per mol of injectant, here B, plotted against the molar ratio of B and STV.

First the whole system was analysed with ITC. Two different concentrations of STV solution, as well as two different concentrations of B solution were used for the ITC experiments, with $c_{STV} = 65$ mg/ml, and $c_{STV} = 109$ mg/ml, all ITC experiments were carried out in a H_2O based buffer. The B solutions were titrated to STV solutions, as shown Fig. *3*. In Fig. *3* (A) it can be observed that the end of binding reactions occurring due to saturation of STV with B is very distinct. To observe the binding stoichiometry Fig. *3* (B) can be used. The change in enthalpy ΔH, is easily determined from these results. In order to calculate ΔS it is necessary to determine the turning point in Fig. *3* (B) to get K_a. It is obvious from Fig. *3* (B) that the turning point cannot be easily determined. As no literature value of K_d or K_a is known for this specific strain of STV it was decided to proceed with a literature value of wild strain STV for $\Delta S_{STV} = -348\mathrm{J/(molK)}$, instead of using its K_a and the ITC experiments. From the different experimental runs taking the statistical fluctuations into account the following result for $\Delta H = -427.1 \pm 47.2$ kJ/mol was calculated.

In this case the ITC specific challenges were solved by compensating with literature values. If a protein ligand system has not been previously characterised it would be necessary to experimentally obtain K_a. By performing the experiments at different STV and B concentrations it was possible to confirm that no concentration effects to the binding reaction or stoichiometry could be observed. The experiments were not repeated in D_2O due to time constraints and the inaccessibility of K_a.

QENS experiments were performed in D_2O based buffers. Due to the small and rigid structure of B, any conformational change on it was ruled out and $\Delta S_{conf,ligand}$ was treated as negligible in this case. For $\Delta S_{trans,rot,ligand}$ literature confirms that the change in entropy caused by desolvation is similar to the entropy contribution from its diffusion, so that any changes even out. Therefore, a deuteration of STV, or B apart from an exchange of free H to D was not necessary. The experiments were carried out on the ps-ns timescale for concentrations in the range of $c_{STV} = 55\ mg/\mathrm{ml}$ - $c_{STV} = 65$ mg/ml. Due to the high binding affinity of this system no oversaturation of STV with B was necessary to ensure complex formation. Therefore, the amount of B necessary to saturate STV was calculated by weight and added to the STV solution. In addition to avoiding the need for an additional measurement of the buffer this system is also sufficiently stable in complex form over time that only the complex was present in solution during the experiments.

Neither DLS nor QENS experiments found a difference in the diffusion between STV and STV+B. In addition the different crystal structures for STV and STV+B did not produce a difference in diffusion coefficient using HYDROPRO [11, 41, 42]. Therefore, it was concluded that $\Delta S_{trans,rot,protein}$ is negligible for this system. The value for $\Delta S_{conf,protein}$ was in good agreement within the error on the different timescales and determined to be $\Delta S_{conf,STV} = -2200 \pm 300\mathrm{J/(molK)}$. The elastic incoherent structure factor (EISF) curves which provided the MSDs for this calculation can be seen in Fig. **4** This shows that the change in conformational entropy is more strongly opposed to the binding than the whole entropy change and illustrates the need for using several complementary methods to obtain each individual component of the total change in entropy.

Due to the experiments being performed in D_2O it was necessary to check that free STV, as well as the STV+B complex were not affected by this change in solvent. This was done by comparing the diffusion in DLS, structure with both SAXS and SANS, as well as CD spectroscopy and later during TDFRS experiments where the ratio of D_2O to H_2O in the buffer was altered. No buffer dependent changes could be observed. In addition, DLS dilution experiments were carried out to check for any changes in diffusive behaviour due to concentration and none were observed.

To gain information about the last entropic component TDRFS experiments were carried out. In accordance with the challenges previously mentioned the buffer used contained up to 50% D_2O, while initial experiments were carried out in H_2O based buffers, the protein's concentration was $c_{STV} = 39\,\frac{\text{mg}}{\text{ml}}$.

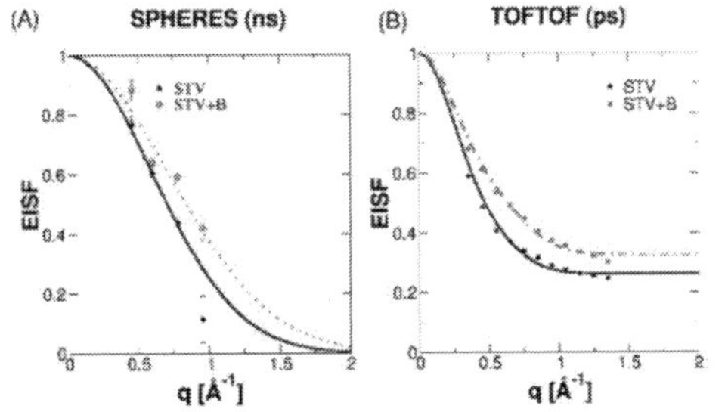

Fig. 4. EISF curves for STV and STV+B on the ns (A), and ps (B) timescale. It can be observed that free STV is more flexible than the STV+B complex [10]. Reprinted with permission from [10] M. Sarter et al., "Strong Adverse Contribution of Conformational Dynamics to Streptavidin-Biotin Binding," J. Phys. Chem. B, vol. 124, no. 2, pp. 324–335, 2020 American Chemical Society.

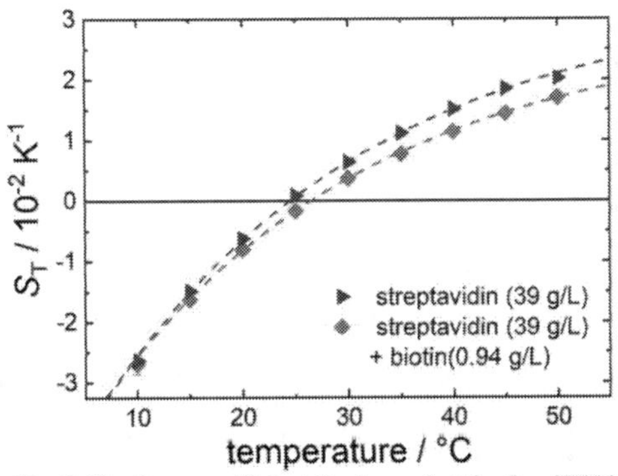

Fig. 5. The Soret coefficient S_T shows that the free STV is more hydrophilic than the STV+B complex. This indicates that the entropy of the hydration layer increases upon complex formation.

From these experiments no changes in diffusion were observed upon ligand binding, while S_T of STV+B is reduced compared to free STV **Fig. 5**. This difference decreases with decreasing T and is almost gone below 298 K. These results indicate that STV is slightly more hydrophilic than STV+B, which agrees with observations made from the structural data and shows that the T dependent change of ΔS_{hydr} relates as follows. $\Delta S_{hydr}^{STV} < \Delta S_{hydr}^{STV}$. From this it can be concluded that an advantageous change in $\Delta S_{surround}$. Increases of D_2O to H_2O ratio did not lead to a change in T sensitivity of S_T.

Regarding $\Delta S_{displaced}$ this could be calculated, as crystal structures show that 5 H are contained in each binding pocket, $\Delta S_{displaced} = 586\ \text{J}/(\text{molK})$.

Using the system of STV+B as an example the advantage of using different complementary methods in addition to QENS to approach the thermodynamic parameters was shown. Several of the potential challenges which come from combining these methods were encountered and the solutions chosen for this system were illustrated. By combining the different methods, it could be confirmed that entropy-entropy compensation occurs upon STV+B complex formation and that this is the reason that the reaction is still thermodynamically favourable, although the conformational contributions would oppose the binding.

5 Conclusion

In order to understand protein ligand systems, it is important to know what the thermodynamics of the ligand binding process are. This proceeding has provided an overview over complementary methods that are available to analyse the entropic component in depths. While there are several challenges in combining different complementary methods, these can be overcome by careful planning and characterisation of the system.

Using the example of the STV+B system it was shown that analysing the different components offers valuable insights about which part of the system influences the binding in what way. This is of special importance in the case of biopharmaceutical applications, where adjusting the binding affinity is often used to improve medications. A better understanding of each components' contribution will allow for more targeted approaches.

Acknowledgement: We thank Bernd W Koenig for fruitful discussion and his help with the ITC measurements.

References

1. C. Sybesma, *An introduction to biophysics, 1st ed.* Acad. Pr., New York (1977)
2. V.B. Teif, Biophys. J. **89**, 2574–2587 (2005)
3. N. Haruta, M. Aki, S.I. Ozaki, et al., Biochemistry **40**, 6956–6963 (2001)
4. A.M. Stadler, E. Knieps-Grünhagen, M. Bocola, et al, Biophys. J. **110**, 1064–1074 (2016)
5. R. Elbert, M. Karplus, J. Am. Chem. Soc. **112**, 9161–9175 (1990)
6. M. Schmidt, K. Nienhaus, R. Pahl, et al., Proc. Natl. Acad. Sci. **102**, 11704–11709 (2005)
7. A.J. Wand, K.A. Sharp, Annu. Rev. Biophys. **47**, 41–61 (2018)
8. T. Lazaridis, A. Masunov, F. Gandolfo, Proteins Struct. Funct. Genet. **47**, 194–208 (2002)
9. B.M. Baker, K.P. Murphy, J. Mol. Biol. **268**, 557–569 (1997)
10. M. Sarter, D. Niether, B.W. Koenig, et al, J. Phys. Chem. B. **124**, 324–335 (2020)
11. M. Sarter *Structural and dynamical propterites in protein ligand interactions,* Thesis (2021)
12. I. Le Trong, Z. Wang, D.E. Hyre, et al., Acta. Crystallogr. Sect. D. Biol. Crystallogr. **67**, 813–821 (2011)
13. P.C. Weber, D.H. Ohlendorf, J.J. Wendoloski, F.R. Salemme, Science **243**, 85–88 (1989)
14. N. Michael Green. Methods. Enzymol. **184**, 51–67 (1990)
15. T. Sano, C.R. Cantor, J. Biol. Chem. **265**, 3369–3373 (1990)

16. E.A. Bayer, T. Kulik, R. Adar, M. Wilchek, BBA - Gene. Struct. Expr. **1263,** 60–66 (1995)
17. D.H. Williams, E. Stephens, D.P. O'Brien, M. Zhou, Angew. Chemie. - Int. Ed. **43,** 6596–6616 (2004)
18. P.C. Weber, J.J. Wendoloski, M.W. Pantoliano, F.R. Salemme, J. Am. Chem. Soc. **114**, 3197–3200 (1992)
19. C.A. K Koppisetty, M. Frank, G.J. L Kemp, P.-G. Nyholm, J. Chem. Inf. Model **53,** 42 (2013)
20. S. Leavitt, E. Freire, Curr. Opin. Struct. Biol. **11,** 560–566 (2001)
21. D.E. Hyre, I. Le Trong, S. Freitag, et al, Protein Sci. **9,** 878–885 (2000)
22. M. Grimaldo, F. Roosen-Runge, F. Zhang, et al., Q. Rev. Biophys. **52,** e7 (2019)
23. F. Gabel, Eur. Biophys. J. **34,**1–12 (2005)
24. L. Yin, *Nanotechnology Research Methods for Foods and Bioproducts* (2012)
25. B. Bagchi, Chem. Rev. **105,**3197–3219 (2005)
26. D. Niether, M. Sarter, B. König, et al., AIP Conf. Proc. **1929,** 1–8 (2018)
27. M. Jerabek-Willemsen, C.J. Wienken, D. Braun, et al., Assay. Drug Dev. Technol. **9**, 342–353 (2011)
28. Microcal. Reading 121 (2004)
29. A. Pérez, E.K. Santamaria, D. Operario, et al., *Neutron Scattering in Biology* (2017)
30. A. Velazquez-Campoy, E. Freire, Nat. Protoc. **1,** 186–91 (2006)
31. M. Bee, *Quasielastic neutron scattering: Principles and applications in solid state chemistry, biology and materials science, I.* Hilger, Bristol (1988)
32. F. Gabel, D. Bicout, U. Lehnert, et al., Q. Rev. Biophys. **35,** 327–367 (2002)
33. J. Fitter. Biophys. J. **84,**3924–3930 (2003)
34. J. Seelig, A. Seelig, Biophys. Reports **2**, 100037 (2022)
35. A.M. Stadler, F. Demmel, J. Ollivier, T. Seydel, Phys. Chem. Chem. Phys. **18**, 21527–21538 (2016)
36. A.M. Stadler, M.M. Koza, J. Fitter, J. Phys. Chem. B **119,** 72–82 (2015)
37. D. Niether, S. Wiegand, J. Phys. Condens. Matter **31,** 503003 (2019)
38. D. Niether, M. Sarter, B.W.B.W. Koenig, et al., Polymers (Basel) **12,** 1–15 (2020)
39. D. Niether, T. Kawaguchi, J. Hovancová, et al., Langmuir **33**, 8483–8492 (2017)
40. S. Wiegand, H. Ning, H. Kriegs, J. Phys. Chem. B **111,** 14169–14174 (2007)
41. A. Ortega, D. Amorós, J. García De La Torre, Biophys. J. **101**, 892–898 (2011)
42. J. García De La Torre, M.L. Huertas, B. Carrasco, Biophys. J. **78,** 719–730 (2000)

Panther — the new thermal neutron time-of-flight spectrometer at the ILL

Björn Fåk[1,*], *Stéphane* Rols[1], *Giuliana* Manzin[1], and *Olivier* Meulien[1]

[1]Institut Laue-Langevin, 71 avenue des Martyrs, F-38042 Grenoble Cedex 9, France

Abstract. Panther is a new high-flux medium-resolution direct-geometry thermal-neutron time-of-flight spectrometer at the Institut Laue-Langevin (ILL). It is designed for inelastic neutron-scattering measurements of excitations in condensed matter using single crystals, polycrystalline samples, and liquids. Panther uses double focusing graphite or Cu monochromators, a Fermi chopper, and position-sensitive ^{3}He detectors covering 2 steradians of solid angle. A system of disc choppers and an optional sapphire filter are used to reduce the epithermal neutron background. Thermal neutron background is reduced by a radial oscillating collimator, a beam dump, and an elaborate set of Cd shielding inside the evacuated detector tank. The outside of the tank is covered by a 0.3 m thick layer of borated high-density polyethylene to reduce ambient and cosmic background. The design and performance of the instrument in its current status are described, as well as planned developments.

1 Introduction

Neutron time-of-flight (TOF) spectrometers are ideal for studies of excitations in condensed matter. In the thermal energy range, from a few up to about hundred meV, typical science applications are crystal field excitations, spin-waves in magnetically ordered materials, magnetic excitations in quantum and/or frustrated magnets, quantum rotations and translations of small molecules in confinement, phonon dispersion relations and density of states, vibrational spectroscopy in the low-energy range, diffusion and relaxation processes in fuel cells, thermoelectric materials, and other energy-related applications. For powder samples, an overview of the excitations can be obtained in a short time, typically a couple of hours on a high-flux spectrometer. With the generalization of large-area highly pixelated detectors on TOF spectrometers, single crystals are increasingly being studied, giving very detailed information on the excitations in four dimensions, provided the sample is rotated during the measurements. Such experiments require longer measuring times, typically a couple of days.

Panther is a completely new thermal neutron TOF spectrometer at the ILL. It replaces IN4C, which was constructed in the mid nineties [1]. Panther uses the same "hybrid" TOF principle with a focusing monochromator and a Fermi chopper operating in time focusing mode. For thermal neutrons, focusing optics using a monochromator is advantageous compared to current supermirror guides in terms of flux. The instrument uses one incoming energy at a time, since the high pulse repetition rate allowed on a continuous neutron source (typically 250–1000 Hz on Panther) is not suitable for so-called repetition-rate multiplication. The main improvements compared to IN4C are that Panther has a huge array of position-sensitive 10-bar ^{3}He detectors allowing single-crystal experiments, a longer secondary flight path to improve resolution, optimized Bragg optics, a larger sample area allowing more complex sample environment, and improved background protection. The incoming energy E_i can be varied between 7.5 and about 150 meV, with the highest flux in the range 15–76 meV. At $E_i = 19$ meV, the flux of monochromatic and pulsed neutrons at the sample position is 5×10^5 n/cm^2/s. The energy resolution varies between 3.8 and 5.7% of the incoming energy for the pyrolytic graphite monochromator, depending on the monochromator take-off angle. The instrument will accommodate a set of five new background choppers operating as a discrete tunable velocity selector. Longitudinal XYZ polarization analysis is also planned.

The design of the Panther spectrometer started in late 2014 when funding was granted from the ILL Endurance program. The IN4C instrument was dismantled in October 2018, after which the floor for Panther was prepared, 200 mm lower than on IN4C to allow for a larger vertical detector coverage. In April 2019, the 8 ton heavy detector tank swung around the whole reactor suspended from the main crane to arrive at its final position. Commissioning started in February 2020 with the graphite monochromator. The first user experiment took place in September 2020 on a powder sample. Regular user operation for powder samples started in early 2021, in parallel with the commissioning of the Cu monochromator and the single-crystal option. The first single-crystal user experiment took place in May 2021, and became routine shortly after.

This paper is organized as follows. The Panther instrument and its different components are described in Section 2. Section 3 discusses the performance and the first user experiments, while planned developments are outlined in Section 4.

*e-mail: fak@ill.fr

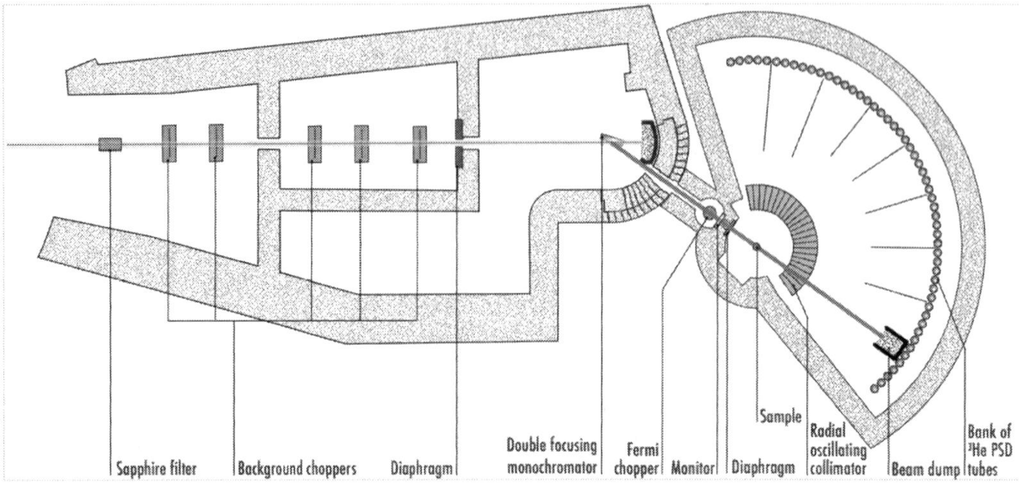

Figure 1. Schematic plan of Panther.

2 Instrument description

A schematic plan of Panther is shown in Fig. 1. The first part of the instrument is enclosed in a heavy concrete casemate, and contains a filter, background choppers, diaphragms, and two monochromators. The detector tank rests on rails in the floor and can rotate around a vertical axis centered at the monochromator in order to change incoming energy. The first part of the mobile detector tank contains a Fermi chopper, a beam monitor, and a set of four independent slits used to adjust the beam size. The sample space is part of the evacuated detector tank, with no windows between the two parts. Typical distances are given in Table 1. The different components will now be described in the order they are intercepted by the neutron beam.

Table 1. Distances L on Panther. The source is taken as the nose (entrance) of the beam tube inside the heavy-water moderator.

Elements	L (m)
Source — Monochromator	12.1
Monochromator — Sample	2.5
Fermi chopper — Sample	0.8
Sample — Detector	2.5

2.1 Beam tube and filter

Panther is installed on the H12 thermal neutron horizontal radial beam tube in the reactor hall, viewing the ambient-temperature heavy-water moderator of the ILL 57-MW high-flux reactor. The beam tube has a circular cross-section of 100 mm diameter and a divergence of 1.27°, both horizontally and vertically.

The first optical element is a single crystal sapphire filter of length 90 mm along the beam, which by Bragg scattering quite efficiently removes high-energy neutrons. The transmission is between 70 and 80% for incoming neutron energies below about 60 meV but drops rapidly below 50% at higher energies. It can be translated out of the beam by remote control.

2.2 Background choppers and heavy slit

For the first years of operation, Panther uses the two old background choppers from IN4C. They consist of steel discs of thickness 100 mm along the beam and diameters of 600 and 640 mm, respectively. The maximum speed is 5000 rpm (83.33 Hz), and to match the higher speed of the Fermi chopper, each disc has eight 60-mm wide openings. In order to reduce the high epithermal neutron flux and gamma radiation in the white beam, the two choppers operate in so-called "static closed" mode, which implies that when one is open, the other is closed. This imposes a phase angle between the two choppers of about 22.5°. The second background chopper is translated along the beam to allow the wanted incoming energy to pass through the chopper system and to block neutrons that otherwise would be reflected as higher and lower orders by the monochromator. The translation covers a distance of 1.5 m, a design criteria from IN4C, which is not sufficiently long for the longer flight distances of Panther. This means that higher- and lower-order contamination is a severe problem for certain incoming energies. This will be solved with the installation of the new background choppers, described in Section 4.1.

A diaphragm ("heavy slit") made from steel of thickness 160 mm along the beam is placed between the background choppers and the monochromator in the casemate. It allows to adapt the beam size both horizontally and vertically to fully illuminate the monochromator and to reduce background.

2.3 Monochromators

There are two monochromators available on Panther. They are mounted on a turntable that brings one of them into the white beam. One monochromator is made of highly oriented pyrolytic graphite (PG), and uses the (002), (004), or (006) Bragg reflection. The graphite plates are 2 mm thick and the mosaic is 0.5–0.6°. The second monochromator uses the (220) Bragg reflection of Cu, but the crystals are cut and aligned such that the (331) Bragg reflection also can be used in non-specular geometry. The crystals are 8 mm thick and the mosaic is 0.35–0.50°. Both monochromators have variable horizontal and variable vertical focusing, used to maximize the flux at the sample position or to produce a less divergent beam for single-crystal measurements. Each monochromator consists of crystal pieces of size 20×20 mm^2 glued onto $^{10}B_4C$ support plates to reduce background. The pieces are mounted in 15 columns and 11 rows and aligned in both horizontal and vertical directions to better than 0.1° and 0.05° with respect to each other for the PG and Cu monochromators, respectively. The total height of the monochromator is 230 mm and the width is 315 mm.

The incoming energy is selected by choosing the PG or Cu monochromator and adjusting its Bragg angle and the take-off angle. The latter is designed to vary between 35 and 70 degrees, and the whole secondary spectrometer (weighting about 40 tons) turns around the monochromator axis. The monochromatic beam leaves the casemate via a system of vertically translated shielding blocks. The incoming energies available, see Table 2, are currently more restricted than the design values because of a misplaced monochromator protection, but also due to the higher- and lower-order contamination related to the use of the old IN4C background choppers as discussed above. Incoming energies above 150 meV are not yet commissioned.

Table 2. Incoming neutron energies for the different monochromators on Panther.

Mono-	E_i (meV)	
chromator	Current	Design
PG(002)	7.5–19	5.5–20
PG(004)	30–76	22–80
PG(006)	67.5–112	50–180
Cu(220)	52–130	38–138
Cu(331)	123–150	91–330

2.4 Fermi chopper

A Fermi chopper is placed 0.8 m in front of the sample and chops the neutron beam in short pulses. It is equipped with magnetic bearings and the maximum rotation speed is 30000 rpm (500 Hz). The chopper is normally operated in time focusing mode to optimize the energy resolution, in which case typical speeds are 8000–18400 rpm. It has a straight slit package and produces thus two neutron bursts per revolution. Typical pulse-repetition rates are 250–1000 Hz, much higher than on pulsed spallation neutron sources. The slit package is 23 mm long, 32 mm wide, and has a divergence of 1.5 degrees (FWHM). It consists of a sandwich of 0.6 mm thick aluminium slits with high neutron transmission and 0.1 mm thick neutron-absorbing slats consisting of $^{10}B_4C$-epoxy.

2.5 Sample area

A ^{10}B beam monitor, inherited from IN4C and operating at 220 V, is mounted after the Fermi chopper close to the entrance to the sample area. Its efficiency and maximal count rate are not adapted for single crystal measurements, and a replacement is foreseen, see Section 4.3.

A set of four independent 2-mm thick sintered $^{10}B_4C$ slits are mounted between the beam monitor and the entrance to the sample area to adapt the beam size to the sample and sample environment. They are positioned with a precision of 0.1 mm by piezoelectric motors and equipped with linear encoders. The nominal sample height is 40 mm and the width is 20 mm.

The sample space has a diameter of 800 mm and is directly connected to the evacuated detector tank without any windows. A motorized elevator allows easy vertical positioning of various standard sample environments from the ILL. Changing the sample environment (from cryostat to furnace, e.g.) requires to break the vacuum of the detector tank. It takes about 50 minutes of pumping to reach a pressure below 0.1 mbar, sufficient for starting measurements. Panther has an orange cryostat specifically designed for the instrument with a 70-mm bore, thinned down aluminum windows with a total thickness in the beam of 1.4 mm, a Gd screen inside the calorimeter, and additional Cd and B_4C shielding on the outside of the cryostat to reduce parasitic scattering. The operation temperature is 1.5–300 K. The cryostat was refurbished in 2021 with a new variable temperature insert equipped with a quadruple heat-exchanger and a 150-W heater for rapid temperature changes. This has reduced cooling-down and heating-up times by a factor of 2–3. It now takes about 50 minutes to cool a typical sample from 300 to 100 K, another 40 minutes down to 1.5 K, and about 20 minutes to warm it up to 100 K again.

A radial oscillating collimator (ROC) is installed in the detector tank just outside the sample area. This positioning implies that the ROC can be used with all sample environments, including large-diameter cryomagnets, high-pressure cells etc., for which background suppression is of outmost importance. The collimator consists of stretched kapton foils of thickness 12.5 μm coated with 85 μm thick Gd_2O_3 layers. The foils are 0.587 m high and 0.2 m long (along the scattered beam), and start outside the sample area at a distance of 0.434 m from the sample center. The angular separation between collimator septa is $2\alpha = 1.25°$, which gives a viewing angle of ±0.69°. This means that a detector sees a region of ±30 mm around the sample. The collimator oscillates over an angle of 1.62° with constant speed at a frequency of 0.02 Hz.

2.6 Detectors and detector tank

Panther has a huge array of position-sensitive detectors located 2.5 m from the sample. They provide a continuous coverage of horizontal angles from -16 to +136 degrees and vertical angles from -13 to +30°, with a total solid angle of 2.0 Sr. The vertical coverage is asymmetric because the beam tube elevation over the reactor floor is quite small. The smallest useful scattering angle depends in principle on the incoming neutron energy and the focusing of the monochromator. In practice, it is about 5 degrees for the whole energy range with the monochromator doubly focused for maximum flux on the sample position. The detector tubes are mounted vertically inside the evacuated detector tank and have an active length of 2.04 m and an outer diameter of 22 mm. The tube walls are made from stainless steel (304L) of 0.4 mm thickness. Neighboring tubes are separated by a gap of 0.6 mm, where 0.5-mm thick Cd spacers reduce cross-talk between tubes. The Cd spacers extend 9 mm beyond the detector surface towards the sample position. The angular separation between tubes is $\Delta\gamma = 0.518°$. The 288 detector tubes are arranged in 9 banks with 32 tubes each. Neighboring banks are separated by a gap of $\Delta\gamma$, corresponding exactly to the dimensions of one tube. In this gap are placed vanes made of Al sheets clad with 0.5 mm thick Cd on each side to reduce parasitic scattering between detector banks. The length of the vanes towards the sample center varies between 0.8 and 1.2 m. The tubes in a bank are interconnected and share a common gas volume, with a partial ^{3}He pressure of 10 bars and an operating high voltage of 2200 V. The detectors are position sensitive with 256 pixels along the tubes. The positions of the pixels have been calibrated using neutrons and a specific Cd mask. An absorbing beam dump is placed in front of the detectors (or pixels) that are in the line of the direct monochromatic beam to avoid saturation.

The detector tank is made of Al alloy plates welded together to form a vacuum-tight (<0.01 mbar) and non-magnetic skin, reinforced by an array of Al profiles to ensure the mechanical strength of the structure. It is covered by 1-mm thick Cd sheets glued to the inside of the tank to reduce parasitic reflections. The outside of the tank is covered on all sides (including top and bottom) by 300-mm thick borated high-density polyethylene (HDPE) to reduce background. Because of the related fire hazard, the instrument zone forms a 500-mm deep pit surrounded by a firewall, which can contain the whole volume of the 21 m^3 (or tons) of HDPE if melted. In addition, all cables in the instrumental area are flame-retardant.

2.7 Data acquisition and reduction

Panther uses the ILL remote control system "nomad" for control of motors, sample environment, and data acquisition. Detected neutron events are stored in a histogram with 288 × 256 = 73728 detector pixels and typically 512 time channels. With the current data acquisition system, the histogram method allows higher count rates than so-called event mode. The data are stored in nexus files for subsequent data reduction using the ILL branch of mantid [2].

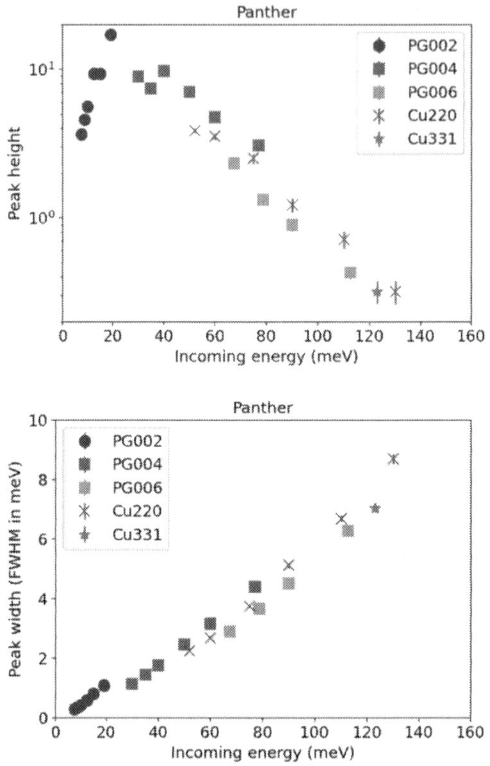

Figure 2. Peak intensity (top) and elastic energy resolution (bottom) on Panther from vanadium measurements.

Single-crystal measurements are typically performed by scanning the sample orientation over 353° around a vertical axis in steps of 1°. Such a scan takes one day for a measuring time of four minutes per sample angle, which is sufficient in most cases. The advantages with scanning (nearly) a full turn are that the measurements can start before the exact sample orientation is known and that no time is lost whilst optimizing the angular scanning range. The data are reduced in mantid [2] and analyzed using horace [3], with which the actual sample orientation can be refined and the data can be symmetrized to improve the statistics, thereby compensating for the (too) wide angular scan range.

3 Performance and user operation

3.1 Flux and resolution

The peak intensity and elastic energy resolution from a vanadium sample are shown in Fig. 2 for different incoming neutron energies and different monochromators. For the pyrolytic graphite monochromator, the energy resolution varies between 3.8 and 5.7% of the incoming energy

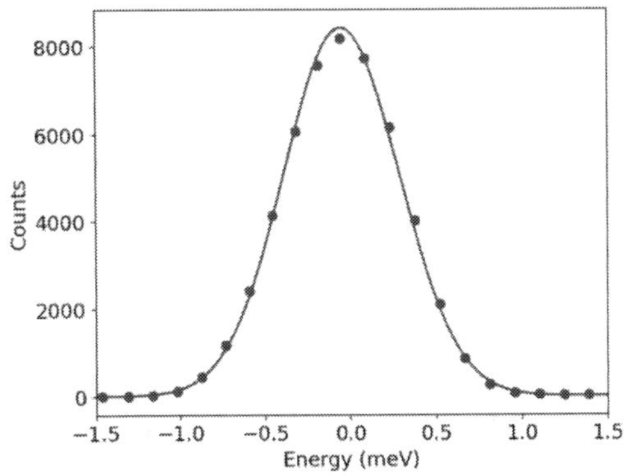

Figure 3. Line shape of the elastic scattering from a vanadium sample at T = 10 K and an incoming neutron energy of 15 meV. The line is a fit by a Gaussian plus a flat background.

depending on the monochromator take-off angle. An advantage with "hybrid" spectrometers on continuous neutron sources is that the energy resolution has a nearly Gaussian line shape, as shown in Fig. 3.

3.2 Comparison

All modern high-performance thermal time-of-flight spectrometers have the same detector solid angle within a factor of 1.5. The flux on Panther is 5×10^5 n/cm^2/s for an incoming energy of 19 meV. For a similar incoming energy (E_i = 19–50 meV) and energy resolution ($\Delta E/E_i$ = 3–5.5%), Panther has about 8 times more flux than MERLIN at ISIS [4], and probably a factor of 1.4–2 times more signal than the thermal TOF spectrometers 4SEASONS at JPARC [5] and ARCS at SNS [6].

3.3 User operation

Panther entered the user program of the ILL in 2020, and the average oversubscription factor since then is in the range 3–4. The first official user experiment on a powder sample concerned a crystal-field study of a geometrically frustrated magnet [7] and the first single-crystal experiment was a study of Dirac magnons in a honeycomb ferromagnet [8]. An example of a constant-energy slice from the latter experiment is shown in Fig. 4. The efficient Bragg focusing optics on Panther allows measurements of small samples. Figure 5 shows an example of the quantized energy levels of single ^{3}He atoms trapped inside C_{60} fullerene cages [9]. The measurement time was 1 hour for a mass of ^{3}He of 3.5 mg.

4 Planned developments

The Panther project comprises several parts that are not yet in operation. These planned upgrades are discussed here.

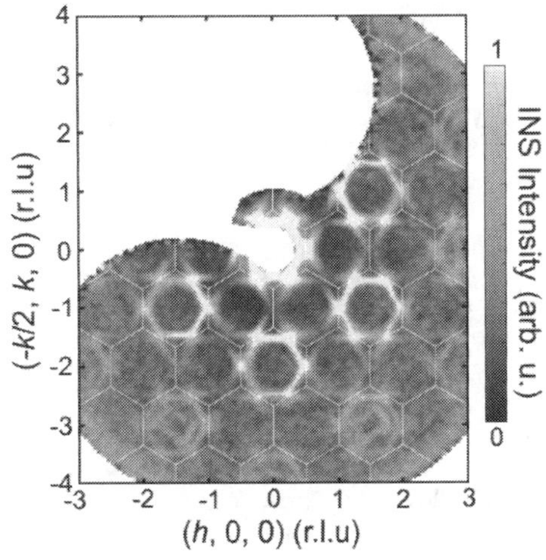

Figure 4. Constant-energy slice of magnon scattering in a honeycomb ferromagnet at T = 1.7 K, measured with an incoming energy of 15 meV. The data are integrated over energy transfers between 4 and 5 meV and for values of the perpendicular Q direction between -5 and +5. The white lines show the Brillouin zone boundaries.

4.1 New background choppers

A set of five new background choppers have been designed, manufactured, and delivered. They are expected to substantially reduce the background and also remove higher- and lower-order contamination from the monochromators. The choppers operate like a discrete velocity selector with variable pitch, obtained by spinning them in the same direction at the same speed but with different phases. Because of the large number of choppers, the "static-closed" condition required for the IN4C background choppers is not necessary.

The five identical disc choppers have a diameter of 750 mm and a thickness (along the beam) of 100 mm. They are made from epoxy-loaded carbon fibre, where the fibre volume content varies in the radial direction of the disc to ensure mechanical stability and that the epoxy content is at least 60% in the region of the neutron beam. It is the hydrogen in the epoxy that reduces the epithermal

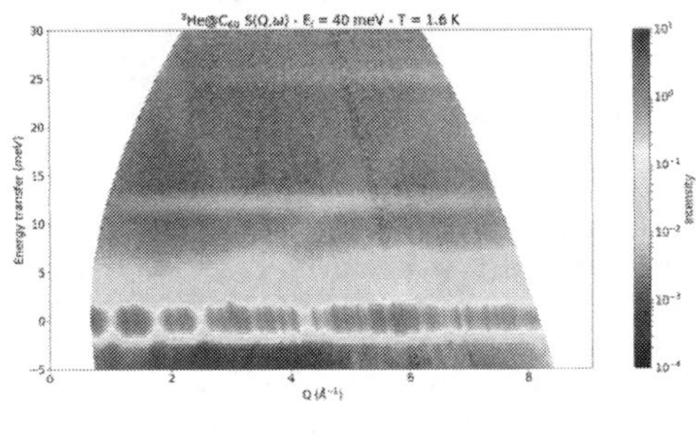

Figure 5. ^{3}He trapped inside C_{60} at T = 1.6 K showing quantum transitions at energies of 11 and 24 meV.

neutron flux by scattering. To absorb thermal neutrons, the discs are coated with a layer of $^{10}B_4C$-epoxy on each side, with a thickness corresponding to a 0.18 mm layer of pure $^{10}B_4C$ per side. The non-metallic construction of the discs leads to a reduced weight, only 52 kg per disc, which allows to run them at a relatively high speed of 10000 rpm (166.67 Hz). To match the speed of the Fermi chopper, limited to 30000 rpm, each background chopper has 6 windows of width 105 mm.

Each chopper has its own evacuated ($2 \cdot 10^{-3}$ mbar) and water-cooled housing made from stainless steel to prevent damages to surrounding equipment in case of a disc failure. The inner part of the housings are coated with a (non-enriched) B_4C layer to reduce neutron background. The neutron beam enters and exits the housings through Al windows of thickness 0.4 mm. Each disc is directly driven by an electrical motor, the bearings are non-magnetic, and the cables are radiation-proofed. The ramp rate is 12 rpm/s for both acceleration and deceleration and the phase stability at constant speed is better than 0.01°. The vacuum levels, water temperature and flow rate as well as the motor temperatures and disc vibration levels are monitored by a control system, which automatically shuts down the choppers in case of anomalies, and also provides information on the chopper status to the interlock system of the instrument.

4.2 Increased energy range

Repositioning of the misplaced monochromator protection will remove the penalizing gap in incoming energies between 19 and 30 meV, which corresponds to the maximum of the Maxwellian flux distribution from the heavy-water moderator. Incoming neutron energies beyond 150 meV are expected to become exploitable with the installation of the new background choppers because of their better higher- and lower-order suppression and also since the sapphire filter may not be needed.

4.3 New beam monitor

A high count-rate ^{3}He beam monitor, required for single crystal measurements, is being developed in-house. It is based on the multitube concept with a maximum count rate of about 100 kHz for an efficiency of $3 \cdot 10^{-4}$ for thermal neutrons. Care is taken to assure that the monitor is fully non-magnetic, as required for the future polarized neutron option.

4.4 Sample environment

Temperature changes between 1.5 and 100 K are very quick using the Panther refurbished cryostat. At higher temperatures, between 150 and 300 K, temperature changes are slower, which may hamper measurements of e.g. the temperature-dependence of phonon density of states. To address this issue, the development of a cryoloop as well as a new cryofurnace is envisaged.

The sample space is designed to accommodate the 10 Tesla magnet developed for IN5, although the scientific case for such low fields (compared to the energy range of Panther) is limited. No particular problems are expected in terms of stray fields or magnetic forces, because the sample area is non-magnetic by design and confirmed by measurements using a three-dimensional magnetic field probe. However, possible interferences with the magnetic bearings and the motor drive of the Fermi chopper remain to be investigated.

4.5 Polarized neutrons

A new device has been developed at the ILL to perform wide-angle XYZ polarization analysis using thermal neutrons, the so-called Pastis-3 concept. It has been successfully tested and brought into operation on the thermal triple-axis spectrometer IN20 at the ILL [10]. The device employs two cells containing highly polarized ^{3}He gas: one to polarize the incoming beam and the second to analyze the polarization of the scattered neutrons over a wide angular range, both horizontally and vertically. The ^{3}He filters are incorporated in a mechanical device where a tilted coil design produces a weak magnetic guide field at the sample. Measurements can be performed for three orthogonal directions of the guide field, XYZ, but more general directions can also be measured. The neutron spin is "flipped" in the first ^{3}He cell by an adiabatic fast-passage technique, where an amplitude-modulated radio-frequency signal inverts the polarization of the ^{3}He gas.

A Pastis-3 device designed to operate in the vacuum environment of Panther is being developed. It fits into the 800-mm diameter non-magnetic sample area of Panther. The angular acceptance range is -14 to +112 degrees horizontally and ±10 degrees vertically. The device will allow completely new types of overview measurements using polarized neutrons for both powder samples and single crystals. The main application will be to separate magnetic from nuclear inelastic scattering in a wide range of scattering angles and energies simultaneously. This separation is particularly important in the thermal energy range (5–60 meV), where phonon excitations often overlap with magnetic scattering from spin waves, crystal field excitations, and spinons. In single crystals, it is also possible to separate e.g. longitudinal from transverse magnetic excitations. A third application is to separate coherent from spin-incoherent nuclear inelastic scattering, which is of interest for studies of polymers and soft matter as well as for hydrogen spectroscopy.

5 Summary

Panther is a new inelastic neutron scattering time-of-flight spectrometer at the Institut Laue-Langevin offering a high flux of thermal neutrons in the energy range 7.5–150 meV at medium energy resolution and a large solid angle coverage of position sensitive detectors. It is expected to become a workhorse for studies of excitations in condensed matter using both powder and single crystalline samples with a high scientific productivity. The development of polarized neutrons will open completely new horizons for separating magnetic from nuclear scattering.

Significant contributions to the planning, design, or construction of Panther were made by: Luc Didier, Baptiste Amoudruz, Joseph Pequeno, Ronan Moreau, David Mezerette, Christophe Vitally, Benoît Jarry, Christophe Gillart, Marc David, Jérôme Nucci, Roland Gandelli, Julien Bonnevaux, Bruno Guérard, Jean-François Clergeau, Jean-Claude Buffet, Sylvain Cuccaro, Fabien Lafont, Martin Platz, Patrick Van Esch, Gilles Pastrello, Pierre Courtois, Benoît Mestrallet, Franck Barneaud, Sandrine Michallat, Florian Philit, Paulo Mutti, Franck Cécillon, Julien Blanc Paques, Franck Rey, Thierry Mary, Emmanuel Courraud, Eric Lampasona, Benjamin Sornin, Eddy Lelièvre-Berna, Yohan Memphis, Jean-Philippe Rey, Fabrice Rencurel, Antti Soininen, Gagik Vardanyan, Alain Filhol, Stéphane Fuard, Emmanuel Farhi, Anton Fossum Hylin, Martin Boehm, Bela Farago, Charles D. Dewhurst, Charles Simon, Mark R. Johnson, and Helmut Schober. The first author has profited from discussions with Rob Bewley and Jacques Ollivier. Ursula Bengaard Hansen, Marek Koza, and Stanislav Nikitin are acknowledged for their help with the commissioning.

References

[1] G. Cicognani, H. Mutka, F. Sacchetti, Physica B **276–278**, 83 (2000)

[2] O. Arnold *et al.*, Nucl. Instrum. Methods Phys. Res., Sect. A **764**, 156 (2014)

[3] R. A. Ewings, A. Buts, M. D. Le, J. van Duijn, I. Bustinduy, T. G. Perring, Nucl. Instrum. Methods Phys. Res., Sect. A **834**, 132 (2016)

[4] R. I. Bewley, R. S. Eccleston, K. A. McEwen, S. M. Hayden, M. T. Dove, S. M. Bennington, J. R. Treadgold, R. L. S. Coleman, Physica B **385–386**, 1029 (2006)

[5] R. Kajimoto, T. Yokoo, K. Nakajima, M. Nakamura, K. Soyama, T. Ino, S. Shamoto, M. Fujita, K. Ohoyama, H. Hiraka, K. Yamada, M. Arai, J. Neutron Res. **15**, 5 (2007)

[6] D. L. Abernathy, M. B. Stone, M. J. Loguillo, M. S. Lucas, O. Delaire, X. Tang, J. Y. Y. Lin, B. Fultz, Rev. Sci. Instrum. **83**, 015114 (2012)

[7] N. Qureshi, A. R. Wildes, C. Ritter, B. Fåk, S. X. M. Riberolles, M. Ciomaga Hatnean, O. A. Petrenko, Phys. Rev. B **103**, 134433 (2021)

[8] S. E. Nikitin, B. Fåk, K. W. Krämer, T. Fennell, B. Normand, A. M. Läuchli, and Ch. Rüegg, Phys. Rev. Lett. **129**, 127201 (2022)

[9] M. Aouane, S. Rols, R. Whitby, M. Levitt, in preparation

[10] D. Jullien, A. Petoukhov, M. Enderle, N. Thiery, P. Mouveau, U. Bengaard Hansen, Ph. Chevalier, P. Courtois, Nucl. Instrum. Methods Phys. Res. A **1010**, 165558 (2021)

Recent Progress on DNA ToF Backscattering Spectrometer in MLF, J-PARC

Yukinobu Kawakita[1,*], *Masato* Matsuura[2], *Taiki* Tominaga[2], *Takeshi* Yamada[2], *Hiromu* Tamatsukuri[1], *Hiroshi* Nakagawa[1,3], and *Keiichi* Ohuchi[2]

[1]J-PARC Center, Japan Atomic Energy Agency (JAEA), 319-1195 Ibaraki, Japan
[2]Neutron Science and Technology Center, Comprehensive Research Organization for Science and Society (CROSS), 319-1106 AQBRC, Ibaraki, Japan
[3]Materials Science Research Center, Japan Atomic Energy Agency (JAEA), 319-1195 Ibaraki, Japan

Abstract. DNA is a time-of-flight backscattering neutron spectrometer installed at the Materials and Life Science Experimental Facility in J-PARC. After opening user program, we have been developing and upgrading DNA. The energy resolution is improved and achieved the designed value of 1.4 μeV by renewing the disks of the pulse shaping chopper that can rotate at max 300 Hz instead of 225 Hz of the previous disks. The installation of Si311 analyser mirrors has been almost accomplished, which extends momentum transfer region up to 3.7 Å^{-1}. We demonstrate the effect of this extension by measuring and analysing methyl rotation dynamics in solid toluene at 10 K and 40 K. We also report variety of sample environment available at DNA and their remote handling.

1 Introduction

DNA is a time-of-flight (tof) backscattering spectrometer [1] installed at BL02 neutron port seeing a coupled moderator with the largest intensity in a cold neutron wavelength among three types moderators [2] of the Materials and Life Science Experimental Facility (MLF) in the Japan Proton Accelerator Research Complex (J-PARC). Apart from the name of DNA biomolecular dynamics spectrometer, it aims at revealing pico to nano second dynamics in a wide field of basic sciences and industrial applications including soft matter, bio-molecules, chemical molecules, battery, catalyst, and magnetism. Since it has been opened to users from March 2012, the fields of publications are distributed as 34 % liquid, 18 % polymer, 16 % biomatter, 11 % crystal chemistry, 10 % magnetism, 3 % ionic protonic conduction, 3 % hard matter (lattice dynamics) together with 5 % instrumentation and sample environment.

DNA has an advantage of flexibility of various energy resolution by utilizing a pulse-shaping chopper, PSC, which has counter-rotating double disks with four slits (three 3-cm slits and one 1-cm slit) [1]. The PSC is located at 7.75 m downstream of the coupled moderator and plays a role of a pseudo pulsed neutron source with the multiple slit-openings and the flexible open-timing, which gives us an interesting and useful function of scanning a wide energy-transfer region with keeping good resolution. [1] The shift of the slit opening phases enables us to scan a wide energy transfer, E, range from -600 to 1500 μeV for Si111 analyzers even with the highest resolution.

DNA has another advantage of extremely good signal-to-noise ratio of more than 10^5 which is attributed by a curved neutron guide cutting high-energy neutron and a back-coating of Gd metal or Gd_2O_3 of the analyzer mirrors.

In this paper, we will report recent progress on DNA which has been achieved after the previous report of the instrumentation [1]. We will introduce an improvement of the energy resolution by re-fabricating the disks for the PSC in §2, and the extension of Si311 analyzer mirrors in §3. In §4, variety of sample environment available in DNA will be introduced along several types of experiment. In §5, near-future plan on the instrumentation for DNA will be explained. All contents of the paper will be summarized in §6.

2 Improvement of Energy Resolution

The new disks for the PSC, which are thicker by 1 mm than the old disks, were re-fabricated to improve mechanical strength, because a large vibration of the old disks limits the maximum rotation speed to 225 Hz instead of the designed 300 Hz. The energy resolution is determined by (1) mirror factor originating from the Bragg reflection angle and the mosaicity of the analyzer mirrors and (2) tof factor defined by time width of the pseudo pulsed neutron source defined by the PSC rotating speed. The contributions depend on neutrons with the final energy determined by Bragg law. The first factor is constant by geometry of Bragg angle of 87.5° and the mosaicity of the mirror material while the second factor becomes smaller with faster rotation of PSC.

* Corresponding author: kawakita.yukinobu@jaea.go.jp

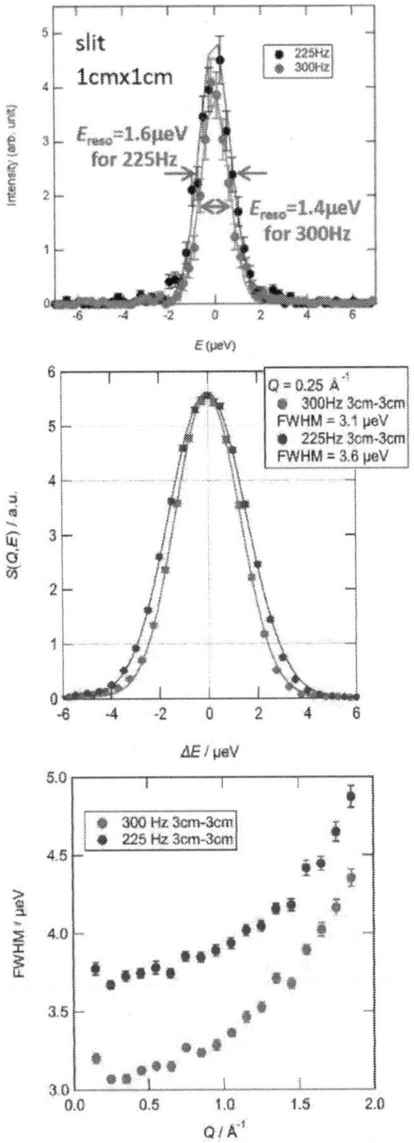

Fig. 1. The energy resolutions of Si111 analyzer mirrors under the highest-resolution condition of the sample size of 1 cm (D) × 1 cm (H) with the slit size of 1 cm × 1 cm (top) and the conventional condition of the sample size of 1.4 cm (D) × 4 cm (H) with the slit size of 3 cm × 3 cm (middle). The *Q*-dependences of the energy resolutions for these conditions are compared in the bottom.

DNA is designed to have both terms balanced at 300 Hz rotation. The 300 Hz PSC operation improved the best performance of the energy resolution for Si111 analyzer mirrors from 1.6 μeV to 1.4 μeV for 1 cm (D) × 1 cm (H) sample size with 1 cm × 1 cm slits, and the conventional resolution from 3.6 μeV to 3.1 μeV for 1.4 cm (D) × 4 cm (H) of sample size with 3 cm × 3 cm slits as shown in Fig. 1. The figures of the quasi-elastic neutron scatterings (QENS) spectra were plotted with an arbitrary vertical scale because of easy comparison for their energy

Fig. 2. The photo of the Si311 analyzer mirrors.

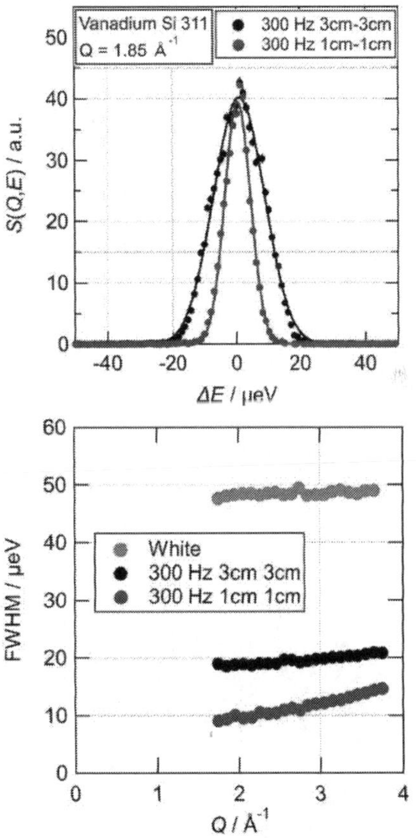

Fig. 3. The energy resolutions of Si311 analyzer mirrors under the conventional condition of the sample size of 1.4 cm (D) × 4 cm (H) with the slit size of 3 cm × 3 cm and 1 cm × 1 cm (Top). The *Q*-dependences of the energy resolutions for the Si311 analyzer is shown in the bottom. "White" (red) corresponds to the condition without operating PSC.

resolutions. The ratio of the integration of the spectrum at 300 Hz to that at 225 Hz is around 52 %, which almost

 https://doi.org/10.1051/epjconf/202227202002

equals to the square of 225/300 derived from the beam slice by the counter-rotating double disks.

As an example that the finest energy resolution would be useful, hyperfine splitting of Nd nuclei in Nd compounds was studied by utilizing DNA[3]. In their study, PSC condition of 225 Hz but small sample size leaded to a relatively fine resolution of around 2 μeV. They were able to distinguish hyperfine contribution aside the elastic peak. To utilize the finest resolution with 300 Hz rotation would give much clearer separation between them. Such very low energy lying excitation could be a target of DNA.

3 Extension of Analyzer Mirrors

In addition of the Si111 analyzer mirrors covering horizontal scattering angles from -30° to 150° (vertical ones from -14° to 21°), Si311 analyzer mirrors covering horizontally from -150° to -55.5° have been installed year by year as shown in Fig. 2. While the Si111 covers momentum transfer, Q, range is from 0.08 to 1.98 Å^{-1}, Si311 extends Q-range to 3.7 Å^{-1}. The energy resolutions of Si311 under various conditions were examined by measuring a vanadium standard sample, as shown in Fig. 3. The highest resolution of 10 μeV is achieved by 1 cm × 1 cm slits with 300 Hz PSC operation. The intensity is two order weaker than that for Si111 with 300 Hz and 3 cm × 3 cm slits. The useful energy resolution of 19 μeV is obtained by 300 Hz with 3 cm × 3 cm slits. Without the PSC operation, it is 50 μeV with a reasonable intensity. The Q-E ranges measurable by Si111 and Si311 are individually shown in Fig. 4. It is noted that simultaneous measurement by using Si111 and Si311 analyzers is impossible because the energy band to be introduced to the sample through PSC and the frame choppers is different between Si111 and Si311 analyzers. Although it is not shown in Fig. 4, Si333, the 3rd harmonics of Si111, has a very weak intensity and limited energy band due to

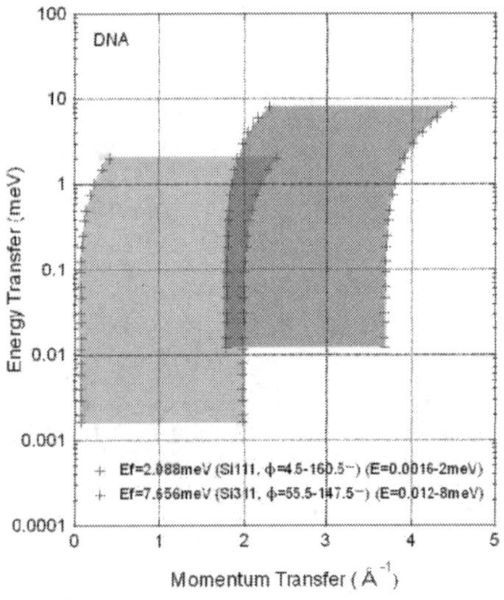

Fig. 4. The wide Q-E region measurable by DNA. The areas for Si111 and Si311 are hatched by orange and purple.

the curved mirror guide cutting high energy neutrons and it is available only for alignment check of single crystal whose QENS or inelastic neutron scatterings (INS) will be measured.

To demonstrate the performance of the extending Q-E region, we measured QENS spectra for methyl rotation in solid toluene at 10 K and 40 K. Fig. 5 shows some combinations of Si111 and Si311 analyzers. At 10 K, tunneling modes of methyl rotation [4] are observed at around ±30 μeV as shown in Fig. 5(a). In such case, the combination of both the high resolution modes of 3.6μeV for Si111 with 225 Hz and 3 cm × 3 cm slits and 10 μeV for Si311 with 300 Hz and 1 cm × 1 cm slits are useful. The measurement time may be a few ten minutes for Si111 and several hours for Si311. Figures 5 (b) and (c) show the spectra at 40 K where only QENS signal is observed with the combinations of the high flux mode of Si111 without PSC (13 μeV in resolution) and (b) the high resolution mode of Si311 with 300 Hz and 3 cm × 3 cm slits (19 μeV in resolution) and (c) the high flux mode of Si311 without PSC. The latter exhibits the widest energy region accessible by DNA.

The QENS profile at 40 K was fitted by a sum of a delta function ($\delta(E)$) representing the elastic component and a Lorentz function ($L(\Gamma,E)$) representing the QENS component convoluted with a resolution function ($R(Q,E)$), with constant background (BG) as follows,

$$S(Q,E) = [A_E\delta(E) + A_L(\Gamma,E)L(\Gamma,E)] \otimes R(Q,E) + BG \quad (1)$$

where the A_E and A_L are coefficients of the delta function and the Lorentz function, respectively. The Γ is a half width at half maximum of the Lorentz function. Figures 6 (a) and (b) show a typical fitting profile at Q = 2.35 Å^{-1} and the Q dependence of the Γ, respectively. The elastic incoherent scattering factor (EISF) was calculated as follows.

$$EISF = \frac{A_E}{A_E + A_L} \quad (2)$$

In Fig. 6(c), the calculated EISF is plotted together with a theoretical curve assuming methyl rotation,

$$EISF = \frac{5}{8} + \frac{3}{8}\left[\frac{1+2j_0(Qr)}{3}\right] \quad (3)$$

where the first term was due to the benzene ring and the second term represents a three site jump of the methyl group[5]. The j_0 and r are the zeroth-order spherical Bessel function and H-H distance (=1.78 Å) in the methyl group. One can understand the effectiveness of QENS measurement in a wider Q region.

Difference in a jump length of the jump diffusion model can be more clearly seen in a high Q region, which means that this combination method has an ability to distinguish an ionic conduction path among possible paths determined by powder diffraction. Coherent QENS measurements can be also a target for the combination method of Si111 and Si311 analyzers.

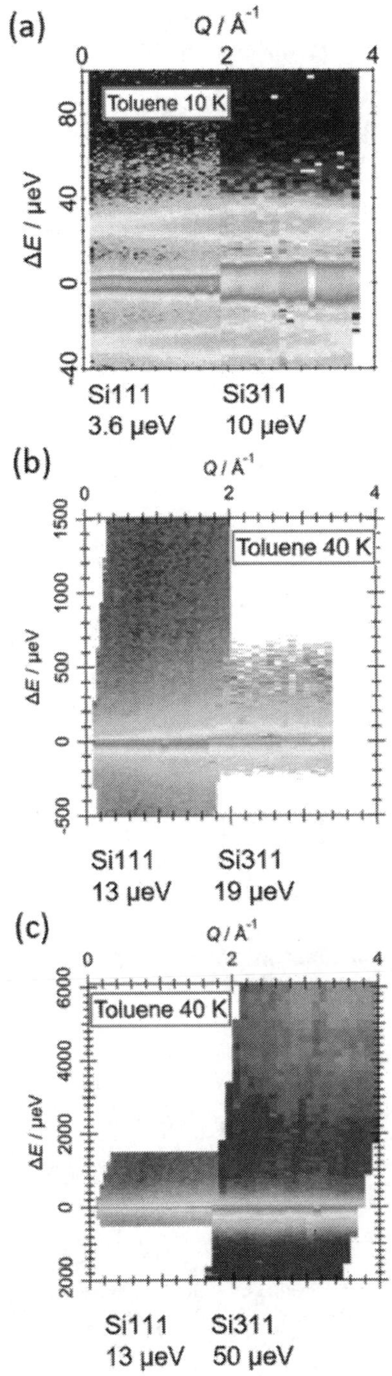

Fig. 5. The QENS spectra for solid toluene at 10 K and 40 K with various energy resolutions obtained with Si111 and Si311. (a) The combination of 3.6 μeV for Si111 (225 Hz, 3 cm × 3 cm slits) and 10 μeV for Si311 (300 Hz, 1 cm × 1 cm slits), (b) the combination of 13 μeV for Si111 (without PSC) and 19 μeV for Si311 (300 Hz, 3 cm × 3 cm slits), and (c) the combination of 13 μeV for Si111 (without PSC) and 50 μeV for Si311 (without PSC)

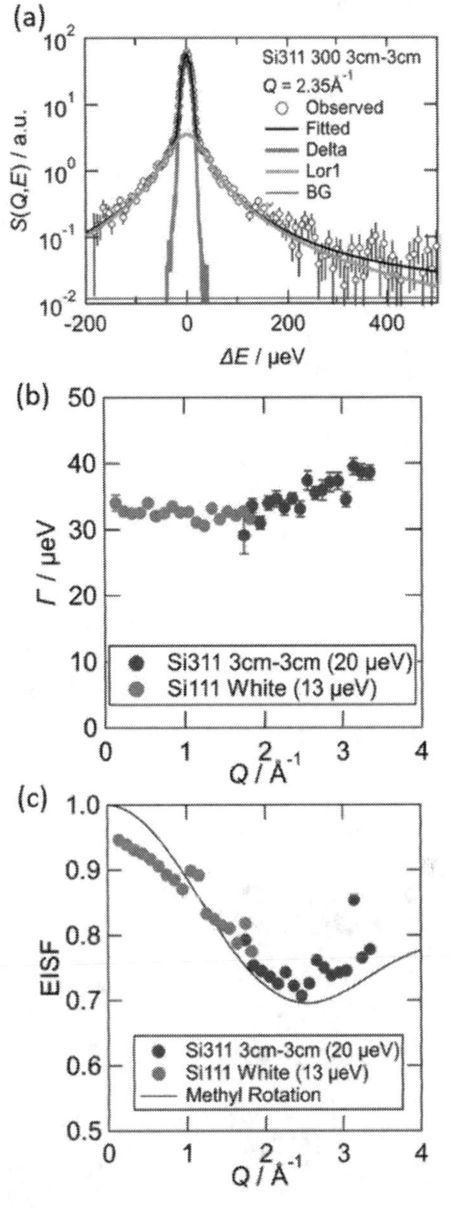

Fig. 6. QENS analysis of toluene at 40K. (a) A curve fitting with a delta plus a Lorentian functions together with a constant background which are convoluted by instrument resolution function. (b) The *Q* dependence of the Lorentzian width. (c) The elastic incoherent structure factor together with a theoretical curve assuming methyl rotation.

One of coauthors (MM) studied ionic diffusion in Li based solid electrolyte [6]. In their study, single crystal was mounted to access the (hhl) horizontal scattering plane and rotated to make scattering trajectories nearly parallel to the three principal axes [110], [111] and [001]. The Q-dependence of the energy width of the QENS signal is different along three principal axes, which makes us determine ionic diffusion path. Since the difference appears more clearly in a high Q region, Si311 would be powerful to see it.

4 Variety of Sample Environment

In DNA, many kinds of sample environment (SE) are available according to wide variety of target sciences. In this section, such SEs are introduced and a remote control and automation are focused.

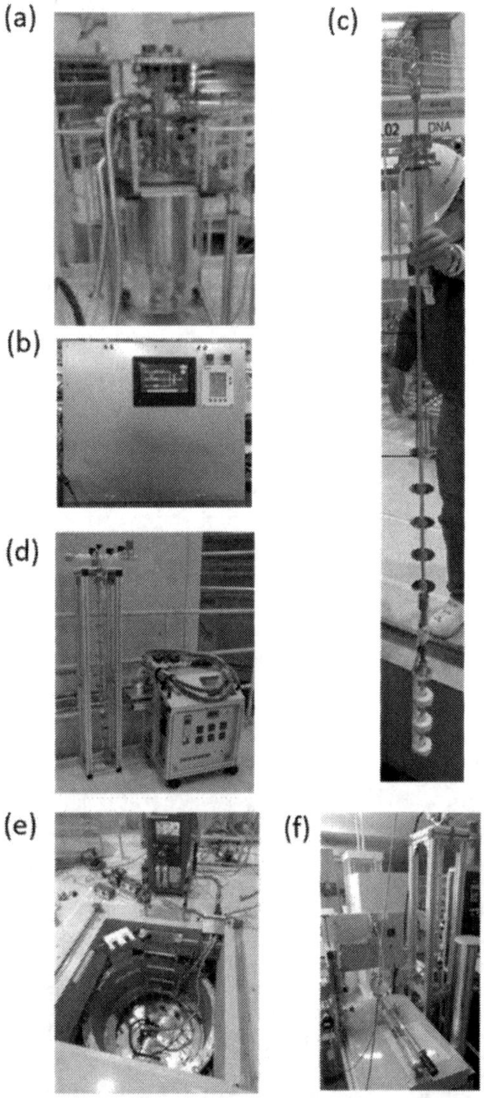

Fig. 7. (a) Standard cryo-furnace at DNA (5 K < T < 700 K), (b)gas handling system DICE, (c)sample stick with automatic sample positioning function for three samples, PEACE. (d)humidity control system, (e)new humidity control system with gas mixing function (common SE), (f) hydraulic press system (0.1 < P < 145 MPa).

4.1 Standard SEs

A standard SE is a cryo-furnace available in a temperature range from 5 K to 700 K (Fig. 7(a)). The cryo-furnace has two types of vacuum chamber, namely an outer vacuum chamber (OVC) which stores a cold head of the closed cycle refrigerator (CCR) and an inner vacuum chamber (IVC) which is a space for installation of a sample stick with an internal heater. The CCR cools IVC and then the sample stick through dilute He gas (approximately 8×10^3 pascals) inserted to IVC, while heating of the sample stick is carried out by the internal heater with thermal isolation from the cool IVC by evacuation. An automated gas-handling system named as DICE [7] developed by the DNA team (Fig. 7(b)) is always auxiliary used together with the cryo-furnace. As well as several standard sample sticks, a stick with sample changer function with positioning three samples vertically arranged (named as PEACE) is frequently utilized in a temperature range below 340 K [7] (Fig. 7(c)).

4.2 SEs for Soft-matter, Bio-molecule and Polymer Sciences

Humidity is an important control parameter for polymer science. DNA has its own humidity control system as shown in Fig. 7(d) and a corresponding sample stick to be installed into the cryo-furnace. Based on this technique a new one which can provide vapor gas mixed H_2O and D_2O at a favorite concentration flexibly has been developed as a common SE device at MLF (Fig. 7(e)). This new machine is also available.

A special sample stick with pressure generator which hydraulic pressure up to 145 MPa [8] can be applied is available at the room temperature (Fig. 7(f)).

4.3 SEs for Energy Science and Solid-State Physics

Two kinds of special purpose sample stick were developed. One can apply electrical current up to 2 A to the sample (Fig. 8(a)), while the other can apply high voltage up to 1.5 kV to the sample (Fig. 8(b)). Measurements under current or voltage may accelerate science of battery electrolyte and multiferroics, pharmaceutical application such as drug delivery. An air-tight sample cell with electrodes is developed for in situ current experiments as shown in Fig. 8(c).

For solid state physics, extremely low temperature is required, which is realized by the combination of liquid He type 2 K cryostat (Blue Cryo) (Fig. 8(d)) and a dilution option (common SE in MLF) (Fig. 8(e)). The Blue Cryo itself is available in the temperature range from 300 K to 1.5 K and with the dilution option 60 mK is achieved.

4.4 Sample cells

Including the sample cell for current and voltage application described above, we developed several sample cells such as standard aluminum cells for cooling and heating, quartz glass cell [9], single crystal sapphire cell, molybdenum cell for high temperature. We also investigated corrosion of aluminum cell by solution [10,11] and proposed surface treatment of aluminum cell adequate for some cases [12]. The details are shown in the references.

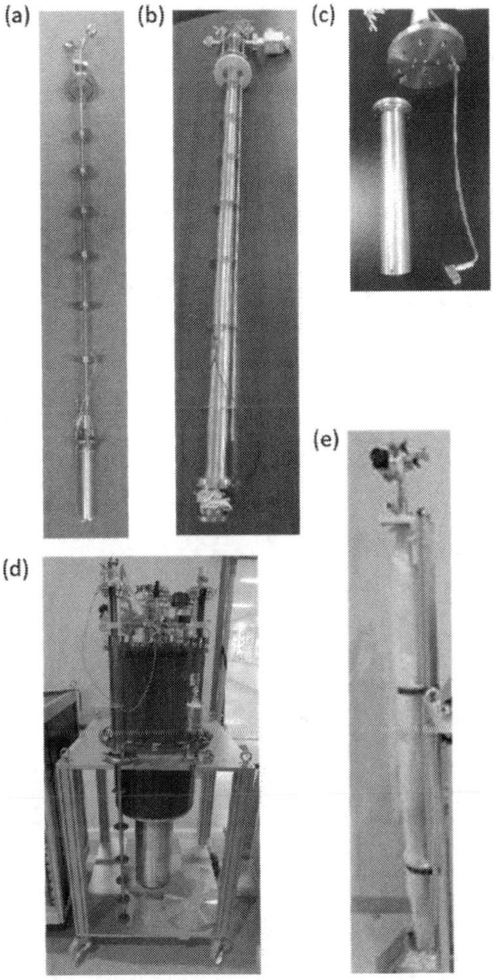

Fig. 8. (a)Sample stick for current application (Max 2 A, T < 500 K), (b)sample stick for voltage application (Max 1.5 kV, T < 500 K), (c)air-tight sample cell with electrodes, (d) liquid He type 2K cryostat (Blue Cryo) (1.5 K < T < 300K), (e) dilution insert (60 mK with the Blue Cryo)

4.5 Automation and Remote Control of SEs

Many of SEs can be controlled from the software platform called as IROHA2 [13] which unifies the data acquisition system (DAQ) and device control. For almost every SE, middlewares controlling its operation parameter such as current, voltage, vacuum level, position encoding etc. and monitoring its environment values such as temperature, pressure, humidity etc. are developed to work on the

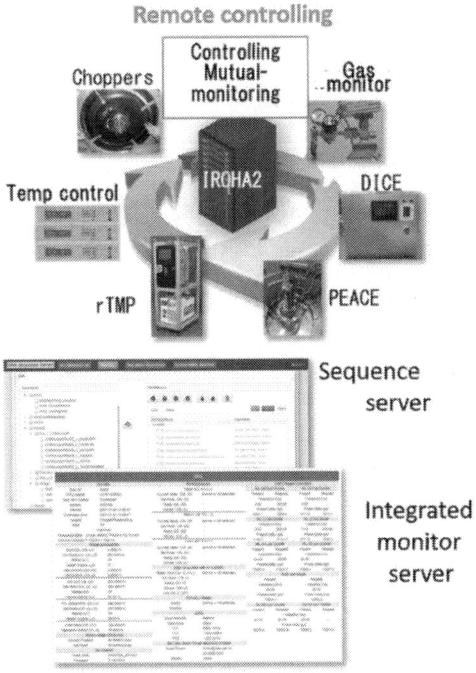

Fig. 9. Remote controlling way at DNA on the software framework IROHA2[13]. From the sequence server, command lists according to timeline of experiment can be conducted remotely. All status of devices is watched by the integrated monitor system.

IROHA2 platform. In the sequencer window, successive commands are arranged to make timelines to be done which involves also ques to start DAQ and end it and therefore we can control all devices and measurement sequence automatically. IROHA2 has also the integrated monitor system showing all status of the instrument which we can watch from outside of the facility. The status of the remote controlling at DNA is schematically shown in Fig. 9. We can watch a residual gas pressure of the gas cylinder such as He gas frequently used for inert gas for IVC of the cryo-furnace and operate a turbo molecular pomp (rTMP) remotely, as well as PEACE, DICE, temperatures of sample sticks, and choppers including PSC.

5 Near Future Plans of DNA

In the summer 2022, we plan to replace the converging neutron guide supermirrors in the vacuum chamber of DNA to new mirrors to improve neutron reflectivity. The guide jacket will be also renewed to non-magnetic materials. Then a small superconductor magnet with max 3.5 Tesla fitted to the small 400A flange of DNA will be commissioned. A high temperature furnace is also planned to be introduced. Two types of oscillatory radial collimator with different collimation degree are fabricated and under commissioning. These will be utilized to the cryo-furnace, the Blue Cryo, the high temperature furnace to eliminate unfavourable scattering from the wall of IVC and OVC, a container for single crystal with relatively large diameter, thermal shields and so on.

6 Summary

In this paper, we report recent progress including near-future plans on DNA. The improvement of the energy resolution was achieved by re-fabricating the disks for the pulse shaping chopper. Si311 analyzer mirrors extend accessible Q region with high energy resolution (~10 μeV). Variety of sample environments are available to user experiments. The auto and/or remote operations have also been developed.

We express special thanks to Dr. Kaoru Shibata, the previous instrument leader of DNA, Dr. Kenji Nakajima and Dr. Masatoshi Arai to their encouragement and kind support, Mr. Wataru Kambara and Mr. Keiichi Inoue to their helps on developments from engineering knowledge, Dr. Yasuhiro Inamura and Dr. Satoshi Kasai to their continuous support for software. The energy resolutions for Si111 and Si311 with 300 Hz operation were checked in the instrument group experiments with proposal No. 2018I0002 and 2020I0002, and neutron experiments for toluene at DNA were performed in the Proposal No. 2021I0002 and 2022I0002 with approval of the registered institute CROSS.

References

1. K. Shibata, N. Takahashi, Y. Kawakita, M. Matsuura, T. Yamada, T. Tominaga, W. Kambara, M. Kobayashi, Y. Inamura, T. Nakatani, K. Nakajima, M. Arai, JPS Conf. Proc. **8**, 036022 (2015)
2. M. Arai, R. Kajimoto, M. Nakamura, Y. Inamura, K. Nakajima, K. Shibata, N. Takahashi, J. Suzuki, S. Takata, T. Yamada, S. Itoh, J. Phys. Soc. Jpn. **82**, SA024 (2013)
3. N. Metoki, K. Shibata, M. Matsuura, H. Kitazawa, H. S. Suzuki, H. Yamuchi, M. Hagihala, M. D. Frontzek, M. Matsuda, J. Phys. Soc. Jpn. **91**, 054710 (2022)
4. M. Prager, A. Heidemann, Chem. Rev. **97**, 2933–2966 (1997)
5. Bée, M. *Quasielastic Neutron Scattering* (Adam Hilger: Bristol, England, 1988)
6. M. Matsuura, Y. Fujiwara, H. Moriwake, K. Ohara, Y. Kawakita, Phys. Rev. B **104**, 094305 (2021)
7. T. Tominaga, M. Kobayashi, T. Yamada, M. Matsuura, Y. Kawakita, S. Kasai, JPS Conf. Proc. **33**, 011095 (2021)
8. T. Yamada, H. Arima, T. Hattori, JPS Conf. Proc. **33**, 011084 (2021)
9. T. Tominaga, Y. Kawakita, H. Nakagawa, T. Yamada, K. Shibata, JPS Conf. Proc. **33**, 011086 (2021)
10. T. Tominaga, M. Sahara, Y. Kawakita, H. Nakagawa, N. Shimamoto, JPS Conf. Proc. **33**, 011094 (2021)
11. T. Tominaga, H. Nakagawa, M. Sahara, T. Oda, R. Inoue, M. Sugiyama, Life **12**, 675 (2022)
12. T. Tominaga, M. Sahara, Y. Kawakita, H. Nakagawa, T. Yamada, J. Appl. Cryst. **54**, 1631, (2021)
13. T. Nakatani, Y. Inamura, T. Ito, T. Otomo, JPS Conf. Proc. **8**, 036013 (2015)

Broadband Wide-Angle VElocity Selector (BWAVES) neutron spectrometer designed for the SNS Second Target Station

Eugene Mamontov[1,*], *Heloisa* N. Bordallo[2], *Olivier* Delaire[3], *Jonathan* Nickels[4], *Judith* Peters[5], *Gerald J.* Schneider[6], *Jeremy C.* Smith[7], and *Alexei P.* Sokolov[8]

[1]Second Target Station Project, Oak Ridge National Laboratory, Oak Ridge, Tennessee 37831, USA
[2]Niels Bohr Institute, University of Copenhagen, Copenhagen, Denmark
[3]Department of Mechanical Engineering and Materials Science, Department of Physics, and Department of Chemistry, Duke University, Durham, North Carolina 27708 USA
[4]Department of Chemical and Environmental Engineering, University of Cincinnati, Ohio 45221, USA
[5]Univ. Grenoble Alpes, CNRS, LiPhy and Institut Universitaire de France, and Institut Laue-Langevin, Grenoble, France
[6] Department of Chemistry and Department of Physics & Astronomy, Louisiana State University, Baton Rouge, Louisiana 70803, USA
[7]UT/ORNL Center for Molecular Biophysics, Oak Ridge National Laboratory, Oak Ridge, Tennessee 37831, USA and Department of Biochemistry & Cellular and Molecular Biology, University of Tennessee, Knoxville, Tennessee 37996, USA
[8]Chemical Sciences Division, Oak Ridge National Laboratory, Oak Ridge, Tennessee 37831, USA and Department of Chemistry, University of Tennessee, Knoxville, Tennessee 37996, USA

Abstract. A recently proposed wide-angle velocity selector (WAVES) device for choosing the velocity of detected neutrons after they have been scattered by the sample paves the way for inverted geometry neutron spectrometers with continuously adjustable final neutron wavelength. BWAVES broadband inverted geometry spectrometer proposed for the Second Target Station at the Spallation Neutron Source at Oak Ridge National Laboratory is designed using WAVES to simultaneously probe dynamic processes spanning 4.5 decades in time (energy transfer). This makes BWAVES a uniquely flexible instrument which can be viewed as either a quasielasitc neutron scattering (QENS) spectrometer with a practically unlimited (overlapping with the vibrational excitations) range of energy transfers, or a broadband inelastic vibrational neutron spectrometer with QENS capabilities, including a range of accessible momentum transfer (Q) and a sufficiently high energy resolution at the elastic line. The new capabilities offered by BWAVES will expand the application of neutron scattering in ways not possible with existing neutron spectrometers

1 Introduction

Inelastic neutron scattering (INS) is a powerful technique for probing atomic and molecular-level dynamics in condensed matter. The type of a neutron spectrometer (e.g., direct vs. inverted geometry) largely defines the scientific areas that the instrument is best capable of addressing. Inverted geometry neutron spectrometers measure neutrons scattered by the sample with a certain fixed "final" momentum and energy, $\mathbf{k}_f$ and E_f. From a range of the momenta and energies in the neutron beam incident on the sample, the "initial" momentum and energy, $\mathbf{k}_i$ and E_i, are found for a particular detected neutron using its time-of-flight TOF from the source to the sample to the detector. This determines the momentum and energy transfer, $Q = |\mathbf{Q}| = |\mathbf{k}_i - \mathbf{k}_f|$ and $E = E_i - E_f$, for each detected neutron scattered by the sample. A sum of all scattering events gives the scattering intensity, I(Q,E) that is proportional to the dynamic structure factor, S(Q,E), $E=\hbar\omega$, which can be interpreted as a double Fourier transformation of the particle's van Hove correlation function, G(r,t) which is classically the probability of finding a particle at position r after time t. INS thus uniquely provides simultaneous information on both the geometries of motions via the Q dependence and the time dependence via E. Further, the simple relation between S(Q,E) and G(r,t) illustrates that INS is a uniquely direct probe of atomic motion, requiring no extra knowledge of electronic, optical, or other properties for its interpretation.

Scattering of neutrons by a sample is comprised of coherent and incoherent parts, and measurements are usually designed to minimize the undesirable contribution from either signal. Inverted geometry neutron spectrometers tend to be utilized for measurements of predominantly incoherently scattering, usually hydrogen-bearing, samples. This is largely because the two prominent classes of such spectrometers, molecular vibrational and backscattering, efficiently utilize the main advantages of the inverted geometry: access to a broad range of energy transfers, even at low temperatures, and a high energy resolution at the elastic line. However, even the recently deployed high energy-resolution backscattering spectrometers [1-4], which excel in probing stochastic

* Corresponding author: mamontove@ornl.gov

dynamics that give rise to quasielastic neutron scattering (QENS), suffer from a limited range of available energy transfers. At the same time, broadband molecular vibrational spectrometers [5,6] lack the energy resolution and Q-coverage to probe QENS-type signal due to stochastic dynamics.

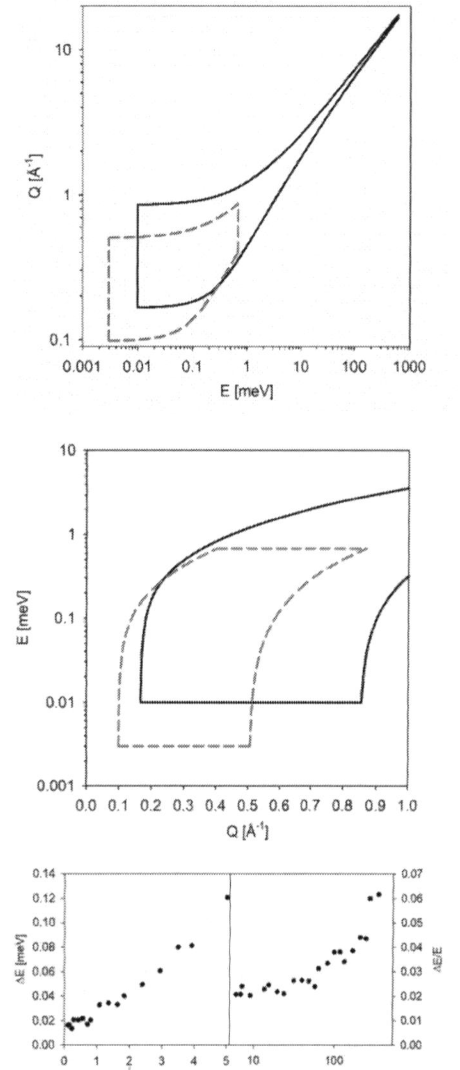

Fig. 1. Top panel: solid lines: the (E,Q) space coverage provided by the BWAVES spectrometer using the default $\lambda_f = 14.50$ Å in the standard operation mode. Also presented (dashed lines) is an example of the BWAVES "booster" mode with $\lambda_f = 24.50$ Å chosen for illustration. Middle panel: zoomed-in and truncated data from the top panel replotted as a function of Q. Bottom panel: the simulated energy resolution of BWAVES. Absolute energy resolution, ΔE (full width at half-maximum), up to 5 meV (left) and relative energy resolution, ΔE/E above 5 meV (right).

To address these shortcomings, a proposal has been recently approved to construct an inverted geometry spectrometer at the Spallation Neutron Source (SNS) Second Target Station (STS). Named BWAVES, which is an acronym for broad-band wide-angle velocity selector spectrometer, the spectrometer utilizes a novel wide-angle velocity selector (WAVES) device [7] to choose the velocity of neutrons after they have been scattered by the sample. BWAVES employs several novel technical solutions (besides the use of WAVES), as described in detail elsewhere [8], to become:

(1) a spectrometer with an unmatched dynamic range covering energy transfers from below 0.010 meV to above 500 meV.

(2) the only inverted geometry spectrometer currently proposed for SNS STS, well suited to map relaxational excitations that gradually emerge as a function of temperature, thereby efficiently deciphering complex dynamic landscapes.

(3) a spectrometer with a small beam size (5 mm by 5 mm), thus enabling studies of small samples and facilitating the use of advanced sample environment equipment.

(4) an open-tabletop sample geometry instrument, providing easy access to the sample position for application of external stimuli and multimodal experimentation, which includes default optical access to the sample.

(5) the gateway (oftentimes, the one-stop) spectrometer for neutron scattering studies of dynamics in materials, which is analogous to the role played by a general-purpose small-angle or broad-angle neutron diffractometer for structural studies.

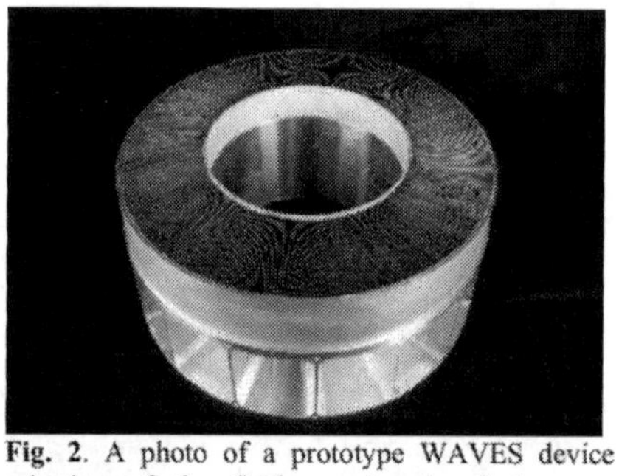

Fig. 2. A photo of a prototype WAVES device for selection of the final wavelength of the neutrons scattered at the sample position.

While each of these features is prominent on its own, the broadband character of BWAVES spectra is perhaps of utmost importance. The remarkable coverage of (E,Q) space, which at BWAVES is achieved within a single neutron frame (before the next proton pulse hits the target), is illustrated in Fig. 1, together with the energy resolution simulated using McStas [9,10] and MCViNE [11,12] packages. The widest dynamic range can be available at a non-constant Q.

A WAVES rotor is designed and built with curved blades of a shape described in polar coordinates as $r = A\varphi$, $A = v_f/(360°f)$, where v_f is the intended velocity of the neutrons transmitted by the rotor, and f is the rotation frequency. A photo of a prototype WAVES rotor is presented in Fig. 2. The WAVES resolution (full width

https://doi.org/10.1051/epjconf/202227202003

at half-maximum) can be evaluated [7] as $\Delta E_f = 2E_f(360°A/n_b(R_{max} - R_{min})) = 2E_f((v_f/f)/n_b(R_{max} - R_{min}))$, where n_b is the number of blades with an internal and an external radius of R_{min} and R_{max} and E_f is the energy of the neutrons transmitted by the rotor. Thus, when designing blades for a WAVES rotor, the maximum rotational frequency is chosen for the target v_f to improve the energy resolution. As a consequence, for a given WAVES rotor with a blade shape of $r = A\varphi$, a longer transmission wavelength, resulting in a finer energy resolution, can be readily chosen by lowering the rotation frequency and, therefore, the transmitted neutron velocity, because for a neutron to be transmitted, the ratio of (v_f/f) must remain constant. In contrast, a higher transmission neutron velocity (a shorter λ_f) cannot be selected readily, because this would require increasing the rotation frequency, which is already chosen to be as high as is technically feasible to ensure the best possible energy resolution in the standard operation mode. A freely selectable, in a continuous fashion, longer transmission wavelength results in finer energy resolution, and such an operation regime is thus dubbed a "booster" mode. An example of the BWAVES coverage in the (E,Q) space in the "booster" mode using an arbitrarily chosen $\lambda_f = 24.50$ Å is also presented in Fig. 1.

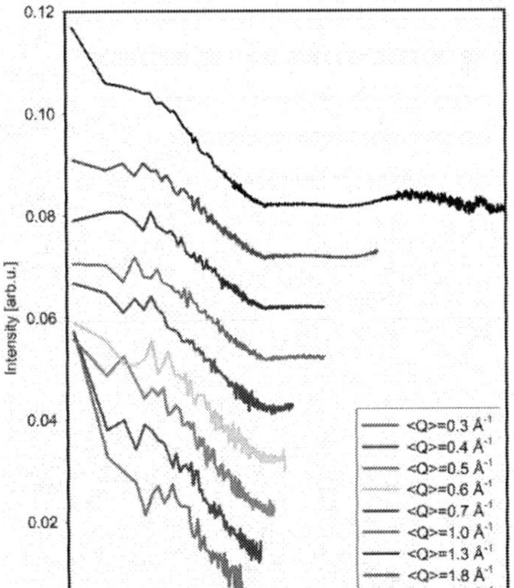

Fig. 3. HIV-1 protease at 120 K, measured at BWAVES using the default $\lambda_f = 14.50$ Å for 5 hours (virtually). The nine average <Q> values refer to the Q-bins of 0.25-0.35 Å^{-1}, 0.35-0.45 Å^{-1}, 0.45-0.55 Å^{-1}, 0.55-0.65 Å^{-1}, 0.65-0.75 Å^{-1}, 0.75-1.25 Å^{-1}, 0.75-1.85 Å^{-1}, 0.75-2.85 Å^{-1}, and 0.15-17.85 Å^{-1}.

Fig. 3 illustrates the output of a virtual BWAVES experiment to measure dynamics of HIV-1 protease. evaluated using McStas [9,10] and MCViNE [11,12] packages. In agreement with the data presented in Fig. 1, as the Q increases, plotting continuous broadband spectra requires the use of progressively broader Q-bins. The maximum possible dynamic range is attained for the data integrated over all scattering angles, similar to molecular vibrational spectrometers [5,6]. The dynamic scattering kernel, S(Q,E) for this system was mainly developed in the course of carrying out modelling for the recently published study revealing dynamical softening in the HIV-1 protease upon binding of an inhibitor darunavir [13], a finding of particular importance in determining the physical factors leading to inhibition of this critical drug target. The virtual neutron scattering experiment at BWAVES was performed using a D_2O-hydrated protein powder sample of HIV-1 protease, loaded in a cylindrical sample holder, 5 mm in height and 5 mm in diameter, with the packing density to give the total incoherent scattering cross-section for the sample of 0.2 cm^{-1} (that is, a 10% incoherent scatterer with the BWAVES design sample geometry). The higher-energy data in Fig. 3 show vibrational spectrum, with the C-H stretch modes visible near 300 meV. The lower-energy part of the spectrum exhibits what appears to be two components, with the positions independent of Q, indicative of the localized relaxational dynamics. It should be noted that development of yet another QENS component, due to the water solvent-driven protein dynamics, could be expected at higher measurement temperatures in this system.

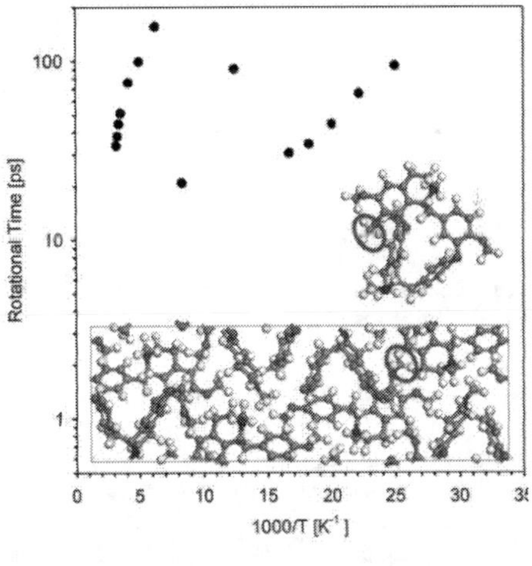

Fig. 4. Arrhenius plot of the measured relaxation times (symbols) for the methyl group rotation in a molecular drug tetrandrine (the data points are adapted from [14]). Insets: molecular structure (top) and crystal structure (bottom) of tetrandrine. The methyl group with the lowest rotational barrier, giving rise to the low-temperature data points, are encircled by an oval. The other methyl groups have higher rotational barriers and give rise to the high-temperature data points.

Another illustration of the advantage of an inverted geometry spectrometer with a broad dynamic range is provided by Fig. 4. Our recent efforts to characterize the dynamics of methyl groups in an antiviral molecular drug tetrandrine [14] were complicated by the fact that there are six methyl groups per tetrandrine molecule (Fig. 4). A series of the temperature dependent QENS measurements carried out at the SNS backscattering spectrometer BASIS (with a ±0.1 meV range of energy transfer) gives a non-monotonic temperature dependence for the methyl rotational jump time (Fig. 4). The reason for this is that, above 60 K, the quasielastic signal from the tetrandrine methyl group with the lowest rotational barrier, which first emerges on warming up from the baseline temperature, becomes too broad for BASIS to measure. However, the other methyl group rotational signals gradually enter (and eventually exit) the spectrometer's accessible range of energy transfers, thereby giving rise to the apparent non-monotonic temperature dependence of the measured rotational jump time. In contrast, measurements at BWAVES, with its wide range of accessible energy transfers, carried out while gradually raising the temperature from the baseline, will show the emergence of the methyl group dynamics with increasing rotational barriers one by one together with the simultaneous shift to the high energy (but not disappearance) of the methyl group dynamics with the lower rotational barriers. Thus, the dynamics of all the methyl groups would be identified, revealing their activation energies, which will also be linked to the vibrational modes simultaneously measured from the INS part of the spectrum. This approach requires access to the full relevant dynamic range, even at low measurement temperatures, not being constrained by detailed balance limitations, which can be achieved with inverted, but not direct, geometry spectrometers. As the only inverted geometry spectrometer currently proposed for SNS STS, BWAVES will be uniquely suited to systematically map the relaxational excitations as they gradually emerge with increasing temperature.

The technical details of BWAVES and its components are thoroughly described elsewhere [8]. This paper concentrates on the relationship between the unusual characteristics of BWAVES and the science it is anticipated to enable.

2 Science enabled by BWAVES

2.1 Molecular dynamics in soft matter

Soft matter is a broad class of materials including, among others, liquids and polymers, colloids and foams, lipid assemblies, and biological systems. Soft materials are used in many current technologies with significant potential for breakthrough future applications. One of the major properties of soft matter is an enormous number of metastable states with only small differences in their free energies, of the order of kT. As a result, soft-matter systems fluctuate rapidly between these states, and such fluctuations and rearrangements define their unique properties. However, the enormous number of metastable states leads to a very high complexity of these materials, with a broad range of relaxation phenomena, dynamic heterogeneities, cooperativity, and ageing of non-equilibrium states. QENS and INS provide critical information on transitions between the metastable states as well as a broad range of collective and individual dynamics in soft matter systems, including liquids, lipid assemblies and polymers. QENS/INS also helps unravel detailed mechanisms of proton/ion transport and elucidate many other aspects of soft matter dynamics. As a typical example, a combination of separately measured scattering spectra from polymer systems is usually necessary for thorough characterization of the polymer dynamics (Fig. 5, [15]), but unfortunately this relies on arbitrary matching of the data from different spectrometers summed over all Q values. In contrast, with its broad dynamic range BWAVES will be an ideal instrument to enable unequivocal analysis of polymer dynamics, including the geometries of molecular motions via the Q dependence of the scattering signal over the broad energy range. BWAVES will be well suited for measurements of dynamics in complex polymer systems, e.g., as presented in recently published studies of bottlebrush polymers [16,17] that had to utilize several neutron spectrometers at different facilities.

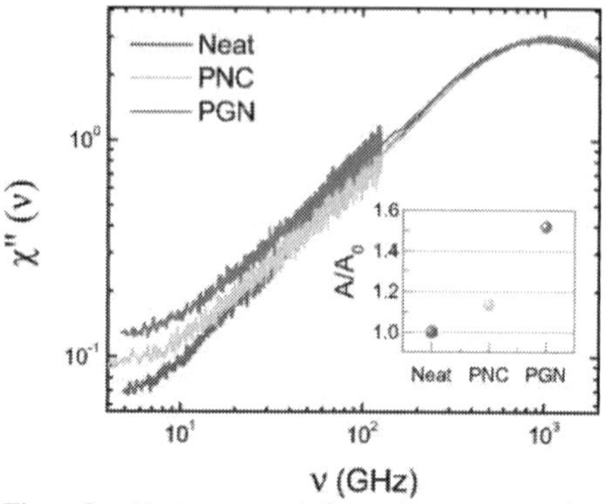

Fig. 5. Neutron scattering data presented as susceptibilities (T = 300 K, summed over all Q values, combined BASIS (SNS) and CNCS (SNS) measurements). The inset shows the normalized QENS area under the curve of a neat polymer vs. the polymer nanocomposite (PNC) and the polymer grafted to nanoparticles (PGN). Reprinted with permission from [15]. Copyright 2017 by the American Physical Society.

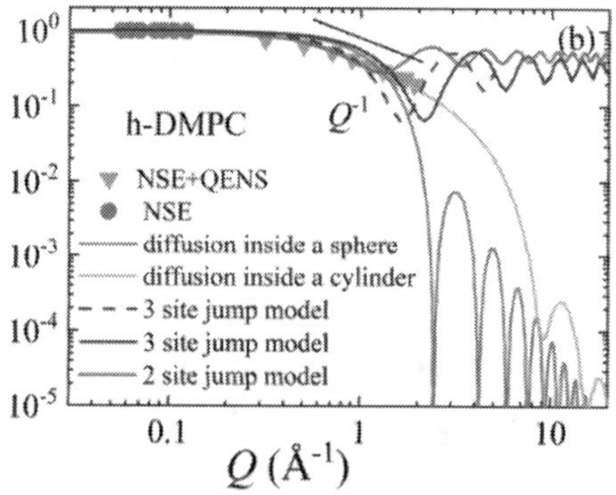

Fig. 6. Elastic incoherent structure factor for h-DMPC lipids obtained from NSE (SNS) and BASIS (SNS) studies. The data are modeled with motions within a sphere, a cylinder, and compared with three and two site jump models. Reprinted with permission from [18]. Copyright 2020 The Royal Society of Chemistry.

In another example, liposomes are an excellent model system for studying non-covalent interactions in complex cell membranes. Neutron spectroscopic studies of liposomes will shed light on effects of drugs,

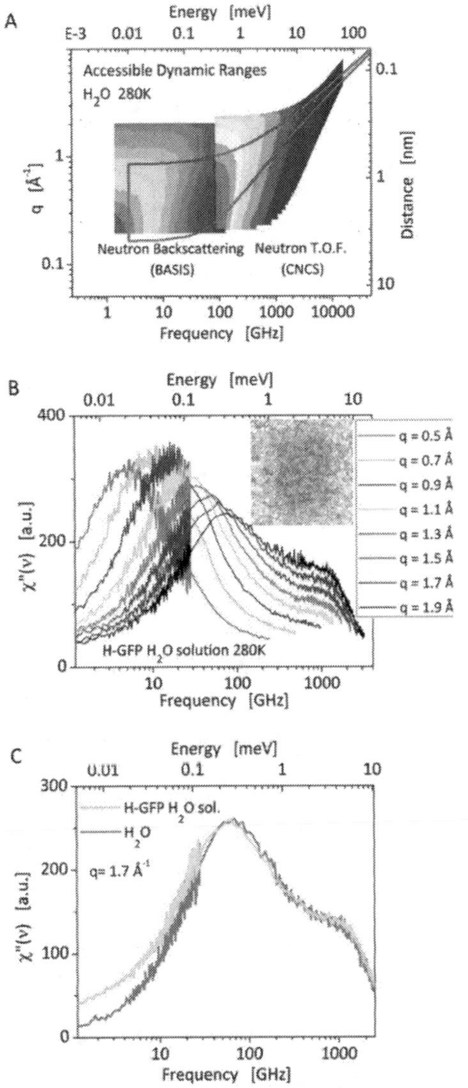

Fig. 7. Broadband neutron scattering capabilities are highly desirable to study the dynamics of water. (A) Recent work studying water hydrating Green Fluorescent Protein was compelled to utilize multiple spectrometers to achieve an experimental window covering the ns-ps dynamics of water. The capabilities of BWAVES would be an ideal fit for such studies. (B) Currently, Q-resolved spectra must be sliced and stitched in a cumbersome process from instruments with mismatched resolution and dynamic window. (C) Ultimately, this window is important to probe different water populations around model biomolecules. The data are adapted from [23].

investigate how viruses enter a cell membrane, and suggest how to harness the properties of cell membranes in technical applications, e.g., with embedded molecules for light harvesting. A recent study [18] combined data collected at the SNS neutron spin-echo (NSE) and BASIS spectrometers to cover a broad Q-range in the time range shorter than 5 ns. Fig. 6 shows the obtained elastic incoherent structure factor (EISF) representing

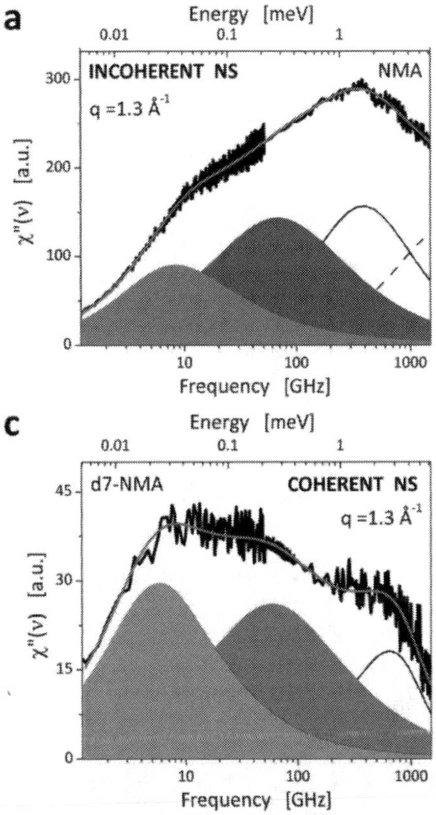

Fig. 8. Fitting of inelastic neutron scattering spectra from n-methylacetamide. Top: incoherent spectra, four components. Bottom: coherent spectra, three components. Reprinted with permission from [26]. Copyright 2017 The Royal Society of Chemistry.

the geometry of the lipid motion. It is likely that the lipid dynamics is best described as motion within a cylinder, yet the gap in the Q-range between the NSE and BASIS data leaves uncertainty regarding the effective cylinder dimensions. BWAVES, due to its capability of selecting long final wavelength (thus, lower Q and higher energy resolution) when operated in a "booster" mode, will be able to close the gap in the Q-range between NSE and BASIS and allow unequivocal determination of the geometry of the lipid motion.

While understanding cooperativity and heterogeneity in the dynamics of liquids and polymers remains a great scientific challenge, the dynamics of water, especially in confined space and at interfaces, is critical for a wide range of industrial and bio-medical applications and is a topic of equally active discussion [19-22]. For example, dielectric spectroscopy detects the main relaxation process in bulk water on the time scale ~10 ps, while light, neutrons and X-ray scattering detect the main structural relaxation on the time scale ~

1 ps [19-22]. BWAVES may resolve this conflict and is positioned to become an invaluable tool in hydration

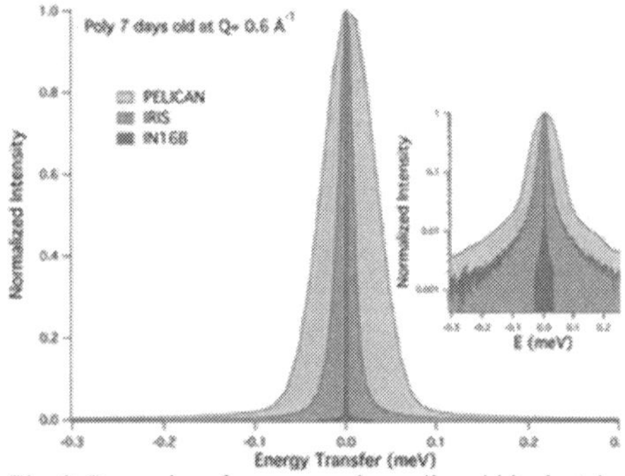

Fig. 9. Dynamics of aqueous polyacrylic acid in dental restorative cements: QENS spectra collected at backscattering and time-of-flight spectrometers. Reprinted with permission from [27]. Copyright 2018 American Chemical Society.

studies, increasing the quality and throughput of experimental studies of water dynamics, with the experimental window overlapping the coupled translational and rotational motions of water on the nanosecond to picosecond timescale as well as with vibrational modes.

The ability to capture the relevant water motions in a single experiment immediately reduces the difficulty of the experiment and ambiguity associated with combining information obtained on multiple spectrometers. For example, Fig. 7 presents the overlap of the current SNS QENS-capable spectrometers (BASIS and the cold neutron chopper spectrometer, CNCS) used for the study of water dynamics [23] in comparison to the proposed capabilities of BWAVES. It should be noted that explicit fits of QENS data collected at BASIS (±0.1 meV only) with more than 2 dynamic components have been infrequent, as such fits are difficult to carry out [24,25]. Another illustrative example of multi-component dynamic data (also collected at BASIS and CNCS), for which BWAVES would become a one-stop spectrometer, is given by a hydrogen-bonding liquid, n-methylacetamide (NMA) which is a model system for the peptide group (Fig. 8, [26]).

Studies of vibrational dynamics in soft matter, such as dental restorative cements [27], are presently championed at SNS by the VISION vibrational spectrometer. In such complex systems functionality is related to relaxational-translational molecular dynamics, and without a spectrometer with the exceptional characteristics of BWAVES, characterization of molecular dynamics again requires the use of many spectrometers. In the example presented in Fig. 9, four spectrometers were used at four different user facilities: in the USA, Australia, France, and the UK. Besides the logistical problems, in this case the traditional approach presents great challenges related to sample ageing, which is not uncommon in soft matter studies. When the experiments are designed to probe the molecular dynamics in the sample as a function of curing time (Fig. 9), the benefits of a single spectrometer capable of measuring relaxational-translational and vibrational dynamics simultaneously, such as BWAVES, are obvious.

2.2 Chemical spectroscopy

Understanding proton and ion dynamics in liquids,

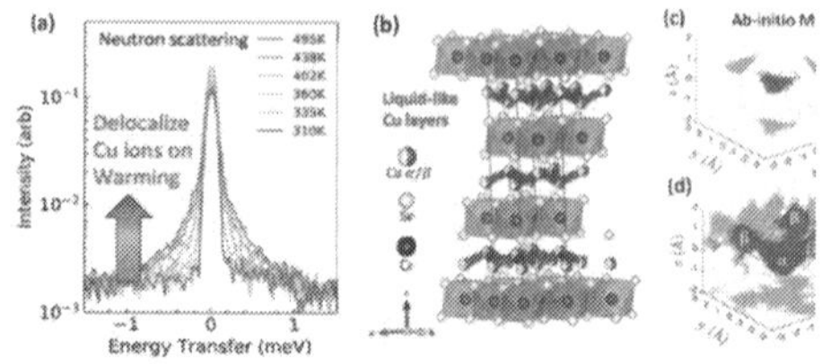

Fig. 10. (a) Quasi-elastic neutron scattering signal showing delocalization of Cu ions with increasing temperature (broadening of central peak). (b) Layered structure of $CuCrSe_2$, showing liquid-like Cu layers in superionic phase above 365 K. (c,d) Ab-initio MD calculations of Cu ion trajectory, showing delocalization across α and β sites at high T (adapted from [58]).

solids and polymers is another example of a scientific

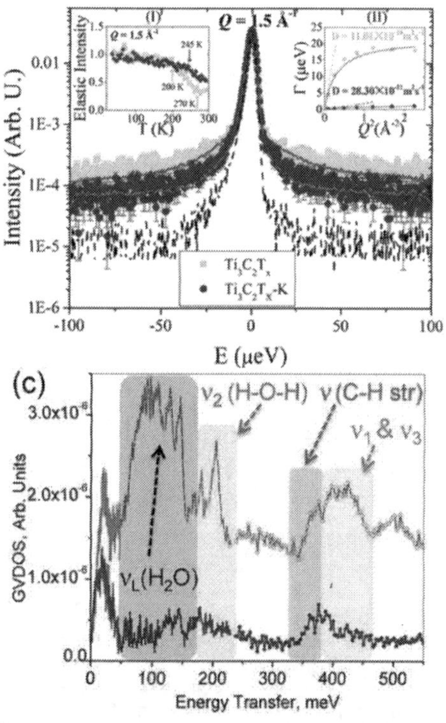

Fig. 11. Top: QENS spectra of MXenes samples measured at BASIS. Reprinted with permission from [77]. Copyright 2016 American Chemical Society. Bottom: the generalized vibrational density of states of MXene samples before (red curve) and after (blue curve) annealing, from the INS spectra measured at SEQUOIA. Reprinted with permission from [78]. Copyright 2014 American Chemical Society.

topic that can benefit tremendously from use of

broadband neutron scattering. For instance, superionic conductors (SICs) are unique solid materials in which a

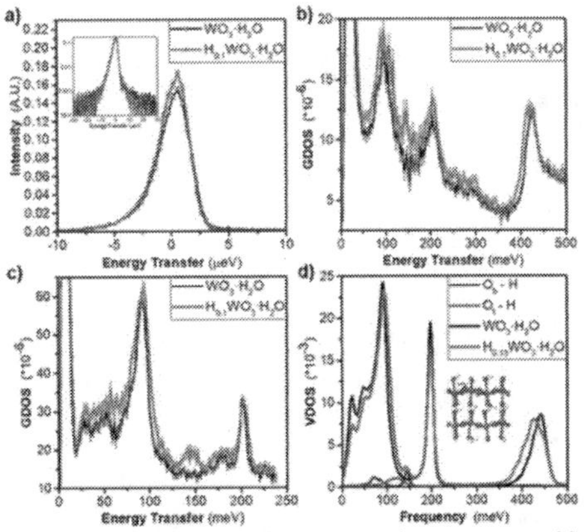

Fig. 12. Confined interlayer water in tungsten oxide hydrates enables fast highly reversible proton intercalation. (a) QENS (BASIS) of an electrode in the pristine, $WO_3{\cdot}H_2O$, black, and intercalated, $H_{0.1}WO_3{\cdot}H_2O$, red, states at T = 300 K. (b-c) INS (SEQUOIA) of the same electrodes at T = 5 K. (d) AIMD simulations of the vibrational density of states show the effect of protons at the terminal (O_t, green) and bridging (O_b, blue) oxygen atoms to the overall spectrum. Reprinted with permission from [79]. Copyright 2019 American Chemical Society.

portion of the structure becomes liquid-like at high temperatures (often above a phase transition), with

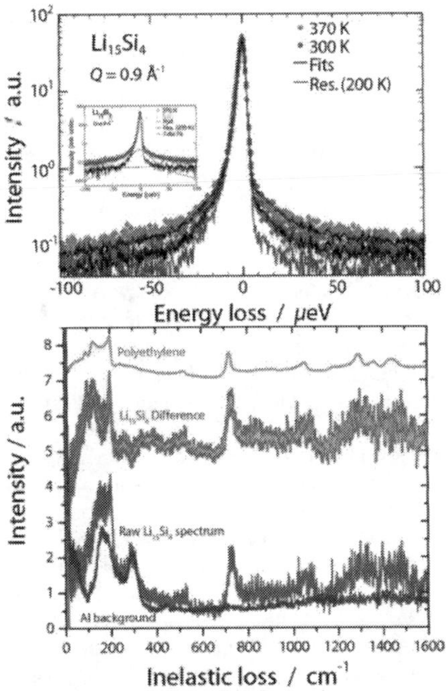

Fig. 13. QENS (top) and INS (bottom) spectra measured from an amorphous Li_xSi anode material. Reprinted with permission from [80]. Copyright 2017 American Chemical Society.

mobile ions acquiring diffusivities similar to liquid water at room temperature (D ~ 10^{-5} cm^2 s^{-1}). Steady interest in these systems [28-35] has spiked recently because of their promising applications as solid-state electrolytes for safer rechargeable batteries [36-46], as promising new thermoelectric materials [47-51], or as solid-oxide fuel-cells [52-54]. However, because of their intermediate nature between the solid and liquid states, SICs exhibit highly complex atomic dynamics involving many intricate processes ranging from solid-like phonons to liquid-like diffusion, which operate across multiple time and length scales [30,42,55,56]. Hence, the thermodynamic and transport properties of SICs remain poorly understood [56]. For example, two recent studies of atomic dynamics in the superionic "phonon-liquid electron-crystal" thermoelectric compounds $AgCrSe_2$ [57] and $CuCrSe_2$ [58] arrived at different conclusions regarding the atomistic processes dominating their thermal transport. These SICs are isostructural and both undergo phase transitions to a superionic state on heating, where Ag/Cu ions become highly mobile. Both exhibit ultra-low thermal conductivities (<1 W m^{-1} K^{-1} above 300 K) [49,59].

In Ref. [57] it was concluded that transverse acoustic phonons disappear in the superionic phase of $AgCrSe_2$, leading to a liquid-like thermal conductivity, while it was shown in [58] that long-wavelength acoustic phonons remain well-defined in the superionic phase of $CuCrSe_2$ but instead disorder and anharmonicity are strong sources of phonons scattering. BWAVES will be highly beneficial for resolving such controversies. The large dynamic range of BWAVES will enable simultaneous measurements of the signals that at present require two or three SNS spectrometers (e.g., BASIS, CNCS, ARCS, SEQUOIA, or VISION) to investigate both the QENS behavior (diffusion) and the lattice dynamics (phonons up to 10-100 meV) critical for understanding solid-state electrolytes for next-generation batteries. Moreover, the QENS signal from non-hydrogenous samples is relatively weak, and systematic studies (e.g., as a function of temperature and compositions) will be facilitated by the higher flux of BWAVES compared to BASIS.

As further example, recent breakthroughs in batteries [45] have achieved solid Li-ion or Na-ion SICs with ionic conductivities rivalling those of conventional organic liquid electrolytes [36,41,60-65]. These could replace liquid electrolytes to improve safety and boost energy density [40,65]. While many structural families of SICs have been uncovered (e.g. argyrodites [43,66], LISICON [67], NASICON [35], garnets [68], lithium nitrides [69,70], halides [71], and perovskites [72,73]), a clear understanding of how superionicity arises in structurally and chemically disparate compounds is still missing [39,45]. QENS is ideally matched to probe the dynamics of fast ions (~GHz), complementing measurements of THz lattice dynamics with INS, and BWAVES will be able to simultaneously measure both regimes in a single instrument setting, a unique feature not accessible with existing neutron spectrometers. Example QENS measurements across the superionic transition in $CuCrSe_2$ are shown in Fig. 10. The broadening of the QENS peak upon warming into the superionic phase (T > 365K) is the hallmark of increasing diffusion. The QENS data analysis will directly estimate the ion diffusion and jump length from the Q dependence of the signal, and these results can be also combined with the analysis of phonons in the same measurements. Thus, BWAVES will open unprecedented opportunities for studying SICs.

The discovery of a novel class of 2D MXene materials [74-76] has revolutionized many applications and lead to an exponential growth in research on MXenes for electrochemical devices. The dynamics, both vibrational and diffusive, of water species in MXenes is critical for their electrochemical performance and long-term operational stability, and strongly depends on the water content and the intercalated ions. To date, MXenes have been studied separately at the SNS BASIS and SEQUOIA spectrometers (Fig. 11, [77,78]) using large-size aluminum sample holders for powders. The combination of QENS and INS capabilities with a small beam size in an open-tabletop sample geometry spectrometer such as BWAVES will facilitate simultaneous studies of vibrational and diffusive dynamics in real-life electrochemical devices, such as MXene-based supercapacitors, under operational condition (e.g., repeated charging-discharging).

There is a strong societal need for energy-related research and development, which requires comprehensive characterization of materials for batteries and supercapacitors. As an example, recent experiments at SNS (Fig. 12, [79]) measured intercalated proton dynamics in tungsten oxide hydrate electrodes. However, the characterization of the vibrational and stochastic energy landscape in these electrode materials relied on disparate measurements at BASIS at 300 K and SEQUOIA at 5 K, and possible soft dynamic modes could not be evaluated at all. The temperature-dependent QENS and INS measurements at BWAVES will provide thorough characterization of the vibrational and stochastic energy landscape in electrode materials, which feature multiple hydrogen-bearing species such as structural water molecules, hydration water molecules, and intercalated protons, each with its own dynamic characteristics. The power of temperature-dependent measurements for elucidating complex energy landscapes, which is specific to the inverted geometry spectrometers such as BWAVES, has been already mentioned above in the context of distinguishing between methyl groups with various rotational barriers (Fig. 4).

Similar to in situ studies of charging-discharging processes in real life supercapacitor devices, the small beam size and an open-tabletop geometry of BWAVES will enable in situ electrochemical studies of batteries under operation conditions. While it is common to use VISION for verification of the proton content in the materials with non-proton charge carriers studied separately at BASIS (e.g., Fig. 13 [80]), BWAVES will provide unprecedented opportunities for the simultaneous monitoring, in real time, of the mobility of Li charge carrier ions and the evolution of the vibrational dynamics in the matrix (e.g., battery anode) associated with the in-operando ion insertion.

2.3 Molecular dynamics in biological systems

Life, as we know it, is based on molecular motions.

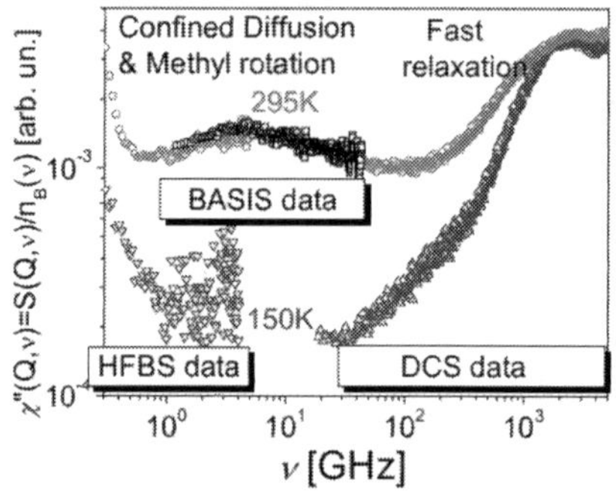

Fig. 14. Susceptibility spectra of protein lysozyme obtained by combination of data from 3 spectrometers, time-of-flight DCS (NIST), backscattering BASIS (SNS) and backscattering HFBS (NIST). Reprinted with permission from [86]. Copyright 2015 The Royal Society of Chemistry.

Biological materials are characterized by enormously complex energy landscapes, and the associated dynamic processes comprise a broad quasi-continuous spectrum of both vibrational and relaxation excitations. Due to the intricate coupling of biomolecules with their aqueous-based environment [81-86], the dynamics of which is already challenging to understand (Fig. 7), the dynamic landscape of hydrated / solvated biomolecules represents even higher level of complexity. Moreover, studies of living systems are challenged with the need to incorporate knowledge on multiple length and time scales. On length scales this involves a continuum from communities, to organisms to organs and down to cells and molecules. On time scales dynamic phenomena in a single protein occur on at least 17 decades of time (fs-10^2s). BWAVES acts in the multiscale research space, connecting phenomena over 4 decades of time, and promises thus to furnish vital information on connected and interdependent multiscale dynamical phenomena and forces at the molecular level of biology.

Neutrons have many desirable and unique properties as a probe of biomolecular dynamics, but the bandwidth limitations of classical neutron spectroscopy are problematic when the broad excitations of diffusive and relaxation motions need to be simultaneously probed. This is true in hard matter systems but is especially detrimental in studies of soft matter and biological systems, where the excitations are broad in energy space and oftentimes overlap with one another. This can, similarly to the above non-biological cases, lead to misinterpretations and ambiguity in analysis of neutron scattering data. As a typical example, analysis of the broad relaxation spectra in a protein required measurements on 3 different spectrometers (Fig. 14, [86]) - a very time and sample consuming process. BWAVES, due to its broad dynamic range, will address this problem. Many experimental techniques can measure longer time dynamics (μs and longer) than BWAVES. However, it remains a great challenge to understand the details of dynamic processes on microscopic time scales (~fs-ns), which are collective, cooperative, and intrinsically heterogeneous. These timescales directly probe the interactions intrinsic in the molecular structure of biomolecules and their assemblies.

Moreover, the internal motions of protein on the timescales accessible to BWAVES are critical to the *in silico* drug discovery procedure of ensemble docking [87,88], in which the dynamic nature of drug binding sites is taken into account in virtual high-throughput screening. To this end, perhaps nothing can illustrate better the importance of neutron spectroscopy for probing the dynamics in complex biosystems than the pandemic caused by COVID-19 virus. The SARS-CoV-2 coronavirus, the causative agent of COVID-19, contains a spike protein, which is a molecular machine for binding to and penetrating human host cells (Fig. 15, [89,90]). Small-molecule drugs binding within the S-protein-human ACE2 interface may prevent this infectious step, and understanding the flexibility of the drug binding sites is essential to the design process. Other SARS-CoV-2 proteins can also be examined, and, to help derive pan-coronavirus treatments for the coming decades, binding site flexibilities of the entire suite of non-structural viral proteins should be derived. INS and QENS measurements at BWAVES will be essential for studies investigating protein-ligand binding.

Further, the mechanics of infection and viral particle assembly requires conformational changes in the protein machines responsible for them. Understanding these steps in the viral lifecycle will require characterization of the anharmonic dynamics responsible for the mechanical soft modes involved. Inelastic neutron scattering spectra ranging from the femtosecond to nanosecond timescale will be invaluable in this regard as they will provide information on the elastic properties of this molecular machine in various states and their perturbations due to the interactions with potential drugs. Moreover, on a fundamental level, the time- and geometry-dependence of motions detectable by BWAVES will be of great help in deciphering the relationship between non-ergodic, ageing, internal protein dynamics [91] in single proteins as determined by molecular dynamics simulation and single-molecule FRET and ensemble characteristics measured by BWAVES. An open question in the studies of biological systems, from single proteins to complex systems as bacteria or cells, concerns the relaxation processes self-similarities over many decades in time, for which there are hints reported in the literature [92]. To obtain information over such broad time ranges, one needs to use spectrometers giving access to complementary time domains, or even better an instrument such as BWAVES, which permits us to see a broad energy-time domain at once, avoiding problems due to different normalizations.

Another biologically important field is the dynamics of hydration water of macromolecules, microbial envelopes, plant cell wall polymers, and many other systems. The structurally and dynamically perturbed hydration shells of proteins and other biomolecules have a strong influence on their function and stability. Many recent studies have focused in great depth on providing a description of hydration water

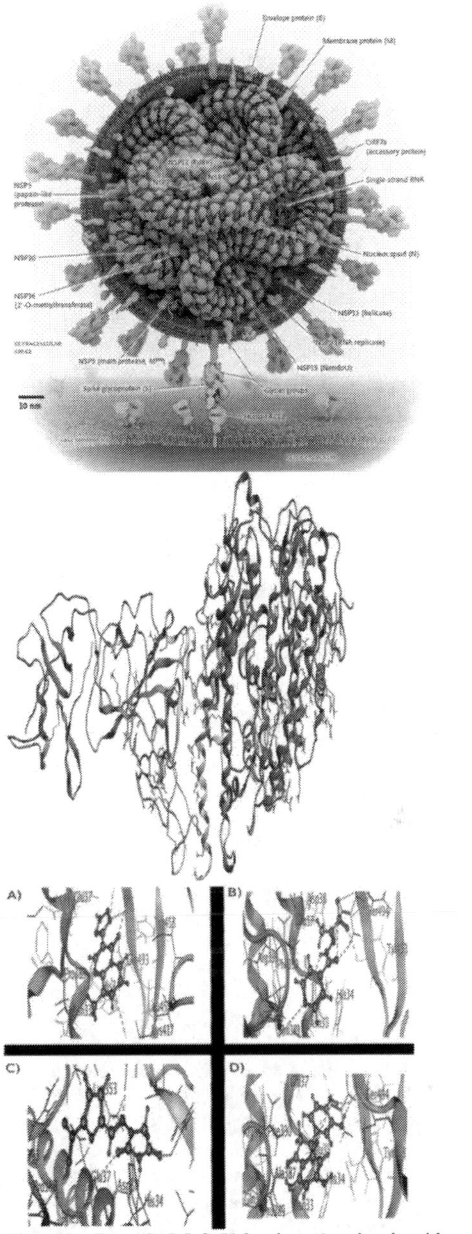

Fig. 15. Top: SARS-CoV-2 virus (reprinted with permission from [89]. Copyright 2020 American Chemical Society). Middle: the S-protein of SARS-CoV-2 and human ACE2 receptor complex. Orange ribbons: the S-protein, purple: the ACE2, yellow: the interface targeted for docking (reproduced from [90]). Bottom: Four of the top scoring small-molecules binding within the S-protein-ACE2 interface. A) pemirolast. B) isoniazid pyruvate. C) Nitrofurantoin. D) Eriodictyol. Orange ribbons: the S-protein, purple: the ACE2 receptor (reproduced from [90]). These calculations required a knowledge of the dynamics of the target proteins on the BWAVES time scales.

dynamics using neutron scattering methods [93-95]. Neutron scattering approaches can quantify the number of molecules perturbed by the presence of a biomolecule and the degree of the perturbation. The recent investigation of water dynamics in crowded solutions of green fluorescent protein (Fig. 7) is a good example [23], where the spectral analysis revealed that slightly less than two shells (~5.5 Å) of water were perturbed by the presence of the protein, with dynamics a factor of 2−10 times slower than bulk water. These results help clarify the seemingly conflicting range of values reported for hydration water retardation as a logical consequence of the different length scales probed by the

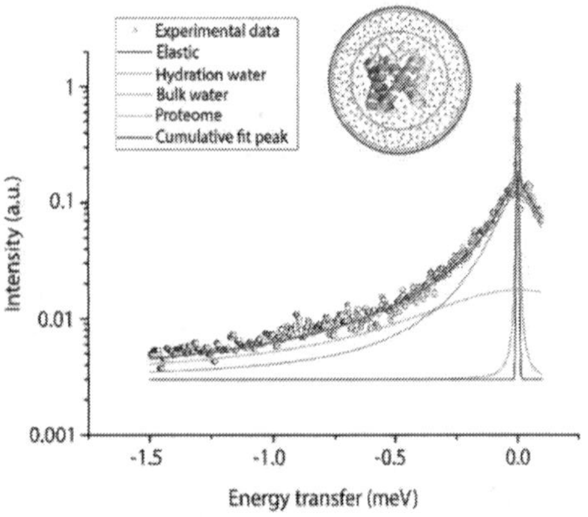

Fig. 16. Normalized QENS spectra and fit for *T. barophilus* at 298 K, 40 MPa and $Q = 0.96$ Å^{-1} on a semi-logarithmic scale. Scheme in the red circle represents the different populations studied, the ribbons represent a protein, and the dots represent water molecules (green for hydration water and blue for bulk water). Reprinted with permission from [105]. Copyright 2019 The Royal Society of Chemistry.

analytical techniques used. Neutrons not only provide a detailed description of water dynamics but do so in a way which adds clarity to the broad community studying the question with other approaches. Even more challenging are studies of water dynamics in cells, where the experimentally determined parameters reveal that hydration water dynamics strongly depend on the confinement environment (size, pH, hydrophilic or hydrophobic surface, etc.) [96]. No single experimental technique can give a full picture of such intricate dynamics, but the broadband capability of BWAVES will provide the most comprehensive information yet.

The study of anticancer drug effects at the cellular level is another fascinating example where the need to measure both vibrational and relaxational phenomena in a single spectrum is critical [97]. BWAVES will make this type of measurement possible. By scrutinizing the effect of distinct chemotherapeutic agents on the water dynamics of different cell lines, the active water role in living organisms over many scales of time and distance will be finally re-evaluated [98]. BWAVES capabilities will offer unparalleled advantages for cellular water investigations under various environmental conditions using neutron scattering (not attainable by NMR [99]) and consolidate the role of neutron scattering in cell-based drug characterization. BWAVES will also play a pivotal role in biotechnology and biomedicine by allowing – with the aid of specialized sample environment [100] – studies of changes in the cellular water dynamics caused by stimuli-responsive drug delivery systems that respond to heat, pH, light, magnetic and/or electrical field. These smart bionanomaterials, where controlled and long-term drug release is enabled by the action of an external stimulus, offer the promise of treatments for chronic diseases. Activating these materials using electrical signals would make this technology easy to generate and control [101,102].

Even though the movement of ions and water dipoles is postulated to be important in the chemical pathway of electroporation (a microbiology technique that uses electrical field to increase the permeability of the cell membrane) and electroosmotic flow (movement of a liquid in a porous material or biological membrane under the influence of an electric field), research gaps still need to be closed to advance this technology. For instance, how are the transport mechanisms of water or ions affected by an applied electric field? Answering this question is key for developing novel hybrid organic-inorganic electric field responsive drug delivery systems. Here, again, the need to measure both vibrational and relaxational phenomena in a single spectrum is crucial [103,104]. The unique ability of

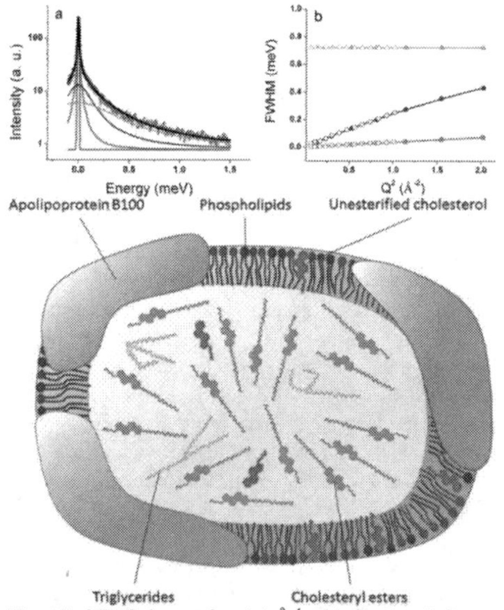

Fig. 17. QENS data at $Q = 1.0$ Å^{-1} with fits (top left) and the associated signal broadening (top right) from lipoprotein particles (bottom) measured at 310 K and 20 bar (reprinted with permission from [106]). The FWHM values of the three Lorentzian distributions as a function of Q-squared as presented in the top right panel were obtained at IN5, ILL (smaller Q-range, open symbols) and IN6, ILL (filled symbols) and fitted using a translational diffusion model (magenta points), a jump-diffusion model (blue points), and a rotational diffusive model (orange points).

neutron scattering to study encapsulated drugs at the molecular level will lead to deeper understanding of their mobility/release dynamics. Being able to predict release profiles of complexed drugs is one of the greatest challenges faced by pharmaceutical technology. Such an understanding would bring to light the possibility of tailoring the drug release rate in accordance with the desired effect, and a combination of density functional theory calculations with incoherent inelastic spectroscopy and state of the art sample environment is a promising approach to overcome this challenge using BWAVES, facilitated by the open sample geometry aiding stimuli application.

The ability of BWAVES to fully characterize the dynamics of a sample in one shot is of critical importance for studies involving stimulus-induced response, because it ensures that the dynamics at all energy transfers is always probed in the identical state of the sample immediately before, during, and after the stimulus application. When carried out at different spectrometers, processes such as in situ hydration/dehydration and drug delivery would require full reproducibility of the external stimulus application that is hard to achieve in practice, thus requiring stitching together data sets collected not necessarily under the identical conditions. This exacerbates the already mentioned issues of the mismatched Q ranges and arbitrary data normalization from different spectrometers. Simultaneous measurements over the full dynamic range, as allowed by BWAVES, will resolve this problem of reproducible stimuli application, which grows more challenging with the increasing complexity of the systems under investigation.

Understanding biological complexity and metabolism in plants and microbes is a further grand challenge. Connecting the molecular scale dynamics of hydrogen atoms or collective fluctuations of biological structures to organism level phenomena requires advances in the tools for assessing molecular motions, such as provided by BWAVES. Rapidly capturing broad spectral snapshots of molecular motions is crucial for understanding differences in molecular motions based on different biological stimuli. Furthermore, studies of complex systems are favored by open tabletop BWAVES design capable of accommodating complex sample environments. A more open sample environment allows for sample interaction with light, added chemicals (salts, metabolites, etc.), and pressure, to name a few. Because biological systems are constantly changing, it is crucial to monitor samples in the course of experiments. Parameters such as optical density, fluorescence, or settling of cells out of solution must be monitored to ensure that changes in dynamics are representative of real phenomena. Thus, easy access to the sample (including access by optical probes) will be a highly beneficial feature of BWAVES design.

Investigation of very complex systems such as prokaryotes can provide information on adaptation mechanisms to extreme external conditions such as high temperature, pressure, and salinity. In recent studies, QENS data were collected on *Thermococcus barophilus*, a microbe living in the deep sea close to hot vents. Fig. 16 [105] shows that, besides the elastic contribution, there are several dynamic components in the cellular dynamics, corresponding to bulk and hydration water and the proteome. These contributions were visible within the rather limited energy transfer range of the spectrometer IN5 (ILL), but the correct interpretation of the two components arising from water required other measurements on IN6 (ILL). Therefore, for these relaxational studies, a broadband range spectrometer is again imperative. In addition, it is desirable to probe low-energy vibrational dynamics simultaneously and on the same samples, as they are very difficult and expensive to produce. Hence a spectrometer that resolves all the relaxational components shown in this spectrum, but also goes much beyond the energy transfer shown here, becomes mandatory. Moreover, modern MD simulations allow a combination of all-atom and coarse-grain approaches, permitting to reasonably model bacteria as *Eschericia coli* and therefore to compare the results from vibrational and local motions to global movements, thus allowing to make progress on both the experimental and simulations sides. Therefore, an instrument as BWAVES will make a dramatic difference in this kind of studies of microorganisms.

Another example where dynamics influences function is human plasma lipoproteins - macromolecular assemblies of lipids and proteins which transport lipids through blood circulation. The main function of low-density lipoproteins (LDLs) is the transportation of

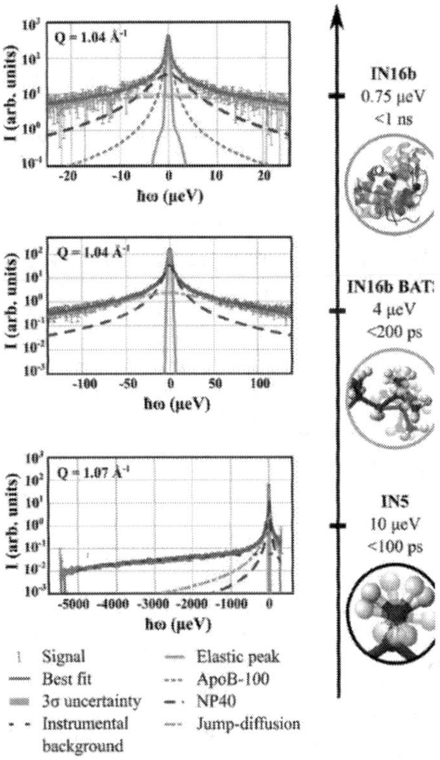

Fig. 18. An example of apo B-100/NP40 data fitting (modified from [107]) on 3 spectrometers (IN16b - top; IN16b BATS - middle; IN5 - bottom) at Q = 1.05 Å^{-1}.

cholesterol from the liver to peripheral tissue. LDL particles (Fig. 17, [106]) comprise a combination of lipids and a single copy of a protein, apolipoprotein

B100 (apoB100). They have a hydrophobic core and an amphiphilic shell. While the core hosts cholesteryl esters, triglycerides, and parts of the unesterified cholesterol, the shell consists of a phospholipid monolayer, unesterified cholesterol, and the amphipathic apoB100 molecule wrapped around the surface. LDL is a key contributor to lipid deposition in the arteries leading to atherosclerosis. Exploration of the dynamical features of the LDL particles with either a normal or triglyceride-rich lipid composition is essential for accurate modeling and understanding of the LDL interactions with various biological environments in vivo. The complexity of the lipid dynamics in the LDL particles has necessitated the use of different neutron spectrometers for the global fit of the QENS spectra [106]. However, because all the dynamic components have been demonstrated to contribute prominently to the QENS spectra measured at Q = 1.0 $Å^{-1}$ and below (Fig. 17), BWAVES will enable single-spectrometer measurements of lipoprotein dynamics, while its open-tabletop geometry will facilitate the in situ control over the physiologically relevant conditions of the lipoprotein's host solutions.

One moiety of LDL, the apolipoprotein B-100 (apo B-100), has been scarcely studied. Due to its amphiphilic nature, it is partly located at the surface of the LDL and partly embedded in its phospholipid monolayer. Therefore, it has characteristics close to classical membrane proteins and is very difficult to extract from LDL, and thus its solubilization is associated with detergent remaining in the sample. Recently, the dynamics of apo B-100 and its detergent, Nonidet P-40 (NP40) were separated using incoherent neutron scattering experiments [107] but through a combination of measurements on three different spectrometers at ILL: IN5, IN16b in BATS mode, and IN16b (see Fig. 18). This approach allowed the motions of the detergent alone to be identified, and, by fixing some of the parameters, to separate the parameters specific to apo B-100. Such methodology can render studies of other membrane proteins achievable. The potential to collect all necessary data on one instrument, such as BWAVES, under the same conditions, would greatly simplify such measurements and improve data reliability.

3 Conclusion

BWAVES will be capable of simultaneous characterization of vibrational and relaxational spectra in an enormous range of soft matter systems. Herein we discussed the benefits of this approach using numerous examples from recent neutron spectroscopy studies in a broad range of systems. We reiterate the main characteristics of BWAVES, which we have discussed in the paper, as follows:

(1) Measurements of continuous spectra covering energy transfers from below 0.010 meV to above 500 meV.

(2) The only currently proposed inverted geometry spectrometer for SNS STS, well suited to map relaxational excitations gradually emerging as a function of temperature.

(3) A small beam size (5 mm by 5 mm).

(4) An open-tabletop sample geometry instrument.

(5) The flexibility to function as the workhorse spectrometer at SNS STS for studies of chemical, soft matter, and biological systems.

The importance of the first three features of BWAVES can be demonstrated using recent studies [14,108,109] of several potential therapeutic agents for COVID-19 treatment, which were carried out at SNS in the Spring of 2020. These studies required the use of BASIS, CNCS, VISION, and SEQUOIA spectrometers for different measurements spanning several weeks. These measurements could have been completed much more readily and efficiently with the use of a single broadband spectrometer such as BWAVES. A broadband inverted geometry spectrometer such as BWAVES would be capable of thorough mapping of the rotational potentials of side groups in molecular compounds such as anti-COVID-19 drugs and other antiviral agents, which proved elusive to achieve with the existing spectrometers (Fig. 4, [14]). Further, it was necessary to bypass several promising drug candidates altogether because they were available only in microgram quantities, which could not be measured using the existing spectrometers with an incident beam size of ca. 10 cm^2 or larger. BWAVES will enable these types of samples to be probed, with its smaller beam size and a high neutron flux.

While the implications of the first three features of BWAVES are well understood and appreciated, the latter points represent novel features of BWAVES that remain to be explored by the community. Never before has a broadband high energy-resolution spectrometer been built with an open sample geometry more resembling that of neutron reflectometers than traditional neutron spectrometers.

An important element of BWAVES operation will be integration with a range of computational and experimental techniques in multimodal studies of important soft matter systems. For example, the convenient and routine application of external stimuli to samples, including optical probes, will greatly broaden the usefulness of the neutron measurements and will and enable the users, both experienced and new, to expand the application of neutron scattering in ways not possible with today's neutron spectrometers. Such expansions may be difficult to appreciate at present. Furthermore, the dynamic range of BWAVES mirrors that accessible to molecular dynamics simulation and developments in the latter utilizing high-performance computing need rigorous experimental testing. BWAVES, by virtue of its direct probing of dynamics on a wide range of timescales, will provide the necessary constraints.

All in all, it may be fair to describe BWAVES as a workhorse spectrometer at SNS STS that will provide a gamut of unprecedented characterization capabilities to a broad user community. Serious consideration needs to be given to the future computational approaches suitable to model both QENS and INS measured signals, not separately, but within the same spectra.

The design of BWAVES has been the result of efforts by many contributors. This research used resources of the Spallation Neutron Source Second Target Station Project at Oak Ridge National Laboratory (ORNL). ORNL is managed by UT-Battelle LLC for DOE's Office of Science, the single largest supporter of basic research in the physical sciences in the United States. The operation and user program at the Spallation Neutron Source is supported by the Scientific User Facilities Division, Office of Basic Energy Sciences, U.S. Department of Energy. OD acknowledges support from the US Department of Energy, Office of Science, Basic Energy Sciences, Materials Sciences and Engineering Division, under Award No. DE-SC0019978. GJS gratefully acknowledges funding by the U.S. Department of Energy under grant DE-SC0019050. The United States Government retains and the publisher, by accepting the article for publication, acknowledges that the United States Government retains a non-exclusive, paid-up, irrevocable, world-wide license to publish or reproduce the published form of this manuscript, or allow others to do so, for the United States Government purposes.

References

1. E. Mamontov, K.W. Herwig, Rev. Sci. Inst. **82**, 085109 (2011)
2. K. Shibata, N. Takahashi, Y. Kawakita, M. Matsuura, T. Yamada, T. Tominaga, W. Kambara, M. Kobayashi, Y. Inamura, T. Nakatani, K. Nakajima, M. Arai, JPS Conf. Proc. **8**, 036022 (2015)
3. B. Frick, E. Mamontov, L. van Eijck, T. Seydel, Z. Phys. Chem. **224**, 33-60 (2010)
4. M. Appel, B. Frick, A. Magerl, Physica B: Condens. Matter **562**, 6-8 (2019)
5. D. Colognesi, M. Celli, F. Cilloco, R. J. Newport, S.F. Parker, V. Rossi-Albertini, F. Sacchetti, J. Tomkinson, M. Zoppi, Appl. Phys. A: Mater. Sci. Process. **74**, S64-S66 (2002)
6. P.A. Seeger, L.L. Daemen, J.Z. Larese, Nucl. Instrum. Methods Phys. Res. A **604**, 719-728 (2009)
7. E. Mamontov, Nucl. Instrum. Methods Phys. Res. A **759**, 83-91 (2014)
8. E. Mamontov, C. Boone, M.J. Frost, K.W. Herwig, T. Huegle, J. Lin, B. McCormick, W. McHargue, A. D. Stoica, P. Torres, W. Turner, Rev. Sci. Inst. **93**, 045101 (2022)
9. P.K. Willendrup, K. Lefmann, J. Neutron Res. **22**, 1-16 (2020)
10. P.K. Willendrup, K. Lefmann, J. Neutron Res. **23**, 7-27 (2021)
11. J.Y.Y Lin, H.L. Smith, G.E. Granroth, D.L. Abernathy, M.D. Lumsden, B. Winn, A.A. Aczel, M. Aivazis, B. Fultz, Nucl. Instrum. Methods Phys. Res. A **810**, 86-99 (2016)
12. J.Y.Y Lin, F. Islam, G. Sala, I. Lumsden, H. Smith, M. Doucet, M.B. Stone, D.L. Abernathy, G. Ehlers, J.F. Ankner, G.E. Granroth, J. Phys. Communications **3**, 085005 (2019)
13. D.W. Kneller, O. Gerlits, L.L. Daemen, A. Pavlova, J.C. Gumbart, Y.Cheng, A. Kovalevsky, Phys. Chem. Chem. Phys. **24**, 3586-3597 (2022)
14. E. Mamontov, Y.Cheng, L.L. Daemen, A.I. Kolesnikov, A.J. Ramirez-Cuesta, M.R. Ryder, M.B. Stone, Chem. Phys. Lett. **777**, 138727 (2021)
15. A.P. Holt, V. Bocharova, S. Cheng, A.M. Kisliuk, G. Ehlers, E. Mamontov, V.N. Novikov, A.P. Sokolov, Phys. Rev. Mater. **1**, 062601 (2017)
16. K.J. Bichler, B. Jacobi, V. Garcia Sakai, A. Klapproth, R.A. Mole, G.J. Schneider, Macromolecules **53**, 9553-9562 (2020)
17. K.J. Bichler, B. Jacobi, V. Garcia Sakai, A. Klapproth, R.A. Mole, G.J. Schneider, Nano Lett. **21**, 4494-4499 (2021)
18. S. Gupta, G.J. Schneider, Soft Matter **16**, 3245-3256 (2020)
19. J.S. Hansen, A. Kisliuk, A.P. Sokolov, C. Gainaru, Phys. Rev. Lett. **116**, 237601 (2016)
20. A. Arbe, P. Malo de Molina, F. Alvarez, B. Frick, J. Colmenero, Phys. Rev. Lett. **117**, 185501 (2016)
21. T. Iwashita, B. Wu, W.R. Chen, S. Tsutsui, A.Q.R. Baron, T. Egami, Sci. Advances **3**, e1603079 (2017)
22. T. Egami, Y. Shinohara, Mol. Phys. **117**, 3227-3231 (2019)
23. S. Perticaroli, G. Ehlers, C.B. Stanley, E. Mamontov, H. O'Neill, Q. Zhang, X.L. Cheng, D.A. Myles, J. Katsaras, J.D. Nickels, J. Am. Chem. Soc. **139**, 1098-1105 (2017)
24. D.V. Wagle, G.A. Baker, E. Mamontov, J. Phys. Chem. Lett. **6**, 2924-2928 (2015)
25. C.Z. Xie, S.M. Chang, E. Mamontov, L.R. Stingaciu, Y.F. Chen, Phys. Rev. E **101**, 012416 (2020)
26. S. Perticaroli, B. Mostofian, G. Ehlers, J.C. Neuefeind, S.O. Diallo, C.B. Stanley, L. Daemen, T. Egami, J. Katsaras, X. Cheng, J.D. Nickels, Phys. Chem. Chem. Phys. **19**, 25859-25869 (2017)
27. M.C. Berg, A.R. Benetti, M.T.F. Telling, T. Seydel, D. Yu, L.L. Daemen, H.N. Bordallo, ACS Appl. Mater. Sci. **10**, 9904-9915 (2018)
28. G.D. Mahan, W.L. Roth, Superionic conductors: [proceedings]. Physics of solids and liquids. 1976, New York: Plenum Press. xvi, 438 p.
29. M.B. Salamon, Physics of superionic conductors. Topics in current physics. 1979, Berlin ; New York: Springer-Verlag. xii, 255 p.
30. J.B. Boyce, B.A. Huberman, Phys. Rep. - Review Section of Physics Letters **51**, 189-265 (1979)
31. W. Dieterich, J. Stat. Phys. **39**, 583-596 (1985)
32. K. Wakamura, K. Hirokawa, H. Shima, K. Takarabe, Solid State Ion. **40-1**, 331-333 (1990)
33. K. Wakamura, F. Miura, A. Kojima, T. Kanashiro, Phys. Rev. B **41**, 2758-2762 (1990)

34. K. Wakamura, K. Hirokawa, K. Orita, J. Phys. Chem. Solids **57**, 75-80 (1996)
35. A. Martinez-Juarez, C. Pecharroman, J.E. Iglesias, J.M. Rojo, J. Phys. Chem. B **102**, 372-375 (1998)
36. N. Kamaya, K. Homma, Y. Yamakawa, M. Hirayama, R. Kanno, M. Yonemura, T. Kamiyama, Y. Kato, S. Hama, K. Kawamoto, A. Mitsui, Nat. Mater. **10**, 682-686 (2011)
37. S.P. Ong, Y.F. Mo, W.D. Richards, L. Miara, H.S. Lee, G. Ceder, Energy Environ. Sci. **6**, 148-156 (2013)
38. Y. Wang, W.D. Richards, S.P. Ong, L.J. Miara, J.C. Kim, Y.F. Mo, G. Ceder, Nat. Mater. **14**,1026 (2015)
39. J.C. Bachman, S. Muy, A. Grimaud, H.H. Chang, N. Pour, S.F. Lux, O. Paschos, F. Maglia, S. Lupart, P. Lamp, L. Giordano, Y. Shao-Horn, Chem. Rev. **116**, 140-162 (2016)
40. J. Janek, W.G. Zeier, Nat. Energy **1**, 16141 (2016)
41. Y. Kato, S. Hori, T. Saito, K. Suzuki, M. Hirayama, A. Mitsui, M. Yonemura, H. Iba, R. Kanno, Nat. Energy **1**, 16030 (2016)
42. X.F. He, Y.Z. Zhu, Y.F. Mo, Nat. Commun. **8**, 15893 (2017).
43. M.A. Kraft, S.P. Culver, M. Calderon, F. Bocher, T. Krauskopf, A. Senyshyn, C. Dietrich, A. Zevalkink, J. Janek, W.G. Zeier, J. Am. Chem. Soc. **139**, 10909-10918 (2017)
44. T. Krauskopf, C. Pompe, M.A. Kraft, W.G. Zeier, Chem. Mater. **29**, 8859-8869 (2017)
45. A. Manthiram, X.W. Yu, S.F. Wang, Nat. Rev. Mater. **2**, 16103 (2017)
46. M.K. Gupta, J. Ding, N.C. Osti, D.L. Abernathy, W. Arnold, H. Wang, Z. Hood, O. Delaire, Energy Environ. Sci. **14**, 6554-6563 (2021)
47. H. Liu, X. Shi, F. Xu, L. Zhang, W. Zhang, L. Chen, Q. Li, C. Uher, T. Day, G.J. Snyder, Nat. Mater. **11**, 422-425 (2012)
48. K.S. Weldert, W.G. Zeier, T.W. Day, M. Panthofer, G.J. Snyder, W. Tremel, J. Am. Chem. Soc. **136**, 12035-12040 (2014)
49. F. Damay, S. Petit, S. Rols, M. Braendlein, R. Daou, E. Elkaim, F. Fauth, F. Gascoin, C. Martin, A. Maignan, Sci. Rep. **6**, 23415 (2016)
50. T.P. Bailey, C. Uher, Current Opinion in Green and Sustainable Chemistry **4**, 58-63 (2017)
51. A.J.E. Rettie, J. Ding, X. Zhou, M.J. Johnson, C.D. Malliakas, N.C. Osti, D. Young Chung, R. Osborn, O. Delaire, S. Rosenkranz, M.G. Kanatzidis, Nat. Mater. **20**, 1683-1688 (2021)
52. A. Villesuzanne, W. Paulus, A. Cousson, S. Hosoya, L. Le Dreau, O. Hernandez, C. Prestipino, M.I. Houchati, J. Schefer, J. Solid State Electrochem. **15**, 357-366 (2011)
53. A. Perrichon, A. Piovano, M. Boehm, M. Zbiri, M. Johnson, H. Schober, M. Ceretti, W. Paulus, J. Phys. Chem. C **119**, 1557-1564 (2015)
54. X.Y. Li, N.A. Benedek, Chem. Mater. **27**, 2647-2652 (2015).
55. R. Hempelmann, Quasielastic neutron scattering and solid state diffusion. Oxford series on neutron scattering in condensed matter. 2000, Oxford: Clarendon Press. xii, 304 p.
56. Y.G. Zhou, S.Y. Xiong, X.L. Zhang, S. Volz, M. Hu, Nat. Commun. **9**, 4712 (2018)
57. B. Li, H. Wang, Y. Kawakita, Q. Zhang, M. Feygenson, H.L. Yu, D. Wu, K. Ohara, T. Kikuchi, K. Shibata, T. Yamada, X.K. Ning, Y. Chen, J.Q. He, D. Vaknin, R.Q. Wu, K. Nakajima, M.G. Kanatzidis, Nat. Mater. **17**, 226 (2018)
58. J.L. Niedziela, D. Bansal, A.F. May, J. Ding, T. Lanigan-Atkins, G. Ehlers, D.L. Abernathy, A. Said, O. Delaire, Nat. Phys. **15**, 73–78 (2019)
59. J. Ding, J.L. Niedziela, D. Bansal, J. Wang, X. He, A.F. May, G. Ehlers, D.L. Abernathy, A. Said, A. Alatas, Y. Ren, G. Arya, O. Delaire, Proc. Natl. Acad. Sci. U. S. A. **117**, 3930-3937 (2020)
60. P. Bron, S. Johansson, K. Zick, J. S. auf der Gunne, S. Dehnen, B. Roling, J. Am. Chem. Soc. **135**, 15694-15697 (2013)
61. A. Kuhn, V. Duppel, B.V. Lotsch, Energy Environ. Sci. **6**, 3548-3552 (2013)
62. A Kuhn, O. Gerbig, C.B. Zhu, F. Falkenberg, J. Maier, B.V. Lotsch, Phys. Chem. Chem. Phys. **16**, 14669-14674 (2014)
63. K. Xu, Chem. Rev. **104**, 4303-4417 (2004)
64. K. Xu, Chem. Rev. **114**, 11503-11618 (2014)
65. M. Armand, J.M. Tarascon, Nature **451**, 652-657 (2008)
66. H.J. Deiseroth, S.T. Kong, H. Eckert, J. Vannahme, C. Reiner, T. Zaiss, M. Schlosser, Angew. Chem.-International Edition **47**, 755-758 (2008)
67. A.R. Rodger, J. Kuwano, A.R. West, Solid State Ion. **15**, 185-198 (1985)
68. V. Thangadurai, S. Narayanan, D. Pinzaru, Chem. Soc. Rev. **43**, 4714-4727 (2014)
69. H. Yamane, S. Kikkawa, M. Koizumi, Solid State Ion. **25**,183-191 (1987)
70. W. Schnick, J. Luecke, Solid State Ion. **38**, 271-273 (1990)
71. W. Schmidt, H.D. Lutz, Berichte Der Bunsen-Gesellschaft – Phys. Chem. Chem. Phys. **88**, 720-723 (1984)
72. O. Bohnke, C. Bohnke, J.L. Fourquet, Solid State Ion. **91**, 21-31 (1996)
73. S. Stramare, V. Thangadurai, W. Weppner, Chem. Mater. **15**, 3974-3990 (2003)
74. M. Naguib, M. Kurtoglu, V. Presser, J. Lu, J. Niu, M. Heon, L. Hultman, Y. Gogotsi, M.W. Barsoum, Adv. Mater. **23**, 4248-4253 (2011)
75. M. Naguib, O. Mashtalir, J. Carle, V. Presser, J. Lu, L. Hultman, Y. Gogotsi, M.W. Barsoum, ACS Nano **6**, 1322-1331 (2012)

76. M. Naguib, V.N. Mochalin, M.W. Barsoum, Y. Gogotsi, Adv. Mater. **26**, 992-1005 (2014)
77. N.C. Osti, M. Naguib, A. Ostadhossein, Y. Xie, P.R.C. Kent, B. Dyatkin, G. Rother, W.T. Heller, A.C.T. van Duin, Y. Gogotsi, E. Mamontov, ACS Appl. Mater. Interfaces **8**, 8859-8863 (2016)
78. Y. Xie, M. Naguib, V.N. Mochalin, M.W. Barsoum, Y. Gogotsi, X. Yu, K.-W. Nam, X.-Q. Yang, A.I. Kolesnikov, P.R.C. Kent, J. Am. Chem. Soc. **136**, 6385-6394 (2014)
79. J.B. Mitchell, N.R. Geise, A.R. Paterson, N.C. Osti, Y. Sun, S. Fleischmann, R. Zhang, L.A. Madsen, M.F. Toney, D. Jiang, A.I. Kolesnikov, E. Mamontov, ACS Energy Lett. **4**, 2805-2812 (2019)
80. R.L. Sacci, M.L. Lehmann, S.O. Diallo, Y.Q. Cheng, L.L. Daemen, J.F. Browning, M. Doucet, N.J. Dudney, G.M. Veith, J. Phys. Chem. C **121**, 11083-11088 (2017)
81. W. Doster, M. Settles, Biochim. Biophys. Act. **1749**, 173-186 (2005)
82. A.P. Sokolov, J.H. Roh, E. Mamontov, V. Garcia Sakai, Chem. Phys. **345**, 212-218 (2008)
83. H. Frauenfelder, G. Chen, J. Berendsen, P.W. Fenimore, H. Jansson, B.H. McMahon, I.R. Stroe, J. Swenson, R.D. Young, Proc. Nat. Acad. Sci. **106**, 5129-5134 (2009)
84. S. Khodadadi, J.H. Roh, A. Kisliuk, E. Mamontov, M. Tyagi, S.A. Woodson, R.M. Briber, A.P. Sokolov, Biophys. J. **98**, 1321-1326 (2010)
85. E. Mamontov, X. Chu, Phys. Chem. Chem. Phys. **14**, 11573-11588 (2012)
86. S. Khodadadi, A.P. Sokolov, Soft Matter **11**, 4984-4998 (2015)
87. J.M. Parks, J.C. Smith, New Eng. J. Med. **382**, 2262-2264 (2020)
88. R.E. Amaro, J. Baudry, J. Chodera, O. Demir, J. A. McCammon, Y. Miao, J.C. Smith, Biophysical J. **114**, 2271-2278 (2018)
89. A. Acharya, R. Agarwal, M.B. Baker, J. Baudry, D. Bhowmik, S. Boehm, K.G. Byler, S.Y. Chen, L. Coates, C.J. Cooper, O. Demerdash, I. Daidone, J.D. Eblen, S. Ellingson, S. Forli, J. Glaser, J.C. Gumbart, J. Gunnels, O. Hernandez, S. Irle, D.W. Kneller, A. Kovalevsky, J. Larkin, T.J. Lawrence, S. LeGrand, S.-H. Liu, J.C. Mitchell, G. Park, J.M. Parks, A. Pavlova, L. Petridis, D. Poole, L. Pouchard, A. Ramanathan, D.M. Rogers, D. Santos-Martins, A. Scheinberg, A. Sedova, Y. Shen, J.C. Smith, M.D. Smith, C. Soto, A. Tsaris, M. Thavappiragasam, A.F. Tillack, J.V. Vermaas, V.Q. Vuong, J. Yin, S. Yoo, M. Zahran, L. Zanetti-Polzi, J. Chem. Inf. Model. **60**, 5832–5852 (2020)
90. M. Smith, J. C. Smith, ChemRxiv, Preprint. https://doi.org/10.26434/chemrxiv.11871402.v4
91. X. Hu, L. Hong, M.D. Smith, T. Neusius, X. Cheng, J.C. Smith, Nat. Phys. **12**, 171-174 (2016)
92. M. Trapp, M. Tehei, M. Trovaslet, F. Nachon, N. Martinez, M.M. Koza, M. Weik, P. Masson, J. Peters, J. R. Soc. Interface **11**, 20140372 (2014)
93. P. Tan, Y. Liang, Q. Xu, E. Mamontov, J. Li, X. Xing, L. Hong, Phys. Rev. Lett. **120**, 248101 (2018)
94. P. Tan, J. Huang, E. Mamontov, V. Garcia Sakai, F. Merzel, Z. Liu, Y. Ye, L. Hong, Phys. Chem. Chem. Phys. **22**, 18132-18140 (2020)
95. S. Li, P. Tan, J. Li, M. Tang, L. Hong, Phys. Rev. Res. **4**, L022003 (2022)
96. E. Persson, B. Halle, Proc. Natl. Acad. Sci. U. S. A. **105**, 6266-6271 (2008)
97. M.L. Martins, A.B. Dinitzen, E. Mamontov, S. Rudić, J.E.M. Pereira, R. Hartmann-Petersen, K.W. Herwig, N.H. Bordallo, Sci. Rep. **9**, 8704 (2019)
98. P. Ball, Proc. Natl. Acad. Sci. U. S. A. **114** 13327-13335 (2017)
99. J. Carvalo, S. Alves, M. Castro, C. Geraldes, J.A. Queiroz, C.P. Fonseca, C. Cruz, J. Pharmacol. Toxicol. Methods **95**, 70-78 (2019)
100. R. Ignazzi, W.P Gates, S.O. Diallo, D. Yu, F. Juranyi, F. Natali, H.N. Bordallo, J. Phys. Chem. C **121**, 23582–23591 (2017)
101. J. Kolosnjaj-Tabi, L. Gibot, I. Fourquaux, M. Golzio, M.-P. Rols, Advanced Drug Delivery Reviews **138**, 56-67 (2019)
102. S. Murdan, J. Control. Release **92**, 1–17 (2003)
103. M.L. Martins, J. Eckert, H. Jacobsen, E.C. dos Santos, R. Ignazzi, D.R. de Araujo, M.-C. Bellissent-Funel, F. Natali, M.M. Koza, A. Matic, E.de Paula, H.N. Bordallo, Int. J. Pharm. **524**, 397–406 (2017)
104. M.L. Martins, R. Ignazzi, J. Eckert, B. Watts, R. Kaneno, W.F. Zambuzzi, L. Daemen, M.J. Saeki, H.N. Bordallo, Sci. Rep. **6**, 22478 (2016)
105. M. Salvador-Castell, M. Golub, N. Martinez, J. Ollivier, J. Peters, P. Oger, Soft Matter **15**, 8381-8391 (2019)
106. M. Golub, B. Lehofer, N. Martinez, J. Ollivier, J. Kohlbrecher, R. Prassl, J. Peters, Sci. Rep. **7**, 46034 (2017)
107. A. Cisse, A.L. Schachner-Nedherer, M. Appel, C. Beck, J. Ollivier, G. Leitinger, R. Prassl, K. Kornmueller, J. Peters, J. Phys. Chem. Lett. **12**, 12402-12410 (2021)
108. E. Mamontov, Y. Cheng, L.L. Daemen, J.K. Keum, A.I. Kolesnikov, D. Pajerowski, A. Podlesnyak, A.J. Ramirez-Cuesta, M.R. Ryder, M.B. Stone, ACS Omega **5**, 21231-21140 (2020)
109. E. Mamontov, Y. Cheng, L.L. Daemen, A.I. Kolesnikov, A.J. Ramirez-Cuesta, M.R. Ryder, M.B. Stone, J. Phys. Chem. Lett. **11**, 10256-10261 (2020)

SHERPA: A Spectrometer with High Energy Resolution and Polarisation Analysis

Kirill Nemkovski[1], *Robert* Bewley[1], *Victoria* García Sakai[1], *Gøran Jan* Nilsen[1], *Adrien* Perrichon[1], and *Ian* Silverwood[1*]

[1]ISIS Neutron and Muon Source, STFC Rutherford Appleton Laboratory, Harwell Science and Innovation Campus, Didcot, Oxfordshire, OX11 0QX, UK

Abstract. SHERPA is a proposed quasielastic neutron spectrometer with polarisation analysis, intended to replace the ageing Iris instrument at the ISIS neutron and muon source. In this paper we present a concept of the instrument along with Monte-Carlo simulations and analysis of possible instrument location. We expect greatly increased count rate compared to Iris (expected from 49 to 660 × Iris) in unpolarised mode and dedicated polarisation analysis capabilities at a more modest count rate increase (~5-70 × Iris). This huge gain in the count rate would be achieved from the combination of three factors: modern neutron guide with high-m coating, and prismatic effect and larger solid angle coverage at the energy analyser. Such an instrument would be the first of its kind and has incredible potential to revolutionise quasielastic neutron scattering technique through the separation of the coherent and incoherent scattering contributions.

1 Introduction

Quasielastic neutron scattering (QENS) is widely used to study dynamics at the atomic and nanoscale in a wide variety of systems including confined materials, [1,2] proteins,[3] polymers,[4,5]ionic liquids [6,7] and solid electrolytes.[8,9] Depending on the scientific problem, it may be important to obtain the purely incoherent or coherent scattering components, which correspond to the individual or collective *and* individual motions of the atoms, respectively. However, since the unpolarised experiment measures the sum of these two contributions, one component can mask the other, thus complicating the analysis of the experimental data.

In many cases, this problem can be effectively solved using neutron polarisation analysis (PA) methods. Recent measurements on direct geometry time-of-flight (ToF) spectrometers such as LET at the ISIS Neutron and Muon Source, UK,[10,11] D7 at the Institute Laue-Langevin, France, [7,12] and the multi-axis crystal spectrometer MACS at the NIST Center for Neutron Research, USA [13] have demonstrated the benefits of PA for this.

On the other hand, many QENS experiments require access to a broad range of energy transfer and a relatively high energy resolution, in order to gain insight into dynamical processes that take place over several decades in time up to nanoseconds. Direct geometry spectrometers have a limit to the resolution they can achieve, with stricter incident monochromation unavoidably leading to lower flux at the sample. Similarly, on triple-axis spectrometers (TAS), a higher energy resolution corresponds to lower values of the incoming/final energies, giving a smaller volume of the resolution ellipsoid and reduced count rate. Counting on single-detector TAS-type instruments is also more time consuming due to the need to scan ***Q***-*E* space point-to-point. Multiple-analyser spectrometers improve the situation, but do not eliminate the issue completely. Polarisation analysis further cuts the count rate and requires more time for measuring different neutron spin orientations. Very high energy resolution combined with some of the advantages of polarized neutrons is offered by the neutron spin echo technique. However, spin echo is also a low count-rate technique.

In contrast, indirect geometry near-backscattering ToF instruments on spallation sources offer benefits in terms of resolution, simplicity, cost, and above all, count rate, but currently there is no such instrument equipped with the ability to exploit polarised neutron analysis.

At ISIS there are two dedicated QENS instruments, Iris [14] and Osiris [15]. In fact, in the period from 2000 to 2005 there was considerable effort to develop PA-QENS on Osiris. However, although polarised diffraction was realised, only proof of concept was obtained with the inelastic mode.[16] Two major upgrades are in progress for Osiris to improve the flux and resolution by replacing the supermirror guide [17,18] and installing a silicon analyser,[19] but neither involve polarisation. In contrast, the future of Iris is still under discussion and the implementation of polarisation analysis is one opportunity being considered.

In this paper we describe a near backscattering spectrometer with dedicated polarisation analysis capabilities as a possible replacement for Iris. It would provide a similar momentum and energy transfer, (***Q***

* Corresponding author: ian.silverwood@stfc.ac.uk

and *E)*, but considerable gains in flux and uniaxial polarisation analysis. The name of the concept is SHERPA: Spectrometer with High Energy Resolution and Polarisation Analysis.

Topically, the main focus of SHERPA is QENS in biological, soft matter and functional materials research. To better understand the requirements for such an instrument from potential users, we organized a dedicated workshop in March 2022. Based on the feedback, we can summarize that the soft matter community is primarily interested in higher energy resolution (< 10 μeV) compared to Iris. Increasing the flux is highly welcome, although not as important as good energy resolution. In opposition, studies of functional materials require rather high flux to be able to work with small samples, while the resolution can stay only moderately high (> 10 μeV). Though not the main target group, interest in high-count rate QENS with PA has also been expressed from the magnetism community. In all cases, it is also important to foresee the provision of sample environment such as furnaces, cryostats, and goniometers to align oriented samples (single crystals, thin films etc.)

2 SHERPA predecessor — Iris

Iris is a near-backscattering indirect geometry ToF spectrometer viewing the hydrogen moderator on Target Station 1 (TS1) at the ISIS Neutron and Muon Facility. It has remained largely unchanged since it was built in 1988.[20] It has a 36 m long neutron guide, which is coated with natural Ni ($m = 1$), ending with a 2.5 m long focusing snout with $m = 2$ coating. The sample position is located 36.5 m away from the moderator. The secondary spectrometer consists of two analyser banks made from pyrolytic graphite (PG) and mica. The corresponding detector banks consist of non-position-sensitive scintillator detectors and cover the scattering angle range of 25-160° for graphite and 25-155° for the mica. The graphite analyser provides an energy resolution of 17.5 μeV with the (002) reflection and is the workhorse setting of the instrument. However, the (004) can be used to access momentum transfers up to 3 Å^{-1} with an energy resolution of 55 μeV. Although mica enables access to higher energy resolution (of 4.5 μeV with 004 and 1 μeV with 002), they are rarely used owing to the much-reduced flux.

3 SHERPA primary spectrometer

The most obvious upgrade to Iris is the complete replacement of the neutron guide with a modern one featuring a high m-value supermirror coating. An initial design has been performed based on recent developments for the "sibling" Osiris instrument.[17,18] Optimization was done via Monte Carlo simulations using the McStas ray-tracing package,[21-23] automated with an in-house MATLAB code. Amongst several guide geometries assessed, a curved guide with elliptical defocusing and focusing sections (see Fig. 1) was found to perform the best. The expected gain in flux is around a factor of 10. Note that the guide also includes the polarising section that will be discussed below.

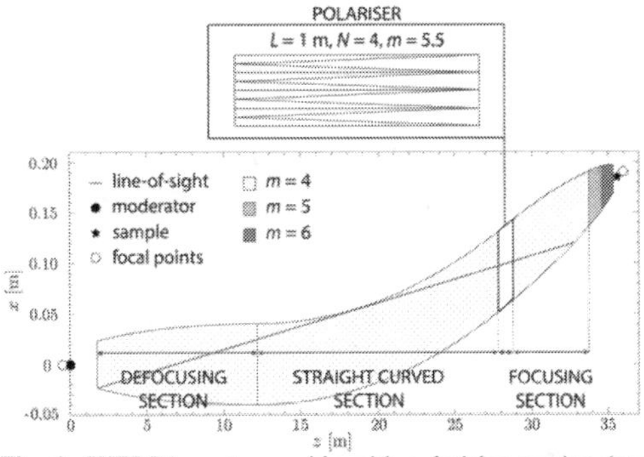

Fig. 1. SHERPA neutron guide with polarising section (see below). Reproduced and adapted from [18]. Model for instrument location on the Iris port.

4 SHERPA secondary spectrometer

As the most used configuration of Iris, we plan to use the (002) reflection from a pyrolytic graphite analyser with a Be filter to suppress higher orders of Bragg scattering.

The detector count rate is proportional to the incoming neutron flux Φ, the solid angle coverage of the analyser $d\Omega$ and wavelength acceptance of the analyser $\Delta\lambda$:

$$I \propto \Phi \cdot d\Omega \cdot \Delta\lambda \quad (1)$$

The wavelength acceptance of the analyser in turn depends on the crystal mosaicity μ, the analyser Bragg wavelength condition λ and $\cot\theta$:

$$\Delta\lambda = \mu \cdot \lambda \cdot \cot\theta \quad (2),$$

where θ is a half of the scattering angle on the analyser.

In backscattering geometry, θ is close to 90° in order to keep $\cot\theta$ very small. To achieve the highest energy resolution. μ should be very small as well. On Iris, θ= 87.5° with $\cot\theta$= 0.04, and μ = 0.8° (see Table 1). However, relaxing the constraint on either of these values would allow for significant gains in intensity. In a conventional backscattering instrument this gain would come alongside a loss in energy resolution. To resolve this, we propose here to use the prismatic effect proposed in the FARO concept,[24] which is already being explored for the future MUSHROOM [25] spectrometer at ISIS. Our initial simulations show that relaxing θ to 75.5° ($\cot\theta$ = 0.26), increasing the PG analysers solid angle by a factor of 3, and changing the mosaicity to 1.5° will achieve a factor 33 gain in intensity. This multiplied by the factor of 10 arising from the guide upgrade, would yield a game-changing total count rate gain of 330, or 660 if PG analysers are placed on both sides of the instrument. We will now demonstrate how this is achieved while simultaneously improving resolution.

The FARO concept is illustrated in Fig. 2(a). It uses the well-known Rowland circle where the sample, analyser and detector all touch the radius of a circle. This focuses neutrons from a point-like sample to a point on the detector with the same final energy while also keeping the final distance travelled almost constant (important for resolution). The large mosaic spread of PG analyser crystals acts just like a prism to light, spatially separating the spread in analysed final energies (or wavelengths), so that as one moves up the position sensitive detector from the bottom the neutron energy is continually changing from higher energy to lower.

Table 1. Comparison (unpolarised) of two SHERPA options (see text) with IRIS and LET. Values for ΔE are given assuming a sample height of 10 mm.

	SHERPA TS1	SHERPA TS2	IRIS	LET
λ (Å)	6.47	6.61	6.64	—
μ (deg)	1.5	1.5	0.8	—
θ	75.5°	82.5°	87.5°	—
$cot\ \theta$	0.26	0.13	0.04	—
$d\Omega$	0.6	0.6	0.2	3.1
Φ	1.0×10^8	2.9×10^7	1.0×10^7	—
Gain	330	49	1	2.5
ΔE (μev)	11.3	5	17.5	21

This effect can clearly be seen in Fig. 2b which shows the McStas simulations; each curve represents the time-of-flight spectrum for a different pixel. Each of these spectra is then analysed separately in the software with its own final energy before being converted to energy transfer, shown in Fig. 3. The FARO approach effectively removes the effect of the large mosaic spread on the energy resolution, as the resolution is dominated now by the angular uncertainties of the sample and pixel sizes, which were 1 cm and 0.5 cm in the simulation. This encoding of the final energy with pixel position allows us to achieve good energy resolution without the usual sacrifice of count rate. Unlike the traditional inverted geometry instrument design, the resolution is however now rather sensitive to the sample geometry. In our example, with 1 cm sample height the simulations show an energy resolution of 11.7 μeV full-width half-maximum (FWHM) (see Table 1, column SHERPA TS1).

The initial simulations considered that SHERPA would replace Iris and would thus be situated on the current beam port viewing the hydrogen moderator on TS1. A possible alternative position is offered by a currently unoccupied beam port on the west side of Target Station 2 (TS2), between the future εMAP and existing IMAT instruments. In this position. SHERPA would take advantage of a much sharper neutron pulse from the solid methane moderator on TS2. This improved resolution and dynamic range, due to the lower repetition rate of TS2, comes at a price, with a reduction in count rate for the TS2 versus the TS1 instrument. Also, to match the improved resolution of the primary spectrometer, the secondary spectrometer needs to be moved further from the sample, from approximately 1 m to 1.3 m, to reduce the sample-size effect on the resolution thus also adding to the cost.

In Fig. 3 we show the resolutions for the TS1 and TS2 options based on our simulations (here we assume that guide transmission is the same for both options). A slight improvement of the line shape of SHERPA on

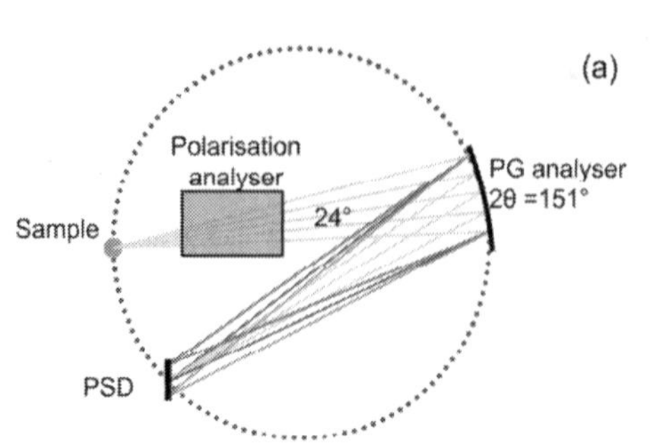

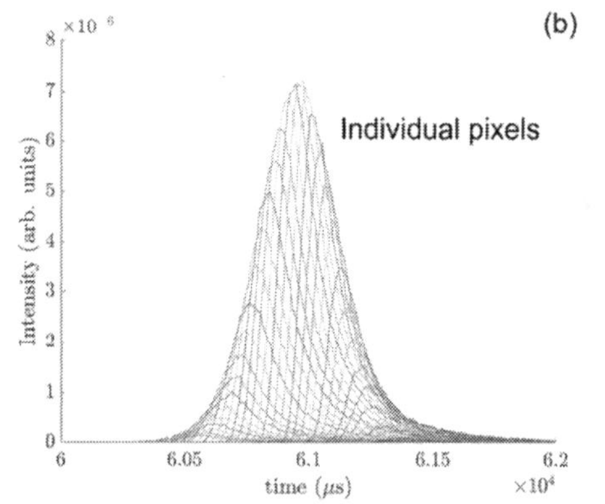

Fig. 2. (a) Prismatic effect on the energy analyser of SHERPA. The rectangular between sample and energy analyser shows schematically the position of the polarisation analyser (discussed in the next section) (b) Spectra for individual pixels.

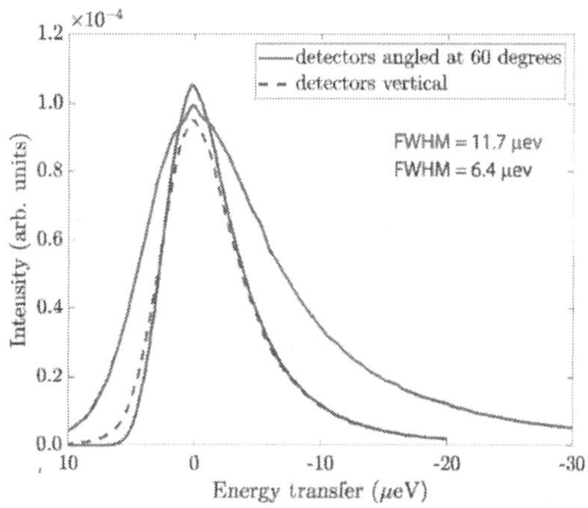

Fig. 3. McStas simulation of SHERPA energy resolution for TS1 and TS2 options, both with a sample at 35 m. The solid red line indicates SHERPA on TS1. The solid blue line is SHERPA on TS2 with detectors tilted 60° towards the Rowland circle, and the dashed blue line is with the detectors vertical.

TS2 can be achieved by tilting the detectors close to the Rowland circle at 60 degrees. The simulated detector intensity in the figure also gives an indication of the relative performance for the same conditions, namely 1 μA proton current on the target. The integral under the TS2 curve is 1.7 smaller and since the current to TS2 is 4 times smaller than to TS1, it means that the SHERPA

 https://doi.org/10.1051/epjconf/202227202004

count rate on TS2 would be approximately 7 times smaller than if installed on TS1.

5 Polarisation analysis at SHERPA

5.1 Polariser

SHERPA is intended to support a uniaxial polarisation analysis mode. Due to its operational simplicity, we have chosen a supermirror polariser, to be installed on a guide changer. The setup is similar to the one used on the LET instrument at ISIS,[26] but an adiabatic fast-passage radio-frequency flipper will be used instead of a precession coil flipper due to its higher performance at long wavelengths and the absence of material in the beam. When SHERPA is use in unpolarised mode, the polarizer will be replaced by a section of regular neutron guide (see Fig. 1.). The cross-section of the polarising section should be exactly the same as a regular one.

We have performed some preliminary optimization of the design using McStas simulations. Results suggest a 1 m long polarising section consisting of 4 channels with absorbing walls in between. Each channel contains a polarising V-cavity, where the polarising supermirrors are coated on both sides with m = 5.5 Fe/Si supermirror. The coating of the outer walls of the device will have $m = 4$, like the non-polarising guide it replaces. The device will be placed in the section of the guide where the divergence is lowest, so as to maximise the wavelength range.

With this set-up, the expected transmission and polarisation for the SHERPA polariser are shown in Fig. 4. In the wavelength range of interest (~6.5 Å, see Table 1) the combined polariser-guide system yields good values for both transmission (~30%) and polarisation (> 95% above the cut-off wavelength).

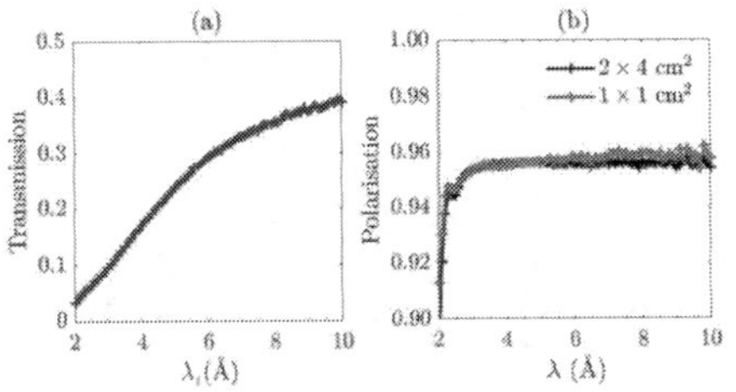

Fig. 4. Transmission of SHERPA polariser and polarisation at the sample position as a function of neutron wavelength. Polarisation is calculated for two characteristic typical sizes: 1x1 cm^2 and 2x4 cm^2 (width × height). The similarity between the curves indicates good homogeneity of the polarisation.

5.2 Analyser

The choice of polarisation analyser is less straightforward: since the final neutron energy will be relatively low (E_f ~ 1.9 meV), both supermirror and ^{3}He-based devices are possible. The latter would either correspond to a replaceable wide-angle cell close to the sample position, as implemented on LET,[7] or an in-situ pumped cell between the analyser and detectors. While the replaceable cell would imply a time-dependent beam polarization and transmission, the in-situ option suffers from the spatial constraints of the secondary spectrometer and limitations on the maximum height of the cell. Regarding supermirror devices, both a multi-channel single-bounce reflection device and multi-channel transmission device are viable.

The former would have the advantage of suppressing unwanted $\lambda/2$ neutrons, although it would require at least one side of the instrument to be permanently polarized and would need to be placed at a large distance from the sample to improve transmission. Besides, with reflection device it is much more challenging to preserve the neutron phase space, whereas for transmission one it is provided naturally. Given these disadvantages, and the inherent operational difficulties of using ^{3}He, a wide-angle transmission-based supermirror device is the current preferred options. This will have similarities with the design presented in [27], although the collimator will be integrated in the analyzer and the channels changed from Z- to V-cavities to suppress spurious multiple reflections (see Fig. 5).

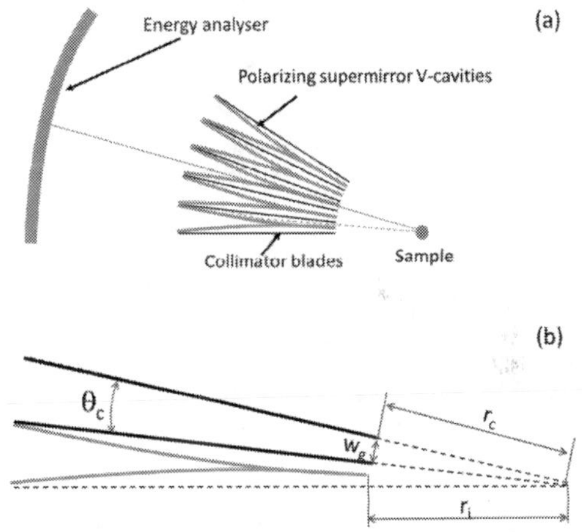

Fig. 5. (a)Schematic layout of the polarisation analyser. Polarizing cavities are shown in green. Black lines in between are collimator blades made from absorbing material. The solid red line indicates the transmitted neutron path from sample to energy analyser. The dashed one shows the path of a neutron with opposite polarization which has been absorbed. (b) Definitions of the analyser geomtertical parameters used in the text and in the next figure: angular opening of the channel θ_c, inner radius of the supermirror assembly r_i, inner radius of the collimator blade r_c. To keep enough space for the notation, in the upper channel we show only collimator blades without the supermirror cavity. Note: r_i and r_c are two different parameters. They not necessarily equal one to another, and in the optimisation calculations are varied independently.

To illustrate the performance of the device, in Fig. 6 (a-c) we plot the dependence of minimum transmitted wavelength from geometrical parameter and sample diameter. Fig.6d shows the relation between channel parameters.

The minimum inner radius is limited by dimensions of standard sample environment to be used (e.g. cryostat). The outer radius is fixed to 0.5 m, which would permit it to be accommodated in the sample area. A possible compromise between polarisation, transmission, and cost is found for a sample diameter of 12.5 mm, inner radius of 0.35 m, and angular opening of 1.3 degrees. The analyser parameters will be fully optimised once a final decision on the analyser type is made. However, it is evident that it will be possible to analyse neutrons at this wavelength even for large sample diameters and compact device dimensions, leaving sufficient space around the sample for sample environment. Irrespective of the final choice of analyser, it is expected that very good performance (a > 0.95, for a 1cm diameter sample, where for analyser efficiency we use the definition of [28]) can be achieved.

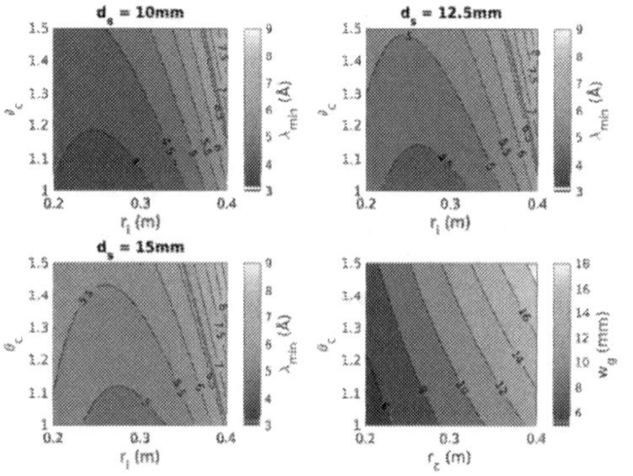

Fig. 6. (a-c) Minimum transmitted neutron wavelength versus the angular opening of the channels (θ_c, see Fig. 5b) and the inner radius of supermirror assembly without radial collimators (r_i) for sample diameter of 10 mm (a), 12.5 mm (b), and 15 mm (c). The final wavelength of the PG (002) analyser is indicated by a solid red line. (d) The gauge width accepted by the device as a function of the angular opening of the channel and the inner radius of the collimating blades.

6 Discussion and concluding remarks

Based on the above-presented simulations, we can summarize that both TS1 and TS2 option provide considerable gain in count rate compared to Iris (see Table 1). Although the gain of 330 at TS1 is particularly impressive, the factor of 49 at TS2 is already enough to get access to the experiments with weaker scattering/smaller samples as requested by the user community. On the other side, energy resolution for TS1 option is just slightly better than the one we have at Iris, while at TS2, SHERPA may have ~2.5 better energy resolution. Besides, positioning SHERPA at TS2 allows to extend the available dynamic range up to 5 meV (instead of 1 meV at current Iris, and the TS1 SHERPA option). All together, these will extend the capabilities of the instrument for the core users of Iris with biological, soft matter and functional materials. Therefore, for the moment we consider building SHERPA at TS2 as the more favourable option.

It is also important to note that completely new functionality is coming with implementation of backscattering PA-QENS. Its increased flux at low energy transfer is of interest to many other research fields. The combination of *routine* PA with ~5-10 μeV-scale resolution makes SHERPA unique and promises to revolutionise the QENS technique.

This work was supported by the Science and Technology Facilities. The authors thank A. Arbe, H. Bordallo, M. Karlsson, A. O'Malley, M. Månsson and M. Nagao for sharing their vision of requirements for polarized QENS instrument, and all participants of the SHERPA workshop and the QENS/WINS conference for fruitful discussion and giving their feedback.

References

1. W.P. Gates, T. Seydel, H. N. Bordallo, Appl. Clay Sci. **201**, 105928 (2021)
2. A. J. O'Malley, M. Sarwar, J. Armstrong, C. R. A. Catlow, I. P. Silverwood. A. P. E. York, I. Hitchcock, Phys. Chem. Chem. Phys. **20**, 11976 (2018)
3. A. Faraone, C. Branca, S. Magazu, G. Maisano, H. D. Middendorf, P. Migliardo, V. Villari, Physica B Condens. Matter **276**, 524 (2000)
4. A. Arbe, F. Alvarez, J. Colmenero, Polymers (Basel) **12**, 3067 (2020)
5. R. J. S. Lima, D. V. Okhrimenko, S. Rudić, M. T. F. Telling, V. García Sakai, D. Hwang, G. Barin, J. Eckert, J-W. Lee, H. N. Bordallo, ACS Appl. Mater. Interfaces **12**, 58161 (2020)
6. F. Nemoto, M. Kofu, M. Nagao, K. Ohishi, S. Takata, J. Suzuki, T. Yamada, K. Shibata, T. Ueki, Y. Kitazawa, M. Watanabe, O. Yamamuro, J. Chem. Phys. **149**, 054502 (2018)
7. T. Burankova, J. F. Mora Cardozo, D. Rauber A. Wildes, J. P. Embs, Sci.Rep. **8**, 16400 (2018)
8. M. Coduri, M. Karlsson, L. Malavasi J. Mater. Chem. A, **10**, 5082 (2022)
9. F. Juranyi, M. Månsson, J. L. Gavilano, M. Mena, E. Pomjakushina, M. Medarde, J. Sugiyama, K. Kamazawa, B. Batlogg, H. R. Ott, T. Seydel, EPJ Web of Conferences **83**, 02008 (2015)
10. A. Arbe, G. J. Nilsen, J. R. Stewart, F. Alvarez, V. García Sakai, J. Colmenero, Phys. Rev. Res. **2**, 022015(R) (2020)
11. G. Cassella, J. R. Stewart, G. M. Paternò, V. García-Sakai, M. Devonport, P. J. Galsworthy, R. I. Bewley, D. J. Voneshen, D. Raspino, G. J. Nilsen, J. Phys. Conf. Ser. **1316**, 012007 (2019)
12. T. Burankova, R. Hempelmann, A. Wildes, J. P. Embs, J. Phys. Chem. B **118**, 14452 (2014)
13. W. Chen, S. Watson, Y. Qiu, J. A. Rodriguez-Rivera, A. Faraone, Phys. B: Condens. Matter **564**, 166 (2019)
14. https://www.isis.stfc.ac.uk/Pages/iris.aspx (accessed 20 October 2022)

15. https://www.isis.stfc.ac.uk/Pages/osiris.aspx (accessed 20 October 2022)
16. M. Telling, private communication
17. A. Perrichon, F. Demmel, Nuclear Inst. and Methods in Physics Research, A 1039, 167014 (2022)
18. A. Perrichon, F. Fernandez-Alonso, M. Wolff, M. Karlsson, F. Demmel, J. Surf. Invest. X-ray, Synchrotron Neutron Tech., **14**, S169 (2020)
19. A. Perrichon, F. Fernandez-Alonso, M. Wolff, M. Karlsson, F. Demmel, Nuclear Inst. and Methods in Physics Research, A **947**, 162740 (2019)
20. C. J. Carlile, M. A. Adams, Physica B **182**, 431 (1992)
21. K. Lefmann, K. Nielsen, Neutron News **10**, 20 (1999)
22. P. Willendrup, E. Fahri, K. Lefmann, Physica B **350**, E735 (2004)
23. P. Willendrup, K. Lefmann, J. Neutron Res. **23**, 7 (2021)
24. R. Bewley, Rev. Sci. Instrum. **90**, 075106 (2019)
25. R. Bewley, Nuclear Inst. and Methods in Physics Research, A **998**, 165077 (2021)
26. J. Košata, G.J. Nilsen, M. Devonport, R.I. Bewley, D.J. Voneshen, P.J. Galsworthy, D. Raspino, J.R. Stewart, Physica B: Condensed Matter **551**, 476 (2017)
27. P. Böni, C. Schanzer, M. Schneider, Nuclear Inst. and Methods in Physics Research, A **966**, 163858 (2020)
28. A. R. Wildes, Neutron News, **17**, 17 (2006)

The correction of Inelastic Neutron Scattering data of organic samples using the Average Functional Group Approximation

Enrico Preziosi[1], *Carla* Andreani[1], *Giovanni* Romanelli[1,*], and *Roberto* Senesi[1]

[1]Physics Department and NAST Centre, Università degli Studi di Roma "Tor Vergata", Via della Ricerca, Scientifica 1, 00133, Roma, Italy

Abstract. The use of the Average Functional Group Approximation for self-shielding corrections at inelastic neutron spectrometers is discussed. By taking triptindane as a case study, we use the above-mentioned approximation to simulate a synthetic dynamic structure factor as measured on an indirect-geometry spectrometer, as well as the related total scattering cross section as a function of incident neutron energy and sample temperature, and the transmission spectra depending on the sample thickness. These quantities, obtained in a consistent way from the Average Functional Group Approximation, are used to calculate the energy-dependent self-shielding correction affecting the sample under investigation. The impact on the intensities of low-energy vibrational modes is discussed, showing that at typical experimental conditions the sample-dependent attenuation factor is about 15% higher compared to the correction at higher energies.

1 Introduction

Inelastic Neutron Scattering (INS) and Quasi-Elastic Neutron Scattering (QENS) are well-established experimental techniques allowing to study the atomic-scale dynamics of condensed-matter systems [1–3]. The investigation of hydrogenous compounds through these techniques has found a naturally successful application, since hydrogen (^{1}H) has the largest bound scattering cross section (82.03 barn) among the elements of the periodic table [4].

Neutron data from hydrogen-containing organic systems collected in INS and, to a lesser extent, QENS experiments, can be subject to non-trivial and sample-dependent attenuation corrections related to the total scattering cross section at thermal neutron energy, also referred to as Thermal neutron Cross Section (TCS). These act as attenuation functions whose intensity can change up to a factor of 4 in the neutron energy range between units and hundreds of meV. As the TCS depends upon temperature as well as upon the structure and dynamics of the system – i.e., the ultimate objectives of the experiment itself - the correction of the experimental spectra becomes, in principle, a very challenging task. The significant change of the TCS affects the INS peak intensities in an energy-dependent way, therefore complicating the accurate comparison with state-of-the-art computer simulations as well as other experiments at different experimental conditions.

Knowing TCSs of the materials of interest is a challenging task, following the dependencies on the molecular structure, dynamics, and temperature. In addition, it is quite uncommon to have the possibility to obtain them from experimental measurements over a broad range of neutron energies, since specific instrumentation (e.g. the VESUVIO beamline at ISIS Neutron and Muon Source, UK [5]) is required. For this reason, experimentally obtained and tabulated TCS are available for just few systems. In order to overcome the impossibility of direct measurements, it is possible to retrieve TCS by calculations from simplified models [6, 7], from molecular dynamics [8], as well as from ab initio calculations [9–13].

In this context, a wide experimental campaign of measurements has been started at the VESUVIO spectrometer [14–16] to investigate TCS of alcohols [17], organic systems [18], water [18, 19] or neutron moderators [20–23]. In particular, in Ref.[24], a new approach for the modelling of TCSs was developed, referred to the Average Functional Group Approximation (AFGA), specifically designed for organic hydrogen-containing systems and based on the incoherent approximation. Within this model, the TCS of large macromolecules and polymers can be estimated, for which a rigorous phonon-based calculation would be prohibitive.

TCS information can be used to correct INS measurements, as presented in [25, 26], via a post-processing procedure referred to as InsCorNorm. This procedure, requiring the knowledge of the energy-dependent TCS of the sample under investigation, allows the correction of the measurements for the experimental conditions, so as to make easier the comparison of the INS experimental spectra with single-molecule or periodic DFT calculations. Moreover, the INSCorNorm algorithm allows to express the experimental INS data on an absolute intensity scale, removing any arbitrary procedure, once the value of the average kinetic energy is known.

Here, we show how the INSCorNorm algorithm in Ref. [25] can be easily adapted so as to take as an in-

*e-mail: giovanni.romanelli@uniroma2.it

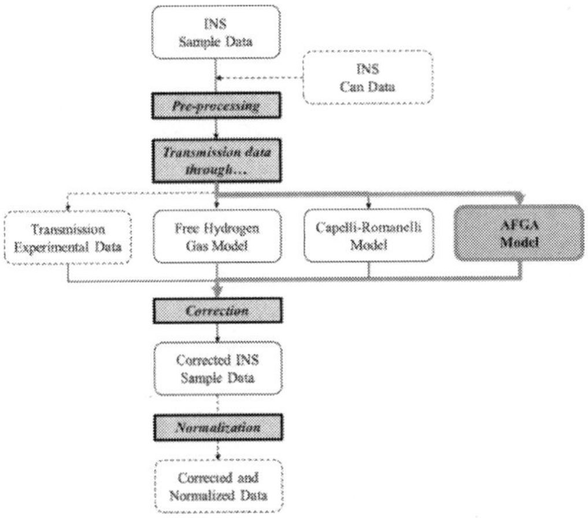

Figure 1. Workflow highlighting the addition of the AFGA method in the INSCorNorm algorithm. Adapted from [25].

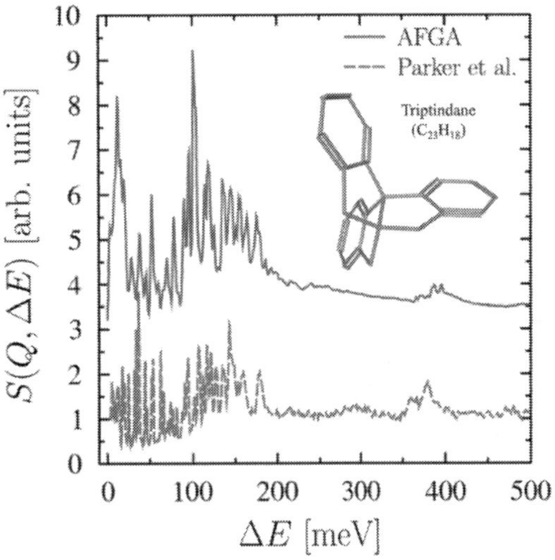

Figure 2. INS spectra of triptindane from Ref[30] (red dashed line) and synthetic spectrum obtained using the AFGA model (green solid line) as a function of the neutron energy transfer. The latter is vertically shifted for the sake of clarity. The figure also shows the structure of the molecule, from [33].

put an AFGA-simulated TCS [24] when the direct measurement of the experimental one is not possible. In Figure 1 the workflow of the InsCorNorm algorithm is integrated with the AFGA method that allows to obtain TCSs of compounds when transmission measurements are not available. With respect to the other methods integrated in the standard INSCorNorm algorithm (namely, Free Hydrogen Gas Model and the model by Capelli and Romanelli [18], see Figure 1), the addition of AFGA Model provides a sample-specific TCS for organic and biological compounds. We take as an example of application triptindane, a molecule at the base of a series of compounds of interest for the industry of hydrogen storage and catalysis [27, 28].

2 Materials and Methods

The triptindane molecule ($C_{23}H_{18}$) was rationalised as composed of 3 CH_2 aliphatic groups and 12 aromatic CH groups. The average hydrogen-projected Vibrational Densities of States (VDoSs), g_H, of a given hydrogen atom in the molecule was calculated as

$$g_H = \frac{1}{18}(6 \cdot g_{CH2} + 12 \cdot g_{CH,aro}) \qquad (1)$$

where g_{CH2} and $g_{CH,aro}$ correspond to the average hydrogen-projected VDoS of a single hydrogen atom in a CH_2 group or aromatic CH group, respectively. The VDoS of specific functional groups as calculated within the AFGA were taken from Ref.[24] (see Supplementary Information therein), and g_H was used to calculate the TCS of the molecule according to the AFGA model. Moreover, we pushed the model a step forward to calculate the double differential scattering cross section as follows. Using such VDoSs as an input and using the multi-phonon expansion (in particular Eqs. 3–7 in Ref.[24]), we calculated the synthetic dynamic scattering function $S(Q, \Delta E)$, as a function of the neutron momentum and energy transfers, Q and ΔE, at a sample temperature of 20 K. We considered an indirect-geometry neutron spectrometer, as well as a scattering angle and final energy equal to those available for the forward-scattering detectors on the TOSCA spectrometer at ISIS [29]. The results, reported in Figure 2, is compared to the experimental data recently measured on TOSCA [30] obtained form the INS TOSCA database [31]. The comparison shows an overall qualitative agreement of the two spectra – a very good result from a model aimed at accurately reproducing total cross sections rather than double-differential ones. However, an interesting feature of the synthetic spectrum is that the low-energy modes seem to be clearly more intense compared to the experimental ones. At the end of our discussion we will relate such difference, at least partially, to the effect of sample self-shielding. It should be noted that a difference in the vibrational intensities, especially at low frequencies, can have an impact on the calculation of the Mean Square Displacement (MSD) and, consequently, on the estimation of the Debye Waller factor that is used in the calculation of the TCS within the multi-phonon expansion. In the present case, at a temperature of 20K, the average MSD from Eq. 1 and the AFGA model is 0.0647 Å^2 for hydrogen atoms. While we could not find an experimental value to compare this result with, we note that in the case of biphenyl investigated in Ref.[32] at room temperature, the relative difference between the calculated and experimental MSDs was found lower than 20%, and the difference should decrease significantly at base temperatures.

Similarly, as anticipated, the total scattering cross section, $\sigma(E)$ for triptindane per formula unit was calculated using the AFGA approximation included within the NCrystal module [34], and it is reported in Figure 3 for

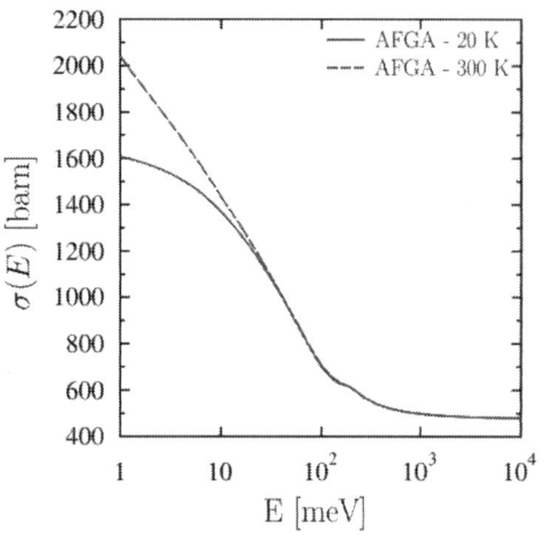

Figure 3. Total cross section per formula unit of triptindane obtained from the AFGA model, corresponding to a sample temperature of 20 K (solid blue line) and 300 K (dashed blue line) as a function of the incident neutron energy.

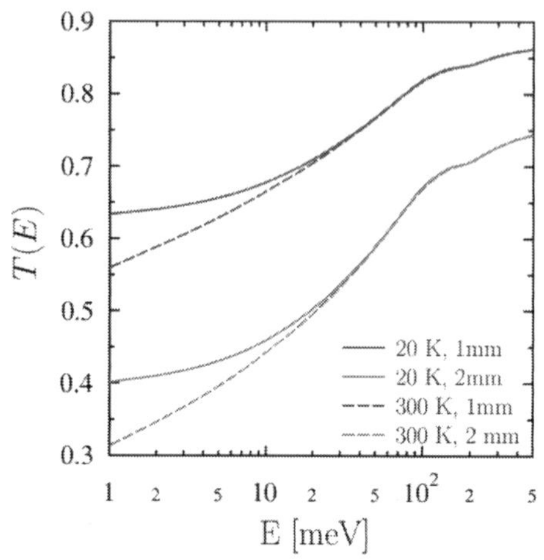

Figure 4. Sample transmission of a sample of triptindane, as obtained from the AFGA model, corresponding to a sample temperature of 20 K (solid lines) and 300 K (dashed lines), and to a sample thickness of 1 mm (blue lines) and 2 mm (gray lines) as a function of the incident neutron energy.

sample temperatures of both 20 K, similar to the one used for experiments on TOSCA, and 300 K. The low-temperature cross section is a clear example of how such function changes drastically as a function of the incident neutron energy, moving from the sum of free scattering cross sections, of hydrogens and carbons in the molecule, at epithermal neutron energies, to the (approximate) sum of bound scattering cross sections, of the same elements, in the cold-neutron energy limit. In the case of the room-temperature sample, it is also worth noting how the low-energy limit of $\sigma(E)$ deviates significantly from the sum of bound scattering cross sections, showing the temperature dependence of the total cross section at enegies below tens of meV.

Finally, sample transmission spectra, $T(E)$, as a function of the incident neutron energy, E, were obtained from the AFGA total cross sections assuming values of the sample thickness, d, of 1 mm and 2 mm, and a bulk sample density of 1.2 g/ml. The results were computed via the Beer-Lambert law,

$$T(E) = \exp\left(-nd\sigma(E)\right), \tag{2}$$

with n the number density, and are shown in Figure 4.

3 Results

Once the AFGA-based synthetic values of $\sigma(E)$, $S(Q, \Delta E)$, and $T(E)$ were generated, they could be used as trial input files for the INSCorNorm algorithm, so as to calculate the self-shielding correction, $f(\theta, E, E_f)$. As discussed in detail in Ref.[25], the self-shielding correction for a flat-geometry sample and a forward-scattering angle θ_f can be approximated as [35]

$$f(\theta, E, E_f) = \frac{T(\theta = 0, E_i) - T(\theta = \theta_f, E_f)}{\log T(\theta = 0, E_i) + \log T(\theta = \theta_f, E_f)}, \tag{3}$$

with

$$T(\theta, E) = \exp\left(\frac{\log T(E)}{\cos\theta}\right). \tag{4}$$

The computed self-shielding corrections for a sample at 20 K and thickness values of 1 mm (blue line) and 2 mm (gray line) are reported in Figure 5. In both cases, the correction causes a marked attenuation of the INS data, enhanced as the sample thickness increases and as the transmission decreases, as one can expect. As a function of the neutron-energy transfer on an indirect-geometry spectrometer, the correction is relatively flat above ca. 200 meV, while it decreases quickly as the energy decreases below ca. 150 meV. Therefore, one should take into consideration two main effects from the sample self-shielding. On the one hand, there is a significant - yet constant - contribution to the attenuation of the INS spectra related to the value of $T(\theta = \theta_f, E_f)$. As the final energy on indirect-geometry instruments is generally around few meV, one can appreciate that the transmission (total cross section) will be approximately at its minimum (maximum) within the dynamic range probed by the instrument. However, being constant in nature, such contribution does not spoil the comparison of the intensities of the experimental VDoS spectra with the ones from computer simulations and models. On the other hand, there is a contribution related to $T(\theta = 0, E)$, related to the incident neutron energy, that may change drastically over the dynamic range available on indirect-geometry INS spectrometers and that is responsible for the drop in the self-shielding correction below ca. 150 meV.

In order to visualise the effect of the self-shielding correction on INS spectra, Figure 6 shows the same synthetic spectrum reported in Figure 2 (dashed green line) together with the same spectrum multiplied by the self-

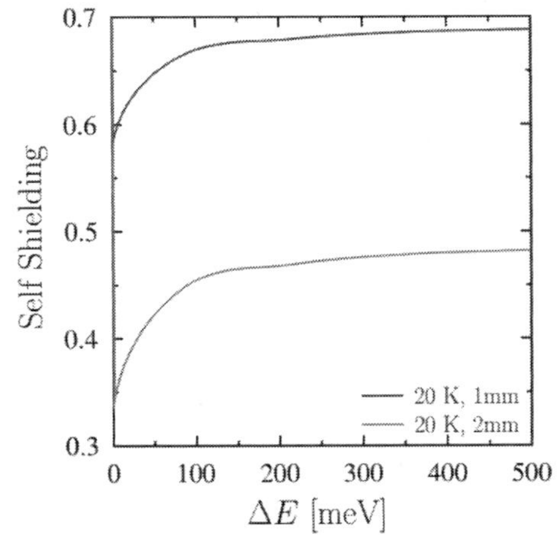

Figure 5. Self shielding correction, as a function of the neutron energy transfer, corresponding to a sample of triptindane, as obtained from the AFGA model, at a temperature of 20 K and for a sample thickness of 1 mm (blue line) and 2 mm (gray line).

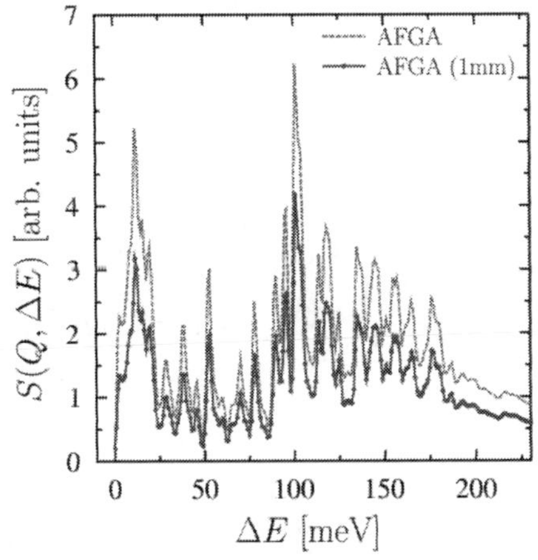

Figure 6. Synthetic INS spectra, as a function of the neutron energy transfer, corresponding to a sample of triptindane as obtained from the AFGA model (green line, the same as in Figure 2), and the same spectrum attenuated for the self shielding correction corresponding to a sample thickness of 1 mm (blue line with circles).

shielding correction for a 1-mm-thick sample at 20 K (blue line with circles). As already mentioned in the previous section, the effect of the attenuation becomes particularly relevant to the intensities of the low-energy modes, which undergo a suppression of about 15% in addition to the almost-constant suppression affecting the intensities at higher neutron energies.

4 Conclusions

In this work we have provided a worked example on how to implement the Average Functional Group Approximation within the INSCorNorm algorithm to obtain an accurate self-shielding correction of organic samples for which an experimental transmission spectrum is not available. We have quantified the effects of self-shielding in the case of tripdindane at 20 K, corresponding, in the case of 1-mm-thick sample, to an additional 15% suppression of the low-energy vibrations to be added to the ca. 30% suppression affecting the vibrations above ca. 200 meV. Moreover, we have discussed how these corrections change depending on the sample thickness and temperature.

In conclusion, the inclusion of accurate sample self-shielding corrections for organic samples, as those available within the AFGA model, is a simple-yet-needed step for the comparison of experimental and theoretical vibrational densities of states, beyond the already successful interpretation of the vibrational energies, towards a quantitative assessment of the vibrational intensities as well.

The authors gratefully acknowledge Regione Lazio (G10795 published by BURL n. 69, 27/08/2019) and University of Rome Tor Vergata for the financial support to ISIS@MACH Regional Project. The Authors gratefully acknowledge the financial supports of the Consiglio Nazionale delle Ricerche - within CNR-STFC Grant Agreement [2014-2020 (n. 3420)] concerning collaboration in scientific research at the ISIS (UK) of STFC - and of the JRU ISIS@MACH ITALIA, Research Infrastructure hub of ISIS (UK) - MUR official registry U. 0008642.28-05-2020.

References

[1] P.C.H. Mitchell, *Vibrational spectroscopy with neutrons: with applications in chemistry, biology, materials science and catalysis*, Vol. 3 (World Scientific, 2005)

[2] F. Fernandez-Alonso, D.L. Price, *Neutron Scattering* (Academic Press, 2013)

[3] K. Drużbicki, M. Gaboardi, F. Fernandez-Alonso, Polymers **13**, 1440 (2021)

[4] V.F. Sears, Neutron news **3**, 26 (1992)

[5] *Isis neutron and muon source*, ISISNeutronandMuonSource, accessed July 2021

[6] J. Granada, Phys. Rev. B **31**, 4167 (1985)

[7] V. Laliena, J. Dawidowski, G. Cuello, J. Campo, Nucl. Instrum. Meth. A **993**, 165071 (2021)

[8] J.M. Damian, D.C. Malaspina, J.R. Granada, J. Chem. Phys. **139**, 024504 (2013)

[9] A. Hawari, I. Al-Qasir, V. Gillette, B. Wehring, T. Zhou, Proc. PHYSOR 2004 pp. 25–29 (2004)

[10] A. Hawari, Nucl. Data Sheets **118**, 172 (2014)

[11] X.X. Cai, E. Klinkby, New J. Phys. **19**, 103027 (2017)

[12] J. Wormald, A.I. Hawari, *Thermal neutron scattering law calculations using ab initio molecular dynamics*,

in *EPJ Web Conf.* (EDP Sciences, 2017), Vol. 146, p. 13002

[13] Y. Cheng, A.J. Ramirez-Cuesta, J. Chem. Theory Comput. **16**, 5212 (2020)

[14] G. Romanelli, M. Krzystyniak, R. Senesi, D. Raspino, J. Boxall, D. Pooley, S. Moorby, E. Schooneveld, N. Rhodes, C. Andreani et al., Meas. Sci. Technol. **28**, 095501 (2017)

[15] J.I. Robledo, J. Dawidowski, J.M. Damián, G. Škoro, C. Bovo, G. Romanelli, Nucl. Instrum. Meth. A **971**, 164096 (2020)

[16] P. Ulpiani, G. Romanelli, D. Onorati, A. Parmentier, G. Festa, E. Schooneveld, C. Cazzaniga, L. Arcidiacono, C. Andreani, R. Senesi, Rev. Sci. Instrum. **90**, 073901 (2019)

[17] L.R. Palomino, J. Dawidowski, J.M. Damián, G.J. Cuello, G. Romanelli, M. Krzystyniak, Nucl. Instrum. Meth. A **870**, 84 (2017)

[18] S.C. Capelli, G. Romanelli, J. Appl. Crystallogr. **52**, 1233 (2019)

[19] C. Andreani, R. Senesi, M. Krzystyniak, G. Romanelli, F. Fernandez-Alonso, Riv. Nuovo Cimento **41**, 291 (2018)

[20] G. Romanelli, S. Rudić, M. Zanetti, C. Andreani, F. Fernandez-Alonso, G. Gorini, M. Krzystyniak, G. Škoro, Nucl. Instrum. Meth. A **888**, 88 (2018)

[21] F. Cantargi, J. Dawidowski, C. Helman, J.I.M. Damian, R.J. Granada, G. Romanelli, J.G. Cuello, G. Skoro, M. Krzystyniak, *Validated scattering kernels for triphenylmethane at cryogenic temperatures*, in *EPJ Web Conf.* (EDP Sciences, 2020), Vol. 239, p. 14002

[22] G. Škoro, G. Romanelli, S. Rudić, S. Lilley, F. Fernandez-Alonso, *Discovery of new neutron-moderating materials at isis neutron and muon source*, in *EPJ Web Conf.* (EDP Sciences, 2020), Vol. 239, p. 17008

[23] L.R. Palomino, J. Dawidowski, C. Helman, J.M. Damián, G. Romanelli, M. Krzystyniak, S. Rudić, G.J. Cuello, Nucl. Instrum. Meth. A **927**, 443 (2019)

[24] G. Romanelli, D. Onorati, P. Ulpiani, S. Cancelli, E. Perelli-Cippo, J.I.M. Damián, S.C. Capelli, G. Croci, A. Muraro, M. Tardocchi et al., J. Phys. Condens. Mat. **33**, 285901 (2021)

[25] C. Scatigno, G. Romanelli, E. Preziosi, M. Zanetti, S.F. Parker, S. Rudic, C. Andreani, R. Senesi, J. Chem. Theory Comput. **16**, 7671 (2020)

[26] D. Colognesi, J. Neutron Res. **19**, 147 (2017)

[27] J. Vile, M. Carta, C.G. Bezzu, B.M. Kariuki, N.B. McKeown, Polymer **55**, 326 (2014)

[28] A.M. Dilmaç, T. Wezeman, R.M. Bär, S. Bräse, Nat. Prod. Rep. **37**, 224 (2020)

[29] R.S. Pinna, S. Rudić, M.J. Capstick, D.J. McPhail, D.E. Pooley, G.D. Howells, G. Gorini, F. Fernandez-Alonso, Nucl. Instrum. Meth. A **870**, 79 (2017)

[30] S.F. Parker, L. Zhong, M. Harig, D. Kuck, Phys. Chem. Chem. Phys. **21**, 4568 (2019)

[31] *Isis ins database*, http://wwwisis2.isis.rl.ac.uk/INSdatabase/Theindex.asp, accessed July 2021

[32] P. Ulpiani, G. Romanelli, D. Onorati, M. Krzystyniak, C. Andreani, R. Senesi, Nuovo Ciment. C **44**, 1 (2021)

[33] *Triptindane, pubchem identifier: Cid 12444533*, https://pubchem.ncbi.nlm.nih.gov/compound/Triptindane, accessed August 2021

[34] K. Ramić, J.I.M. Damian, T. Kittelmann, D.D. Di Julio, D. Campi, M. Bernasconi, G. Gorini, V. Santoro, Nucl. Instrum. Meth. A **1027**, 166227 (2022)

[35] A.K. Agrawal, Phys. Rev. A **4**, 1560 (1971)

A sample holder for simultaneous neutron and dielectric spectroscopy -

dielectric tests with glycerol, glycerol-water, water and phosphoric acid

Bernhard Frick[1,*], *Margarita* Fomina[1,2], *David* Noirat[1,3], *Henriette W* Hansen[3], *Markus* Appel[1,**], and *Kristine* Niss[3,***]

[1]Institut Laue-Langevin (ILL), CS 20156, F - 38042 Grenoble Cedex 9, France
[2]Jülich Centre for Neutron Science (JCNS) at Heinz Maier-Leibnitz Zentrum (MLZ), Forschungszentrum Jülich GmbH, Garching, Germany
[3]Glass and Time, IMFUFA, Department of Science and Environment, Roskilde University, Postbox 260, DK-4000 Roskilde, Denmark

Abstract. We report on dielectric test measurements of a rectangular flat sample holder which serves as capacitor and which is aimed for simultaneous neutron and dielectric (n-DE) spectroscopy of acidic liquid samples. We describe technical details of the sample holder assembly and the dielectric and neutron equipment as well as the sample preparation procedure of the air sensitive acidic samples. The sample holder was characterised off-line from the neutron spectrometer by dielectric spectroscopy, but using the standard IN16B cryofurnace with a dielectric sample stick with 4-wire connection and a Novocontrol - equipment, previously setup by a collaborative effort between ILL and Roskilde University. Temperature-dependent dielectric scans on standard samples (glycerol, glycerol-water, and Milli-Q water) were measured in the frequency range between 0.27 Hz and 1 MHz. Step-like temperature changes allowed to probe the temperature equilibrium conditions and continuous temperature changes were made to mimic typical IN16B backscattering neutron fixed window scans. Both type of scans were carried out in cooling and in heating. The standard samples show that our dielectric setup with flat sample holder is well suited for simultaneous n-DE-experiments. On the other hand, the dielectric scan on phosphoric acid reveals the limitations of our setup in case of high sample conductivities, but also shows that the DC-conductivity can still be accessed in a sufficiently wide low temperature range where the onset of conductivity can be simultaneously probed with the change in proton dynamics as seen by neutron spectroscopy.

1 Introduction

Quasielastic neutron scattering (QENS) [1] and dielectric spectroscopy (DES) [2] are complementary techniques and the application of both is of interest for many areas of science. QENS offers both time and space resolution for characterising molecular motions, whereas DES probes the time- or frequency dependence of electrical dipoles. QENS thus has the potential to inform about the type and time scale of atomic or molecular motions, like e.g. on the diffusional behaviour of ionic or non-ionic species. Regarding the dynamics of ionic species it is known that the QENS signal is not always straightforward related to charge displacements. In contrast DES can probe at lower frequency the DC-conductivity and at higher frequency the transition to AC-conductivity arising from the motion of ionic species. Thus it is obvious that a combination of the two techniques does give deeper insight and this has been widely used by comparing results from dedicated experiments applying these complementary techniques.

Standard broad band DES covers an extremely wide frequency range of more than 9 orders of magnitude in a single experimental setup, like from the μHz to above-MHz region, whereas standard QENS offers typically only 2-3 decades between energy resolution and maximum energy transfer, thus requiring often a combination of different instruments or experimental setups. Even though these methods are probing different time scales, i.e. hundreds of seconds to 10^{-7}s for standard DES and 10^{-7}s to sub-picoseconds for QENS, it is in some cases of interest to carry out experiments simultaneously, in spite of evidence that dedicated individual experiments may deliver superior data quality. Scientific cases where simultaneous neutron and dielectric (n-DE) measurements are expected to give more information include i) non-equilibrium systems in general, ii) systems where a connection between low frequency dipole dynamics, probed by DES, and high frequency atomic motion, probed by QENS, is suspected, such as the hypothesis about a general connection between fast and slow dynamics [3], or more specific, iii) samples with crystallisation tendency or in metastable states which depend on the sample history, where it would be difficult to prepare the samples under exactly the same conditions in two separate experiments. Other more practical reasons could be parameter scans, like temperature or pressure scans, where kinetic delays between the measured

*e-mail: frick@ill.eu
**e-mail: appel@ill.eu
***e-mail: kniss@ruk.dk

parameter value and the actual intrinsic sample parameter value may differ and cause a non-equilibrium situation or samples with crystallisation tendency for which the interest is to measure with both methods exactly under the same conditions.

Simultaneous QENS and DES (called hereafter 'n-DE') experiments were started at the Institut Laue-Langevin some years ago by the group of K. Niss [4], mainly for n-DE experiments under pressure with a n-DE pressure cell allowing to access P=400MPa in a broad temperature range [5] and with Roskilde-owned DE-equipment (standard LCR meter 4284A from Agilent) from which emerged several publications (e.g. [6],[3],[7]). Recently, in continuation of this Roskilde-ILL collaboration, ILL has acquired a Novocontrol Alpha Analyser [8] which is now integrated into ILL's NOMAD data acquisition chain [9] and has built a sample stick with connectors for four-wire DES measurements fitting into the standard 50 mm diameter cryofurnaces and cryostats. Here we are reporting on tests of a flat sample holder, usable for ambient pressure experiments on acidic samples in a wide temperature range between T = 2 K and 340 K. We describe the sample holder assembly and off-line dielectric temperature scans, carried out under the same conditions and with the same equipment as found normally in fixed window scans [10] on IN16B [11]. As test samples we have chosen the well known (non-acidic) samples glycerol, glycerol-water and Milli-Q water, background reference samples as well as phosphoric acid water mixtures (PA-water) on which lies our main focus of future simultaneous n-DE experiments. We briefly describe the sample preparation procedure for the very air sensitive samples of PA and the practicality for user experiments. Dielectric scans were taken in a frequency range from 0.27 - 1 MHz and at temperatures between 2 and 340 K. Our tests show that our sample holder design is well suited for dielectric measurements at low temperatures as long as the sample conductivity and electrode polarisation is not too high. Both reduce the frequency and temperature range for measuring the conductivity with our dielectric setup. As our main interest of future work is to test at low temperatures for phosphoric acid - water mixtures a possible connection between the onset of conductivity (from the DC conductivity of DE spectra) and high frequency dynamics of proton displacement (from neutron fixed window scans, e.g. on IN16B or IN13), this restriction is acceptable.

2 Methods

2.1 Dielectric equipment and measured quantities

The dielectric equipment, which is available at ILL for simultaneous n-DE experiments [12], is based on a Novocontrol Alpha-AT Analyser, controlled by a PC running WINDETA software [8] and which can be triggered by the ILL instrument software NOMAD [9] to run simultaneous with the instrument data acquisition. With dedicated dielectric equipment the Novocontrol analyser covers a wide frequency range in the μHz- to above-MHz region. Our test measurements were carried out in a restricted frequency range from 0.27 Hz to 1 MHz. The DE-analyser is connected via a dispatch box to an ILL built sample stick, which is more than 1 m high and is usable with standard ILL top loading 'Orange cryostats' having a 50 mm diameter bore. For our tests we have used the standard IN16B cryofurnace. The length of the sample stick and the sample orientation is adjustable and serves for placing the sample in the centre of the neutron beam. The cables for 4-wire measurements are fed through the central tube of the sample stick and are at their end welded in pairs to two pin connectors. These can then be plugged onto the pins fixed on the electrodes of any type of sample holder. The sample holder is connected with an intermediate electrically insulating PEEK adapter via an M8 thread to the end of the sample stick and is thus located just below the place where the sample temperature is measured.

A detailed description of broad band dielectric spectroscopy can be found in the book by Kremer and Schönhals [2]. DES derives the electrical properties from a measurement of the complex impedance Z* of a sample placed between electrodes. From the measured impedance one can find the complex capacity C* which is related to the geometry of the sample cell by $C \propto \frac{A}{d}$, with A being the sample area and d the sample thickness. Thus for DES a large sample area A and a small sample thickness d result in a large measurable signal. From the complex capacity C* of the sample cell one can derive the sample permittivity $\epsilon^* = \frac{C^*}{C_0}$, where C_0 refers to the empty sample cell capacity, and the complex dielectric permittivity is given by $\epsilon^* = \epsilon' - i\epsilon''$. From the complex permittivity one can then calculate the complex conductivity $\sigma^* = \sigma' + i\sigma''$, which is related to the complex permittivity by $\sigma^* = i\omega\epsilon_0\epsilon^*$ (ϵ_0 is the dielectric permittivity of vacuum).

2.2 Sample holder and requirements for simultaneous n-DE experiments

For neutron scattering basics we refer to standard books, e.g. [13], [14] or [1] and discuss only the requirements for a neutron sample holder. For maximum intensity the projection of the sample holder area A should cover the incoming neutron beam area, which on the instruments used is typically about 10 -30 mm wide and 20 - 50 mm high. The sample thickness should be chosen thin enough to favour single over multiple scattering processes. For that reason a rule of thumb is to aim for a scattering probability of less than 10% , equal to a transmission larger than 90% if absorption is neglected. For protonated materials, due to the high incoherent cross section from Hydrogen, a sample thickness d for straight neutron transmission should therefore be in the order of 150 μm and can be thicker for other samples. Both, A and d required for n-scattering, are thus compatible with the requirements for DES described above.

Two major sample geometries are used for QENS, a flat sample holder which can be oriented with respect to the incident and scattered neutron beam and thus with respect to the scattering vector **Q** and a hollow cylinder geometry for which the neutrons pass twice through the annular sample thickness. There are standard designs for both types at ILL (and similar design elsewhere), which allow to adapt easily to different sample thickness.

We limit ourselves to the flat sample holder design which we have modified to suit simultaneous n-DE-spectroscopy. A cylindrical sample holder for n-DE spectroscopy will be described elsewhere [15]. For completeness we should mention an earlier sample cell development which was aimed for simultaneous neutron diffraction (different requirements than for QENS) and dielectric spectroscopy [16].

For adjusting to different sample thickness, one part of the sample holder remains usually unchanged and the counter part is slightly changing its dimensions: for flat sample holders it is the base of the holder which remains unchanged and top cover plates of different depth are used which dive into the base plate sample holder (see Fig.1) and leave a gap defining the desired sample thickness between 0 - 3 mm. The standard flat sample holders are made from Aluminium (small scattering cross section for wavelength below the Bragg cut-off) and can be sealed with Indium wire, squeezed in between the outer more massive frames which are held together by Aluminium screws. For dielectric spectroscopy these two parts of the sample holder (base and top) serve as electrodes.

Modifications which were made to this type of flat sample holder are the following 1: i) We have added by cold fitting tiny electrical connector pins on the upper edge of the two sample holder pieces. After screwing the sample holder to the sample stick, these allow to connect the wires for the DE- measurement. ii) We isolate electrically the electrodes (top and base) from each other by thin Teflon sheets on which the central part is cut out in order to leave most of the neutron window free. When closing the sample holder, the top cover pulls the thin teflon foils into the base plate so that it is lining equally the side walls of the sample volume. We have used 2 or 3 such very thin window foils of only $10\mu m$ thickness. This minimises the risk of shearing off the teflon foil when closing the sample holder. It also minimises the neutron background arising from the coherent scattering (Bragg peaks) of Teflon (PTFE), which is important because minor parts of the foil near the border of the neutron window might still reach into the neutron beam. iii) For sealing liquid samples we squeeze an Indium wire, placed in between the foils around the sample volume. The PTFE foils also protect the Indium from contact with the sample as we aim for measuring acidic liquids. At first Aluminium screws are used for sealing the sample cell (squeezing the Indium wire), which after tight closing are then exchanged one by one by electrically insulating Nylon screws. iv) Our aim is to measure acidic samples which would corrode the Aluminium quite fast. Therefore the neutron windows with their side walls were coated on the inside of the sample holder with a very thin Au layer ($0.34\mu m$) in order to minimise neutron activation of the sample holder. v) As the major source for potential neutron background stems from the nylon screws (protonated carbon chain) we attach, for the neutron scattering experiments only, a Cadmium mask on top of the sample holder facing the neutron beam, taking care not to create electrical contact between the electrodes.

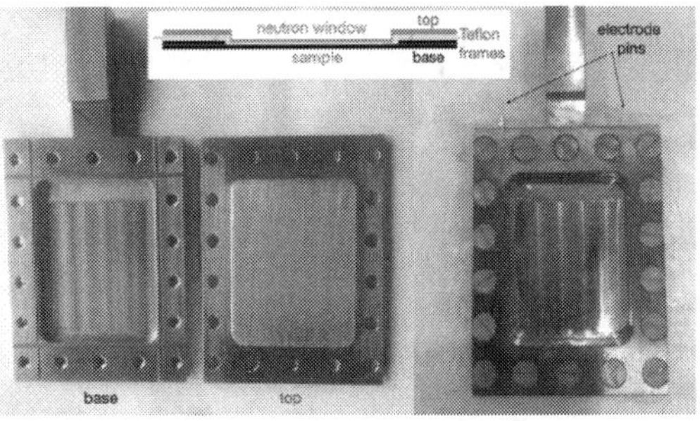

Figure 1. left photo: gold coated sample holder: bottom part (l.h.s.) and top cover (r.h.s.); right photo: filled sample holder with Nylon screws and electrode pins on top. Inset: schematic view how the sample is squeezed between base and top plate and how the 3 thin Teflon foils are placed.

2.3 Sample preparation

All air sensitive samples were handled within a glove bag under Argon flow, including in a few cases melting the crystalline samples on a heat plate, weighing the appropriate sample amount and pipetting it into the sample holder. Sample handling took place only after a low humidity level had been reached (typically 2.5% RH at about 20C). RH and temperature were registered continuously close to the sample with a Sensirion humidity sensor. The exposure time of the samples to the dry atmosphere was minimised to several ten seconds until the sample holder was closed and initially kept together with a special tool until it was definitively closed with screws outside of the Argon bag.

Critical and delicate phases are the preparation of the sample holder with the very thin Teflon foils and the Indium wires in between, the transfer of this assembly to the glove bag, and the closing of the sample holder within the bag, where the foils are hardly anymore visible. During closing the sample holder with screws, care had to be taken not to pull the Teflon foils with the screws into the threads which could rupture the foils and create electrical contact. Punching holes into the very thin foils before assembly seemed difficult, therefore we have perforated the foils with a needle in the center of each screw hole after assembly and then placed the Aluminium screws. In some cases a short circuit was detected after the sample assembly and then the described tedious procedure had to be repeated. In that respect there is room for improving the sample holder design and handling.

3 Dielectric tests on standard samples

In order to test our sample holder design we have carried out dielectric spectroscopy on well known samples and on empty sample holders in the IN16B standard cryofurnace and with the sample stick described above. As test samples we have used Milli-Q-water, glycerol and a glycerol-water mixture. Finally we report on a phosphorous acid - water mixture, belonging to the group of samples which meet closest our main objective, namely measuring the low temperature conductivity of PA-water systems simultaneously with neutron spectroscopy. Systematic measurements on this series will be reported elsewhere. Here we show that our experimental setup is well suited for dielectric measurements on standard samples. But we also point out the experimental limitations of our setup for samples with extremely high conductivity similar to that of PA.

3.1 Tests on glycerol

As an example for the DE-data quality achievable with our n-DE-setup we show in Fig.2 dielectric spectra measured in a frequency range from 2.710^{-1} – 10^6 Hz on glycerol between 209 K and 240 K at temperatures similar to the ones reported in literature (e.g. [17]) which show the α-relaxation peak in the imaginary part of the dielectric permittivity ϵ''.

For two temperatures (T = 209 K and 219 K) we show in Fig.3 fits of the α-relaxation peak observed in the imaginary part of the dielectric permittivity ϵ'' of glycerol. We have fitted ϵ'' applying a single Havriliak-Negami (HN) model fit function with an additional conductivity term, resulting in HN shape parameters $\beta = 0.97$ and $\gamma = 0.48$. β is found to be very similar to the value given in Ref.[2] (β = 0.95), but the smaller γ found by us (γ= 0.6 in Ref.[2]) indicates a lower slope in the high frequency wing. We attribute this to the fact that we have not fitted an additional HN-function to the high frequency wing as it was done in Ref.[2].

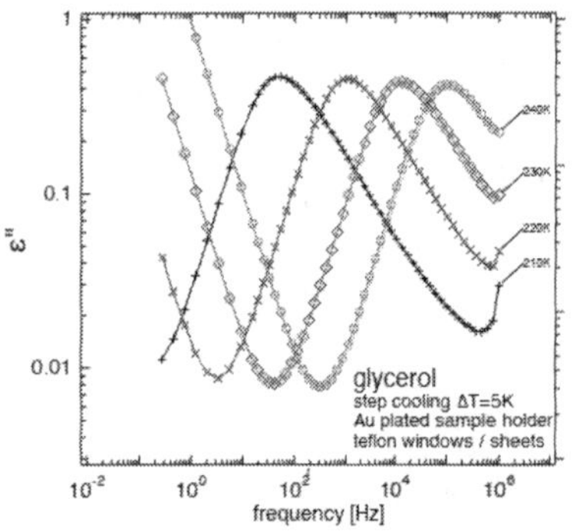

Figure 2. Frequency dependence of the imaginary permittivity of a glycerol sample, measured in a gold coated flat sample holder, placed in an IN16B cryofurnace, suited for simultaneous neutron scattering.

In neutron spectroscopy a powerful way to get an overview over the onset of different relaxation processes as a function of the sample temperature or pressure are so-called (elastic or inelastic) fixed window scans (EFWS, IFWS, FWS) [10]. Such FWS are most efficiently carried out on reactor backscattering spectrometers using a linear-motor Doppler drive, where the incident neutron

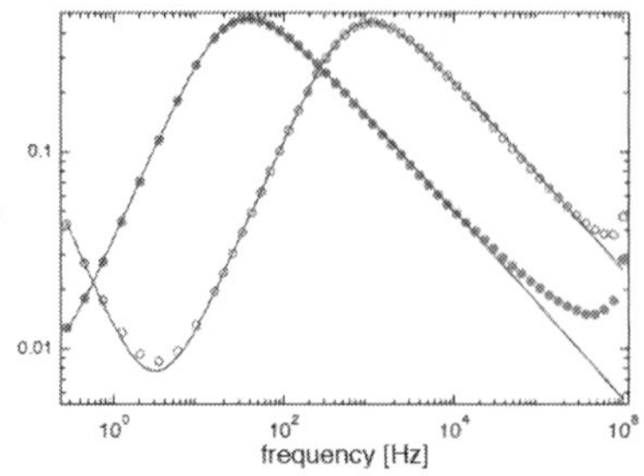

Figure 3. Test fits with a HN-function including conductivity of the α-relaxation peak of glycerol at T = 209 K and T = 219 K.

wavelength $\lambda_{inc}(t) = \lambda_M \pm \Delta\lambda_D = \lambda_M \mp \frac{h}{m_n}\frac{1}{v_D(t)}$ can be flexibly set offset with respect to the fixed wavelength $\lambda_M = \lambda_A$ of the static monochromator and the analysers by choosing an appropriate periodic Doppler velocity profile $v_D(t) \approx \pm const.$ for the backscattering monochromator which, when discarding the regions near the reversal points, is constant [10].

This way most of the neutron detection time is spent in the corresponding energy transfer channel $E = \frac{\hbar^2}{2m_n}((1/\lambda_A)^2 - (1/\lambda_{inc})^2)$ which is probed with sub-μeV energy resolution and thus with the longest relaxation times currently available on neutron spectrometers operating in frequency space. In view of the short neutron measuring times and the superior statistics of FWS compared to QENS, carrying out the fast DE-spectroscopy simultaneously with FWS might be of highest interest (e.g.[18]).

In order to prepare for such n-DE-experiments we have carried out off-line temperature scans on glycerol in cooling from 330 K to 15 K with a continuous cooling ramp of -0.5 K/min (see Fig.4a), followed by continuous heating (Fig.4b) with the same ramp speed of 0.5 K/min using a gold plated flat sample holder and the IN16B standard cryofurnace. In a ramp the sample temperature set point was adapted continuously every 6 seconds. A frequency sweep for the dielectric spectra was chosen to last about 2 min in the range 0.2 to 1 MHz, followed by a waiting time of 1 min, which should be compared with neutron measuring times (not reported here) of about 30 s for an EFWS and 60-180 s for an IFWS step. The time for a DE frequency sweep would become much longer and would no longer match a temperature step of the FWS if lower frequencies were chosen (e.g. one period at 1 μHz equals 1000 s).

For both sets of scans it can be seen in Fig.4 how the -relaxation peak moves in the intermediate T-range through the dielectric frequency window. With increasing temperature the spectra are dominated towards lower frequency by strong DC conductivity leading to $\epsilon'' \propto 1/\omega$. The spectra in Fig.4 show also the limitations of our experimental setup at low temperatures and at frequencies above 10^5 Hz where the systematic errors increase (see also empty can measurements). Some temperatures are highlighted by colours to show their different behaviour.

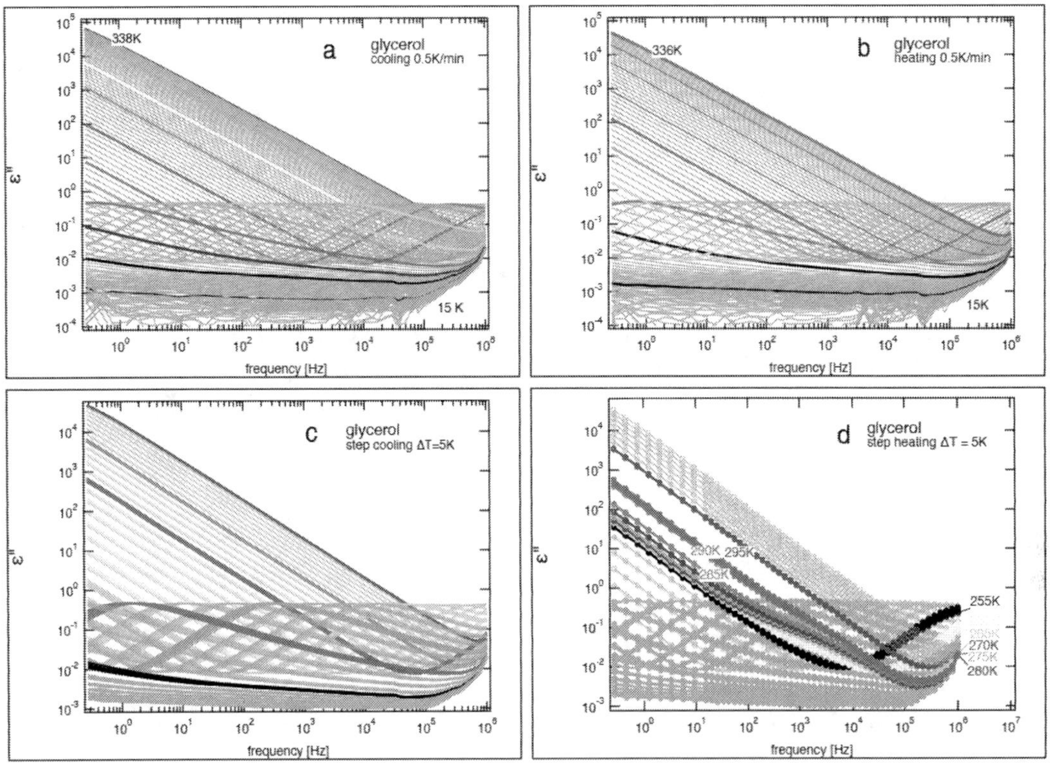

Figure 4. Frequency dependent dielectric permittivity at temperatures measured in: (a) continuous cooling scan with a ramp of 0.5 K/min, (b) continuous heating with the same rate, (c) in step cooling with ΔT=5K, followed by scans during temperature equilibration and (d) in step heating, followed by scans during temperature equilibration. With the slower rate in step heating crystallisation of glycerol is observed.

Depending on the heating or cooling rate a temperature equilibrium at the sample may not always be guaranteed. This depends on the details of the experimental setup and in our case the thermal equilibration is prolongated with respect to normal FWS conditions on IN16B because an electrically and thermally insulating PEEK adapter is placed between the sample and the sample stick (see experimental details). In order to get information on temperature equilibration we have also carried out measurements in step cooling and heating with much longer temperature equilibration times. We judge the equilibration time from the dielectric sample response which is recorded for a nominally identical sample temperature. Such step scans, which followed without sample change the continuous temperature scans (Fig.4a,b), are shown in Fig.4c and Fig.4d. Here the temperature set point was lowered or increased in steps of ±5 K. After a waiting time of 10 min a frequency sweep of the dielectric signal was recorded which took about 2 min and was repeated ten times before changing to the next sample set point temperature.

From Fig.4c) it seems that the sample reaches thermal equilibrium faster at higher temperatures than at low temperatures near the glass transition of glycerol ($T_g \approx 185K$), which could be a kind of aging effect. It might, however, as well reflect that at low temperatures, when approaching the glass transition, an equivalent sample temperature change causes a larger frequency shift of the ϵ'' signal than at higher temperature. In fact the data points in the temperature plot of Fig. 5 below are more spread in the α-peak region where the most pronounced temperature shift is expected on ϵ''.

During the step heating scans of glycerol (Fig.4d) a somewhat different behaviour was found, which is highlighted by the coloured curves. A nearly temperature independent shoulder developed near 260 K in the frequency range $10^2 - 10^5$ Hz, followed by a major rapid increase of ϵ'' in the whole frequency range near 290 K. We ascribe these features to (partial) crystallisation phenomena, followed by melting at higher temperatures in heating. The equilibrium melting temperature for glycerol from literature is about $T_m = 291K$. In cooling or in the faster continuous heating glycerol doesn't show any sign of crystallisation in our measurements.

A detailed analysis of the crystallisation kinetics of glycerol by DE-spectroscopy with dedicated equipment can be found elsewhere (see e.g. Ref.[19],[20] and references therein) and is not the purpose of the present paper. We don't offer a quantitative analysis of the glyc-

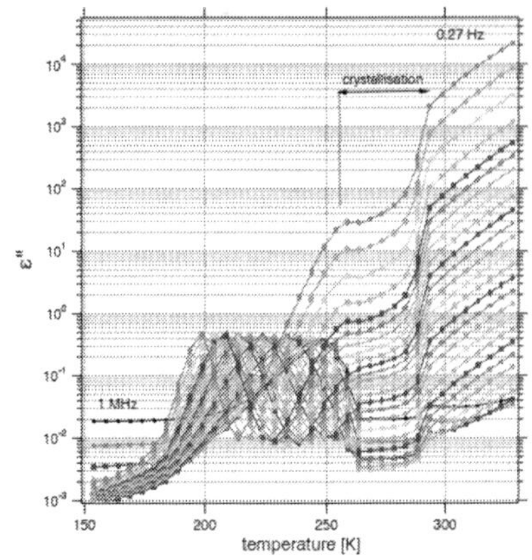

Figure 5. Temperature dependence of the dielectric permittivity of glycerol at selected frequencies measured in step heating.

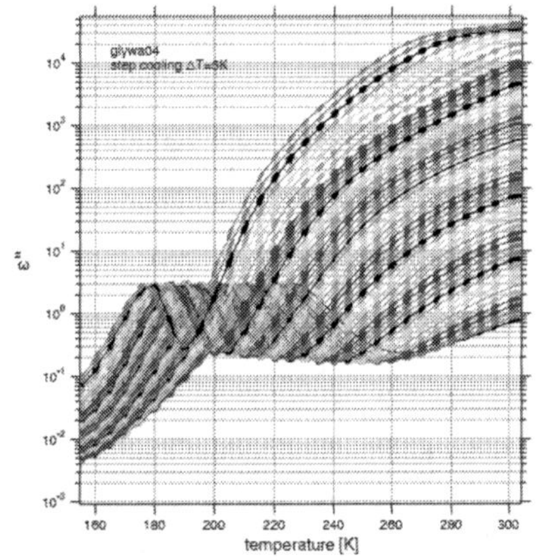

Figure 6. Temperature dependence of the dielectric permittivity of a glycerol-water mixture with a 0.4 molar fraction of glycerol as measured in step cooling.

erol signal but only strengthen our hypothesis that we observe (partial) crystallisation by plotting the temperature dependence at different constant frequencies between 0.27 Hz and 1 MHz (Fig.5). The temperature at which the α-relaxation peak maximum is observed in heating is strongly frequency dependent as expected; it is located near 200K for 0.27 Hz and near 250 K for 1 MHz. With the thermal protocol applied in our study glycerol crystallises near 260 K indicated by a frequency independent feature in Fig.5, a temperature region in which ϵ'' increases only moderately compared to the temperature increase below 260 K. At the end of this crystallisation region, near the melting temperature T_m = 291 K, ϵ'' jumps back to a higher value which can be considered as the extrapolation of the ϵ'' values below the crystallisation region.

3.2 Tests on glycerol-water

A detailed study of the glycerol-water system by DE and neutron spectroscopy as a function of both temperature and pressure is part of a PhD thesis by D.N. [15]. For testing our setup we show here the measurement on a glycerol-water sample with 0.4 molar fraction of glycerol in our flat sample holder, following the same protocol as described for neat glycerol described above. No crystallisation was observed in any of the cooling or heating scans. Fig.6 shows the corresponding temperature dependence for all measured frequencies between 0.27 Hz and 1 MHz and for step cooling with $\Delta T = 5$ K.

3.3 Influence of Teflon

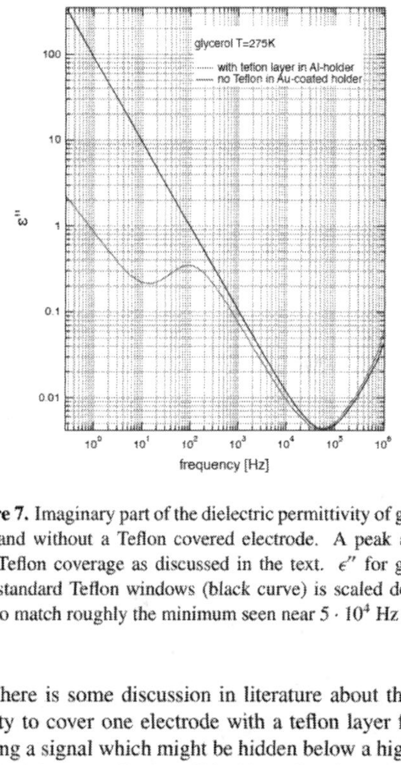

Figure 7. Imaginary part of the dielectric permittivity of glycerol with and without a Teflon covered electrode. A peak appears with Teflon coverage as discussed in the text. ϵ'' for glycerol with standard Teflon windows (black curve) is scaled down by 0.55 to match roughly the minimum seen near $5 \cdot 10^4$ Hz

There is some discussion in literature about the possibility to cover one electrode with a teflon layer for extracting a signal which might be hidden below a high DC-conductivity contribution ('blocking electrode') as published for glycerol by Bergman [17] and commented by Ref.[21] with reply by Ref.[22]. We are not aiming to contribute to this controversial discussion. But for our n-DE-sample holder we used Teflon foil frames for electrical insulation between the electrodes (2 or mostly 3 foils, thus 20 or 30 μm in total) with nominally no Teflon in the neutron window, but only sample (see inset in Fig.1). However, assuming that the 3 foils may not always be perfectly superposed, there could be a possibility that a little part near the edge of the neutron window (capacitor) of our sample holder might have a 10 μm teflon with some sample on one or both sides. Thus we have tested by purpose to increase the electrode coverage with teflon by decreasing drastically the window size which we cut into a single

10 μm thick teflon foil (in the order of 1 cm^2 window). We show in Fig.7 that in that case we detect a peak with the partially teflon covered electrode, which is absent if we use our standard Teflon frames with proper opening as described in ch.2.2. This check and the absence of such a peak is important for our data measured on PA-water, which we will discuss below.

3.4 Tests on Milli-Q water

As another characterisation of our nDS setup we have measured pure Millipore water (see Fig.8). The imaginary permittivity versus frequency, measured in continuous cooling with 0.5 K/min, changes abruptly from a smooth frequency dependence at temperatures above a registered sample temperature of 266.6 K to curves exhibiting at least one peak, which moves with further cooling to lower frequencies. We assign this observation to a freezing of the supercooled milliQ water in this type of cooling scan and at lower temperatures the observation of relaxation peak(s) in ice. In heating with 0.1K/min the melting is found near T = 273K as shown in Fig.8, r.h.s.

Similar to glycerol we have carried out in a limited temperature range on MilliQ-water DE-scans in step cooling (Fig.8c, bottom, l.h.s.) and also during much slower continuous heating with 0.1 K/min (Fig.8c, bottom, r.h.s.)), which confirm the presence of relaxation peaks in the crystal. As the empty sample holder does not show such peaks and as we didn't see relaxation peaks with teflon in the normal setup on glycerol, we can be fairly sure that the small peaks seen at low temperatures in the ϵ''-signal stems from ice water.

3.5 Comparison of the empty holder with glycerol

In order to get an indication of how much the empty sample holder may contribute to the measured ϵ''-signal we have measured an empty sample holder prepared in the same way as the filled samples, but on air. In these measurements the ϵ''-signal of the empty cell (lines in Fig.9) was smaller than 10^{-3} and thus significantly lower than the permittivity of glycerol (symbols in Fig.9) measured under similar conditions. The empty cell signal decreases with decreasing temperature. At high frequencies we see for all temperatures an increase of ϵ'' towards 1 MHz, the upper limit of the frequency range we decided to probe, ending in a similarly high ϵ'' value.

3.6 Tests on PA-water

All DE-measurements reported in the previous sections were carried out on standard samples and suggest that our sample holder design should be suited for simultaneous neutron and dielectric spectroscopy. As our project aim is to measure on phosphoric acid - water (PA-w) mixtures simultaneously conductivity spectra by DE-spectroscopy and proton dynamics by n-spectroscopy, we show here an example of the DE-scans taken on PA with our n-DE-setup. We refer the discussion concerning the physics of PA and PA-w-mixtures to a later publication and discuss

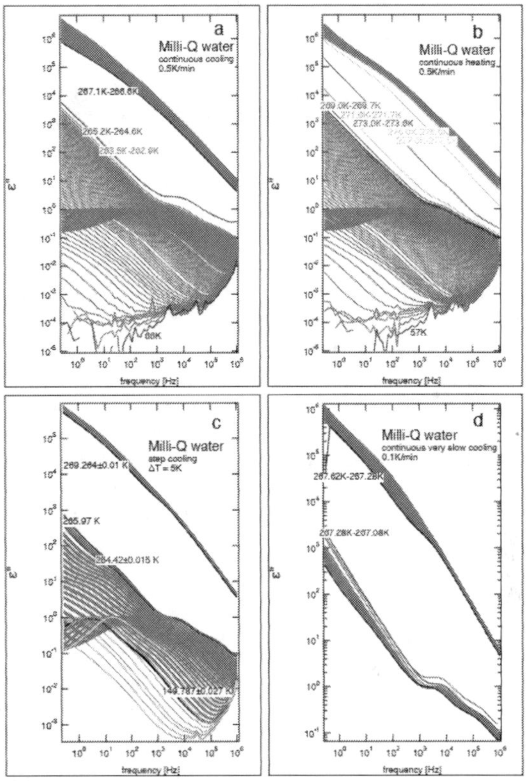

Figure 8. Imaginary part of the dielectric permittivity of Milli-Q water, measured during temperature scan in continuous cooling and heating with rates of ± 0.5 K/min (top), in cooling with steps of $\Delta T = 5K$ followed by equilibration (bottom, left) and around the melting / crystallisation with a slower cooling ramp of 0.1K/min (bottom/right).

here the limitations of our setup which we found on theses systems at high temperatures and high sample conductivity. As a note in passing we mention that phosphoric acid (PA) is known to exhibit an extremely high intrinsic proton conductivity, which can be further enhanced only by the addition of water (see e.g. [23], [24], [25]).

In Fig.10 we show dielectric spectra taken on dry PA. We observe for PA and PA-water mixtures (not shown here) at high temperatures a sharp decrease of the imaginary part of the permittivity ϵ'' in the high frequency range (highlighted within the dashed frame in Fig.10), which is not present at lower temperatures. In Fig.10 and the following we explain why the apparent peak formed is most probably an artefact. This is also suggested by the real part ϵ', which decreases sharply and which takes on negative values at high frequency and high temperatures (for frequencies above the vertical dashed line). This becomes more clear by plotting for 271 K additionally the modulus of ϵ', shown by the dots in the upper part of the figure.

To underline this we add a figure with the measured impedances Z, where Z_s' and Z_s'' corresponds to the serial real and imaginary part of the impedance as it is output by

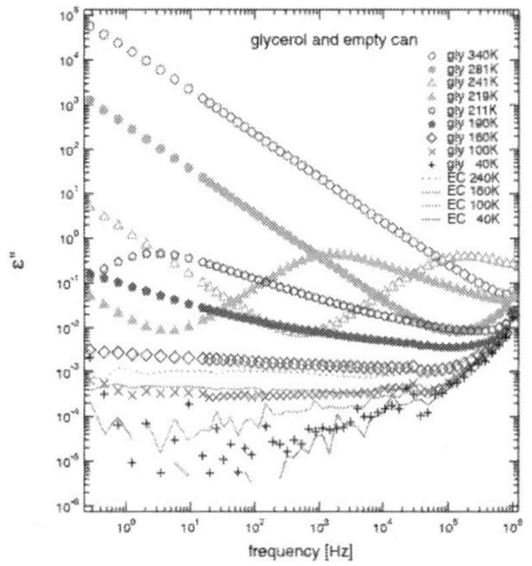

Figure 9. Dielectric scan on empty can (lines) and glycerol (symbols) for comparison are shown at several temperatures (see legend; same colour indicates same temperature).

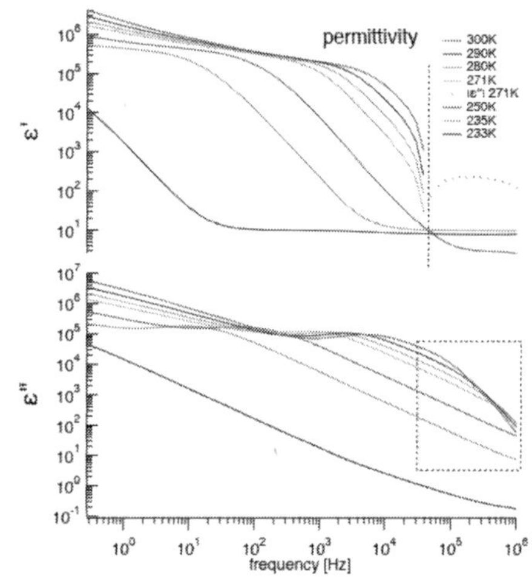

Figure 10. Real and imaginary part of the dielectric permittivity of phosphoric acid (PA), measured in cooling with the setup for simultaneous n-DE-spectroscopy and showing anomalies for high temperatures at high frequency. For ϵ'' we find at high frequency with increasing temperature an increasingly steeper decay (lower figure; dashed square region). In the same temperature-frequency region the real part of the permittivity ϵ' turns negative (upper part: r.h.s. of the dashed line). As the data are plotted on a log-log scale, we have highlighted this by adding for 271 K the absolute values of ϵ' (dots). Note also that crystallisation has appeared between T = 235 K and 233 K.

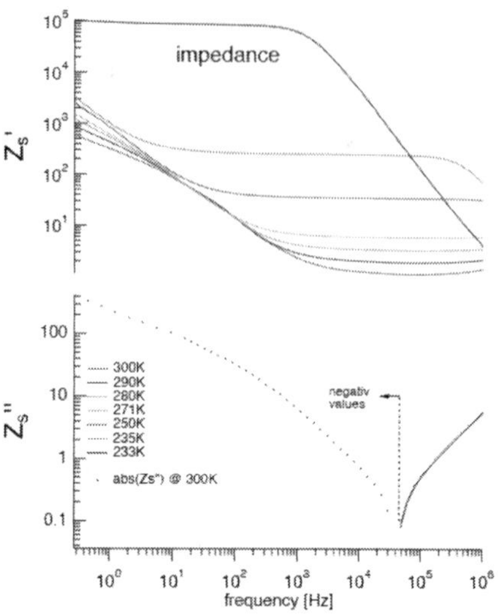

Figure 11. Impedances Z as measured on phosphoric acid (PA) with the setup for simultaneous n-DE-spectroscopy. Upper figure: real part of the serial impedance Z_s'. Lower part of the figure: imaginary counterpart Z_s'' which changes sign at the frequency indicated by a dashed vertical line (positive at high and negative at low frequencies). Dots in the lower figure correspond to the absolute value of Z_s'' shown for T = 300 K only.

the Novocontrol WINDETA [8]. Z_s'' changes from negative to positive sign above this frequency as shown in the lower part of Fig.11.

As the onset of the ϵ''-decrease with frequency varies systematically with conductivity, i.e.with increasing temperature and with increasing water content, we suspect that due to the extremely high conductivity of PA-water induction contributions from the cables might become important, in spite of the 4-wire connection. It is known that it may be difficult to measure highly conducting liquids by dielectric spectroscopy, especially at higher frequency, if the measuring cell is not suited [8]. For DE-spectroscopy a much larger spacing between the electrodes can then be chosen, which however would conflict, at least for protonated samples, with the requirement of simultaneous neutron spectroscopy for which the thickness should be chosen to minimise multiple scattering.

Another possibility which we can't exclude completely as explanation of this artefact could be that our Teflon frames could produce a peak in ϵ' as discussed above in section 3.3 and in Ref.[21]. However, as the temperature scans were measured in cooling, we can expect that the sample geometry (rearrangement of electrodes/Teflon/sample which might depend on viscosity) does not change when cooling to lower temperature and therefore the same feature would be visible as well at lower temperature. These described artefacts unfortunately limit the useful frequency range at high temperatures and for high conductivies for our planned PA-water measurements.

Nevertheless, as can be seen by calculating the conductivity σ' in Fig.12, we are able to determine well the DC-conductivity plateau level with our simultaneous n-DE-setup. Thus we will be able to deduce valuable information by comparing the temperature and water concentration dependence of the DC-conductivity plateau with the neutron scan data, even though it would have been better to see the transition from DC- to AC- conductivity in

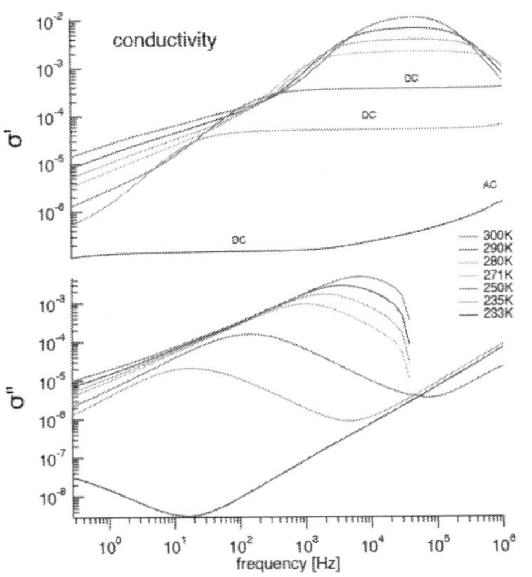

Figure 12. Conductivities as measured on phosphoric acid (PA) with the setup for simultaneous n-DE-spectroscopy. Upper figure: real part of the conductivity σ' and lower part of the figure the imaginary part σ''

a wide frequency range. Such simultaneous comparison is important as we have shown above by step cooling that temperature equilibration can be long and that temperatures recorded in the continuous cooling or heating FWS scans are no equilibrium sample temperatures.

4 Conclusion

We have presented a design for a flat sample holder which can serve as a sample cell (capacitor) for dielectric spectroscopy and in which the same sample volume permits scattering neutrons, thus allowing for simultaneous neutron spectroscopy. The presented type of sample holder allows for measuring normal, air sensitive and acidic liquids, i.e. it is a sealed sample cell which we assembled for some samples in a dry Argon atmosphere - noting that this operation is delicate. The upper temperature range is limited by the Indium seal to about 340K. Dielectric test scans in a wide frequency (0.27 Hz - 1 MHz) and temperature range (2 K - 340 K) on several standard samples (glycerol, water, glycerol-water) and under conditions similar to neutron fixed window scans (FWS) show that our flat cell design, in connection with ILL's DE-equipment, is working well, albeit not being competitive with a dedicated DE-spectroscopy or a dedicated neutron scattering setup, which is what we had expected.

We have pointed out some limitations of our n-DE-setup comprising the empty cell, like systematic errors which become important at higher frequencies. For samples with very high conductivity, like the PA sample tested in this work, another severe restriction in the high frequency range is observed at high temperatures. We ascribe this to the inductivity of cables becoming important for high sample conductivity.

Another limitation which we should stress is that absolute values of the dielectric permittivity or the conductivity are difficult to obtain with our setup. This can be judged e.g. from Fig.2 and Fig.3 were the permittivity of glycerol is found to be lower than regularly reported in literature. Although somehow expected, taking into account that we don't use a dedicated dielectric spectroscopy equipment, it is instructive to learn about the most probable reasons. Certainly the sample holder, originally designed for neutron scattering only, adds a non-negligible contribution as we have shown. Another important contribution comes from imperfect sample filling, which is nearly unavoidable with the present design but which could have been further optimised for the glycerol data shown. The sample is filled in the liquid state near room temperature and the needed sample quantity is calculated from the nominal sample holder dimensions. However, mechanical precision and the necessary play between the top and bottom pieces, uncertainties in the sample thickness due to the squeezed Indium seals and Teflon or minor deformation of the thin Aluminium neutron windows will contribute mostly to underestimate the sample quantity needed. Thus some void or gaz filled volume and an imperfect sample geometry are difficult to avoid and cause more problems for highly viscous samples. Thus one should be aware of the uncertain absolute value of the dielectric permittivity and envisage a more quantitative determination of the deviations in cases where it seems necessary.

In summary, the presented flat sample cell and setup for n-DE-spectroscopy will allow us tackling problems where simultaneous spectroscopy is of interest and we hope that the tests which we report here will help to develop a further improved generation of sample cells for simultaneous n-DE-spectroscopy.

We would like to thank Jérôme Rimet, IN16B ILL, the sample environment group SANE of ILL and the ILL for technical and financial support. We are thankful to Anette Fuchs, MPI Stuttgart, for the sample preparation of phosphoric acid (PA) and KDD Kreuer, MPI Stuttgart and A. Alegria, San Sebastian for discussing the possible origin of a high frequency artefact presented in this paper for PA.

References

[1] M. Bée, *Quasielastic neutron scattering: principles and applications in solid state chemistry, biology and materials science* (Adam Hilger, 1988)

[2] F. Kremer, A. Schönhals, *Theory of Dielectric Relaxation Spectra in Broadband Dielectric Spectroscopy* (Springer: Berlin, 2002)

[3] H.W. Hansen, A. Sanz, K. Adrjanowicz, B. Frick, K. Niss, Nature Communications **9**, 518 (2018)

[4] K. Niss et al., 10.5291/ILL-DATA.LTP-6-7 (2015)

[5] A. Sanz, H.W. Hansen, B. Jakobsen, I.H. Pedersen, S. Capaccioli, K. Adrjanowicz, M. Paluch, J. Gonthier, B. Frick, E. Lelièvre-Berna et al., Review Of Scientific Instruments **89**, 023904 (2018)

[6] H.W. Hansen, B. Frick, S. Capaccioli, A. Sanz, K. Niss, The Journal of Chemical Physics **149**, 214503 (2018)

[7] H.W. Hansen, F. Lundin, K. Adrjanowicz, B. Frick, A. Matic, K. Niss, Physical Chemistry Chemical Physics **137**, 080901 (2020)

[8] *WinDETA 6.02 - Program's Manual*, Novocontrol Technologies Germany (2017)

[9] *https://www.ill.eu/users/support-labs-infrastructure/instrument-control-service/remote-instrument-control* (last viewed 2022)

[10] B. Frick., J. Combet, L. van Eijck, Nuclear Instruments & Methods In Physics Research Section A-Accelerators Spectrometers Detectors And Associated Equipment **669**, 7 (2012)

[11] *https://www.ill.eu/in16b* (last viewed 2022)

[12] K. Niss et al (to be published)

[13] G. Squires, *Introduction to the Theory of Thermal Neutron Scattering* (Cambridge University Press, 1978)

[14] S. Lovesey, *Theory of neutron scattering from condensed matter* (Clarendon Press: United Kingdom, 1984)

[15] D. Noirat, *PhD thesis Roskilde University* (in progress)

[16] M. Jiménez-Ruiz, A. Sanz, A. Nogales, T. Ezquerra, Review of Scientific Instruments **76**, 043901 (2005)

[17] R. Bergman, H. Jansson, J. Swenson, The Journal of Chemical Physics **132**, 044504 (2010)

[18] H.W. Hansen, B. Frick, T. Hecksher, J.C. Dyre, K. Niss, Physical Review B **95**, 104202 (2017)

[19] A. Sanz, K. Niss, Crystal Growth & Design **17**, 4628 (2017)

[20] A. Sanz, K. Niss, The Journal of Chemical Physics **146** (2017)

[21] M. Paluch, S. Pawlus, K. Kaminski, The Journal of Chemical Physics **134**, 037101 (2011)

[22] R. Bergman, H. Jansson, J. Swenson, The Journal of Chemical Physics **134**, 037102 (2011)

[23] K.D. Kreuer, Chemistry Of Materials **17**, 610 (1996)

[24] J-P. Melchior, K-D. Kreuer, J. Maier, Physical Chemistry Chemical Physics **19**, 587 (2016)

[25] J-P. Melchior, B. Frick, Physical Chemistry Chemical Physics **19**, 28540 (2017)

Possible future upgrades of the direct-geometry chopper spectrometer 4SEASONS

Ryoichi Kajimoto[1,*], *Mitsutaka* Nakamura[1], *Kazuya* Kamazawa[2], *Yasuhiro* Inamura[1], *Kazuki* Iida[2], *Kazuhiko* Ikeuchi[2], and *Motoyuki* Ishikado[2]

[1]Materials and Life Science Division, J-PARC Center, Japan Atomic Energy Agency, Tokai, Ibaraki 319-1195, Japan
[2]Neutron Science and Technology Center, Comprehensive Research Organization for Science and Society, Tokai, Ibaraki 319-1106, Japan

Abstract. 4SEASONS is a direct geometry time-of-flight spectrometer installed in the Materials and Life Science Experimental Facility, the Japan Proton Accelerator Research Complex. It is used to study atomic and spin dynamics in the energy range of 10^0 meV to 10^2 meV. Since more than a decade has crossed after the first inelastic scattering experiment, it is essential to consider upgrading the instrument to improve its flexibility and performance. In this paper, we discuss the possible medium-term upgrades of key components of the instrument like the chopper system, which are achievable with the current technology and at reasonable cost. Herein, we demonstrated that 4SEASONS can improve the energy resolution by a factor of two, remove frame overlap of adjacent incident energies, significantly improve the asymmetry in the pulse shape, and increase the flux by a factor of ~1.5, without major technical difficulties.

1 Introduction

4SEASONS, also known as *SIKI*, is one of the inelastic neutron scattering instruments in the Materials and Life Science Experimental Facility (MLF) in the Japan Proton Accelerator Research Complex (J-PARC). It is a direct geometry time-of-flight spectrometer that utilizes a Fermi chopper as a monochromator to study atomic and spin dynamics in the energy range of 10^0–10^2 meV. This instrument was originally constructed for examining high-critical-temperature superconductors and was associated with high intensity and high efficiency [1]. Thus, the instrument is now considered as a middle-resolution and high-intensity spectrometer that covers the middle momentum-energy region in the inelastic/quasielastic neutron spectrometers in MLF [2–4].

The first inelastic scattering experiment using 4SEASONS was conducted in 2009 [5], and since then, its research target has been expanded to include quantum magnets, frustrated magnets, itinerant magnets, multiferroic materials, thermoelectric materials, glasses, and catalysis. Hence, this calls for serious consideration of upgrading the instrument to improve its flexibility and performance. We have made several upgrades to the instrument, such as minimizing background scattering, introducing new sample environments, filling the detector vacancy, replacing malfunctioning or aged equipment, etc. [6–11]. These short-term upgrades were undertaken as per the requirement to ensure the instrument provides its designed performance. In the present study, we discuss the possibilities of medium-term upgrades of key components of the instrument such as the chopper system. Here, we discuss the upgrades achievable using the current technology and within reasonable cost. We also focus on the effectiveness and feasibility of these upgrades in improving the performance and flexibility of the instrument. We set aside long-term upgrades requiring the redesign of the instrument from scratch or new technology currently unavailable, which will be considered in the future accompanying an upgrade of the whole MLF or construction of the second target station.

2 Instrument layout

Figure 1 shows the schematic top and side views of 4SEASONS [1]. 4SEASONS is located at the beamline 1 (BL01) viewing the coupled moderator [12]. The sample is located at $L_1 = 18$ m from the moderator. The supermirror-coated guide is placed between the 2.3 m and 15.8 m positions to minimize the loss of neutrons during the beam transport. It has an elliptically converging shape with one of the focal points being located at the sample. The incident neutron beam transported by the neutron guide is monochromatized by the Fermi chopper. It is located at $L_3 = 1.7$ m upstream of the sample and rotates at 600 Hz at maximum. The scattered neutrons are detected at 3/4-in-diameter and 2.5-m-long position-sensitive detectors (PSDs) surrounding the sample cylindrically at a distance of $L_2 = 2.5$ m from −35° to +130° with respect to the incident beam. The PSDs are ^{3}He tubes with a ^{3}He partial pressure of 16 atm. In addition to the Fermi chopper, 4SEASONS has a T_0 chopper and two disk choppers. The T_0 chopper rotates at 100 Hz at maximum and is placed

*e-mail: ryoichi.kajimoto@j-parc.jp

 https://doi.org/10.1051/epjconf/202227202007

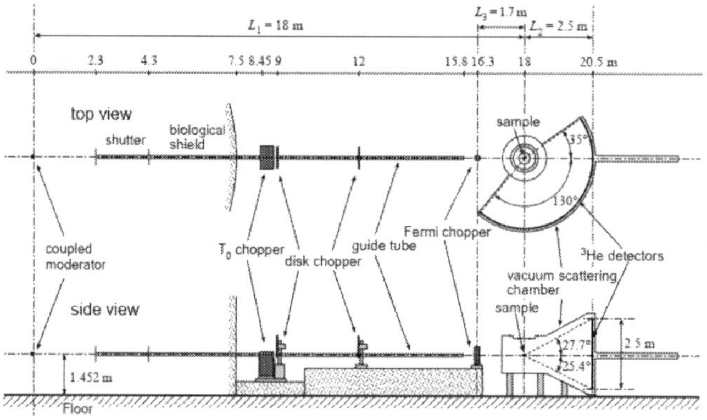

Figure 1. Layout of 4SEASONS [1].

at 8.45 m from the moderator. It is used to suppress the prompt pulse. On the other hand, the two disk choppers are installed at 9 m and 12 m and rotate at 12.5 Hz or 25 Hz. These two choppers are used to define the usable incident energy band and to eliminate the frame overlaps of very low energy neutrons.

3 Candidates of upgrade items

3.1 Fermi chopper and optimum resolution

The Fermi chopper of 4SEASONS utilizes straight-slit package. This package is advantageous in multi-E_i measurements using the repetition rate multiplication (RRM) [5, 13–15] because of the equal transmissions at the $2n\pi$ and $(2n+1)\pi$ rotations (n is integer). The ratio of the width (w) to the length (D) of each slit is 0.02 [16]. The parameters of the slit package were designed to produce approximately 5% energy resolution at elastic scattering under the "optimum" condition in which the chopper opening time Δt_{ch} is equal to the moderator pulse width Δt_m, i.e., $\Delta t_{ch} = \Delta t_m$ [1, 17, 18] [Fig. 2(a)]. The above condition can be fulfilled at E_i below 500 meV by considering that the minimum of Δt_{ch} is 5.3 μs in full width at half maximum (FWHM) at 600 Hz rotation [Fig. 2(b)]. This E_i is sufficient to cover the target dynamic range of the instrument.

However, the energy resolution at low energy transfers may be insufficient with the above condition. This is because the contribution of the Fermi chopper to energy resolution increases substantially at low energy transfers [19, 20]. This further limits the usability of the instrument, even though the low energy transfer region can be covered by lower E_is available via the RRM. It is likely that the above condition was too naïve for 4SEASONS, and we should reconsider the optimum condition for low energy transfers based on the idea that the Fermi chopper contributes to the energy resolution similarly to others [21].

The resolution of the energy transfer relative to E_i is given by [22]:

$$\left(\frac{\Delta E}{E_i}\right)^2 = \left\{2\frac{\Delta t_{ch}}{t_{ch}}\left[1 + \frac{L_1}{L_2}\left(\frac{E_f}{E_i}\right)^{\frac{3}{2}}\right]\right\}^2 + \left\{2\frac{\Delta t_m}{t_{ch}}\left[1 + \frac{L_3}{L_2}\left(\frac{E_f}{E_i}\right)^{\frac{3}{2}}\right]\right\}^2 + \left[2\frac{\Delta L_2}{L_2}\left(\frac{E_f}{E_i}\right)\right]^2, \quad (1)$$

where ΔL_2 is the uncertainty of L_2 because of the sample and the detector sizes. Let us neglect the contribution from ΔL_2 for simplicity because it is independent of Δt_{ch} and Δt_m. Equating the Δt_{ch} and Δt_m terms in Eq. (1) leads the following optimum condition as shown in literature [21, 23],

$$\frac{\Delta t_{ch}}{\Delta t_m} = 1 - \frac{L_{ch}}{L_1 + L_2(E_i/E_f)^{3/2}}, \quad (2)$$

where L_{ch} is the distance between the moderator and the Fermi chopper ($L_{ch} = L_1 - L_3$). The condition $\Delta t_{ch} = \Delta t_m$ corresponds to the extreme case when E is its maximum, E_i. However, for general cases, Eq. (2) is less than unity, and it becomes

$$\Delta t_{ch} = \frac{L_2 + L_3}{L_1 + L_2}\Delta t_m, \quad (3)$$

when $E = 0$. Eq. (3) becomes $0.20\Delta t_m$ with $L_1 = 18$ m, $L_2 = 2.5$ m, and $L_3 = 1.7$ m.

Figure 2(a) shows energy resolutions as a function of E_i calculated using the moderator pulse width of BL01 [24] and Eq. (1) by varying Δt_{ch} from Δt_m to $0.1\Delta t_m$. The increase below $E_i \sim 50$ meV is due to the broadening of the pulse width of the coupled moderator, while $\Delta E/E_i$ is almost constant above this energy because Δt_m is proportional to $E_i^{-1/2}$ [25]. The energy resolution for $\Delta t_{ch} = \Delta t_m$ becomes 5%–6% above ~50 meV as designed. By decreasing Δt_{ch}, the energy resolution smoothly decreases. With $\Delta t_{ch} = 0.2\Delta t_m$, we can obtain less than 3% resolution above 20 meV, and 4% resolution even at 10 meV. Such values of resolutions are comparable with those of

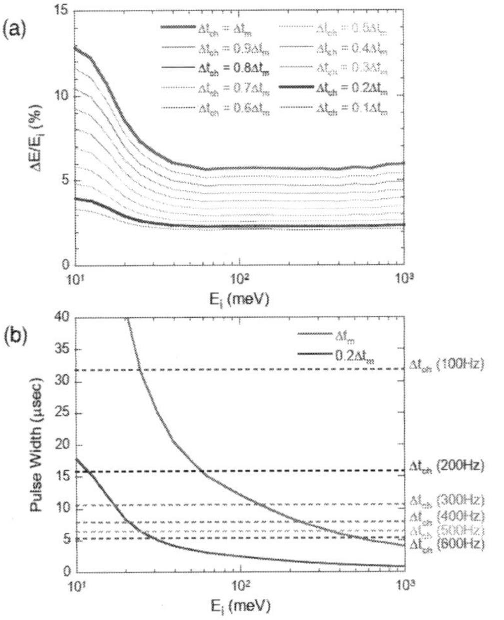

Figure 2. (a) Calculated energy resolutions as a function of E_i when Δt_{ch} varies from Δt_m to $0.1\Delta t_m$. The cases of $\Delta t_{ch} = \Delta t_m$ and $0.2\Delta t_m$ are indicated by thick lines. The sample size is 20 mm in diameter. (b) The moderator pulse width of BL01 Δt_m and $0.2\Delta t_m$ as a function of E_i. Broken lines indicate the chopper pulse width for the 4SEASONS Fermi chopper rotated at 100–600 Hz.

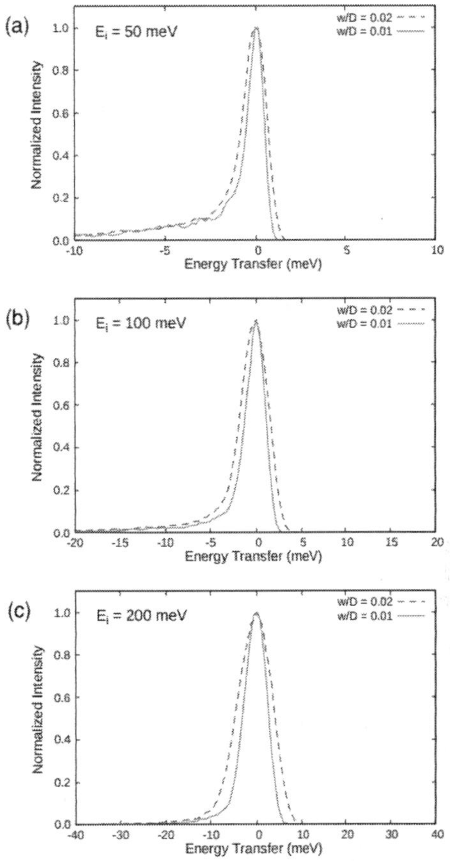

Figure 3. Energy-transfer spectra for $w/D = 0.02$ (broken lines) and 0.01 (solid lines) obtained by Monte Carlo simulations for (a) $E_i = 50$ meV, (b) 100 meV, and (c) 200 meV. The Fermi chopper with $D = 20$ mm is rotated at 600 Hz. The sample is a 1-mm thick hollow cylindrical vanadium that is 20 mm in diameter and 20 mm in height. The peak intensities are normalized to unities.

existing chopper spectrometers and are good enough for a middle-resolution and high-intensity spectrometer like 4SEASONS. It should be noted that the improvement in the resolution slows down with decreasing Δt_{ch} and becomes almost saturated at $\Delta t_{ch} < 0.2\Delta t_m$. This is because the resolution is dominated by contributions other than the Fermi chopper at such a small Δt_{ch}.

Figure 2(b) shows the moderator pulse width in FWHM and the one multiplied by 0.2 together with the chopper pulse widths obtained from the present Fermi chopper for several rotation speeds. The condition of $\Delta t_{ch} = 0.2\Delta t_m$ can be achieved only below $E_i \sim 30$ meV using the present Fermi chopper whose maximum rotation speed is 600 Hz. Since it may be technically difficult to increase the maximum rotation speed, we should introduce a new Fermi chopper with a finer slit package to overcome the limitation of Δt_{ch}. For example, if we employ the slit package whose w/D is half of the present value, i.e., $w/D = 0.01$, the condition of $\Delta t_{ch} = 0.2\Delta t_m$ is available up to about 100 meV. This value of w/D is feasible because Fermi choppers with a similar value [26] or even a finer value of w/D [27] are actually used. Figure 3 shows how the fine slit package improves the energy resolution. The energy-transfer spectra are compared with $w/D = 0.02$ and 0.01 when the Fermi chopper is rotated at its maximum speed of 600 Hz. The data were obtained by Monte Carlo simulation using the ray-trace simulation package McStas [28–32]. $\Delta E/E_i$ estimated by the FWHMs are 3.1%, 3.6%, and 4.4% for $E_i = 50$, 100, and 200 meV, respectively, when $w/D = 0.02$. When $w/D = 0.01$, the estimated $\Delta E/E_i$ reduce to 2.1%, 2.5%, and 2.9%, respectively. These results are consistent with the analytical estimation in Fig. 2. About 70% reduction of the resolution is significant, but the use of fine slits and high rotation speed of the Fermi chopper reduce the neutron transmission. This loss in neutron flux can be compensated by adopting curved slits. The curved slits will be less efficient in multi-E_i measurements because of low transmissions at the $(2n+1)\pi$ positions. Thus, the combination of the present sloppy straight-slit Fermi chopper and a new fine curved-slit Fermi chopper should be considered to be used in 4SEASONS to improve its flexibility.

3.2 Suppression of frame overlaps of incident pulses

One of the most important features of 4SEASONS is the simultaneous utilization of multiple E_is via RRM by the Fermi chopper. In recent years, this feature has become the standard for direct-geometry chopper spectrometers [23, 33–40]. However, the overlap of scattering from the adjacent E_i may limit the usable energy-transfer range, particularly at low E_is. Suppose we choose $E_i = 200$ meV and subsidiary E_is as typical experimental condition at 4SEASONS while rotating the Fermi chopper at 300 Hz. Then, this results in a series of $E_i = 200$, 75.1, 39.0, 23.9, and 16.1 meV. If we calculate the maximum energy transfer E_{clean} below which scattering from the next lower E_i never overlaps [5], E_{clean}/E_i becomes 95%, 87%, 75%, 59%, and 40%, respectively. It is known that the intensity of up-scattering from the next lower E_i drastically decreases as its energy transfer decreases toward $-\infty$ following the Bose population factor $n(E,T) = [\exp(E/k_BT) - 1]^{-1}$, where k_B and T are the Boltzmann constant and temperature, respectively. Thus, the usable energy transfer range can be enlarged, particularly at low temperatures. However, the above example has clearly shown that the overlaps of incident pulses becomes a serious problem for $E_i \sim 10$ meV or lower. To mitigate this problem, the new direct-geometry chopper spectrometers are equipped with pulse-removal choppers to control spacings between incident pulses. These pulse-removal choppers are usually disk choppers that are placed either adjacent to [23, 35–37] or away from the monochromatizing choppers [38–40]. However, 4SEASONS, which is one of the first chopper spectrometers to implement the RRM capability [5], does not have choppers dedicated for such a purpose, though it can select a single E_i by combining the opening times of the two band choppers.

In this study, we briefly discuss the feasibility of adding a pulse-removal chopper to 4SEASONS. Since the pulse-removal chopper thins out the incident pulses defined by the monochromatizing chopper, it has more advantages when placed closer to the Fermi chopper. However, the spatial constraint and the possibility of using the chopper as a pulse-shaping chopper, which will be discussed in the next subsection, need to be taken into consideration. Thus, we suggest the chopper to be placed at half of the distance between the moderator and the Fermi chopper. Figure 4 shows the cutaway view of a part of 4SEASONS near the T_0 chopper and the first disk chopper. The position at $L_{ch}/2 = 8.15$ m is marked by a down arrow. A chopper can be installed at this position without disturbing the existing choppers if the short section of the neutron guide is removed or replaced. In our case, the pulse-removal chopper can remove each second pulse if its opening period τ is the same as that of the Fermi chopper. For the straight-slit Fermi chopper rotating at 600 Hz, the required minimum opening period is 1/1200 s. The maximum rotating speed of a typical disk chopper is about 300 Hz. So, $\tau = 1/1200$ s can be achieved if we use a disk with four equally spaced apertures. On the other hand, a disk chopper with unequally spaced apertures [36, 40] or that with two disks rotating at different speeds [35] provides uneven elimination of E_is. The latter type of disk chopper also needs to be considered in the case of selectively removing low E_is with more serious frame overlap.

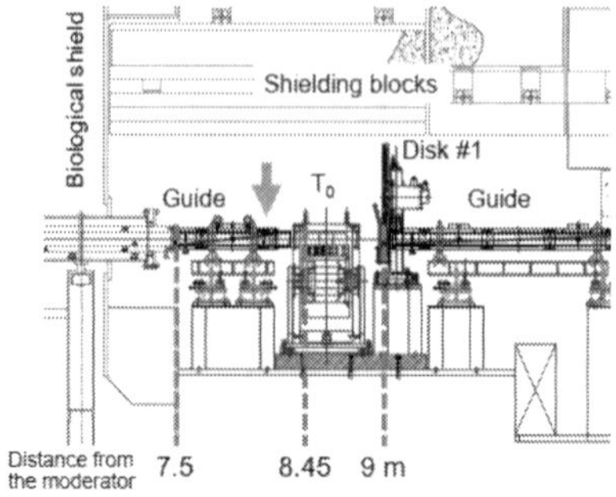

Figure 4. Cutaway view of 4SEASONS after the biological shield. "Guide," "T_0," and "Disk #1" denote parts of the neutron guide, the T_0 chopper, and the first disk chopper, respectively. The arrow indicates the candidate position of the pulse-removal chopper.

3.3 Pulse shaping

We have confirmed the feasibility of adding a chopper at the $L_{ch}/2$ position. Next, we discuss whether this chopper can also be used for pulse shaping. The pulse-shaping chopper is mainly used for two purposes. One purpose is to obtain a finer energy resolution by sharpening the neutron pulse width prior to the monochromatization with the Fermi chopper. The other purpose is to cut the pulse tail to improve the symmetry in the pulse shape.

In order to obtain a finer energy resolution, the opening time of the pulse-shaping chopper Δt_{ps} must be in the same order as that of the moderator pulse width. In quantitative terms, the time width at the moderator defined by the pulse-shaping chopper and the monochromatizing chopper is

$$\Delta t_m^{ch+ps} = \frac{L_{ch}}{L_{ch} - L_{ps}} \Delta t_{ps} + \frac{L_{ps}}{L_{ch} - L_{ps}} \Delta t_{ch},$$

where L_{ps} is the distance of the pulse-shaping chopper from the moderator [33]. The Δt_m^{ch+ps} should be similar to or less than Δt_m to effectively control the energy resolution by the pulse-shaping chopper. If $L_{ps} = L_{ch}/2$, Δt_{ps} must be less than $\Delta t_m/2$ even for the limiting case of $\Delta t_{ch} = 0$. The beam width at the candidate position of the pulse-shaping chopper ($L_{ps} = 8.15$ m) is ~80 mm. In the case of a typical fast-rotating disk chopper having 300 mm radius at the beam center, 80 mm slit width, and 300 Hz rotation speed, Δt_{ps} of this chopper becomes 141 μs. If we use a disk chopper with two counter-rotating disks, Δt_{ps} becomes 70 μs at most. In any case, Δt_{ps} is larger than Δt_m in most of the energy range of 4SEASONS [see Fig. 2(b)]. Therefore, This type of the pulse-shaping chopper cannot be used to obtain finer energy resolution. A narrow-slit disk chopper or a Fermi chopper can achieve sharper Δt_{ps}. However, such choppers are associated with significant loss in neutron flux. 4SEASONS can potentially achieve 2.5%–3% resolution, as discussed in §3.1, without the pulse-shaping chopper. Higher resolution should be left to other high-resolution spectrometers in MLF, such as HRC and AMATERAS.

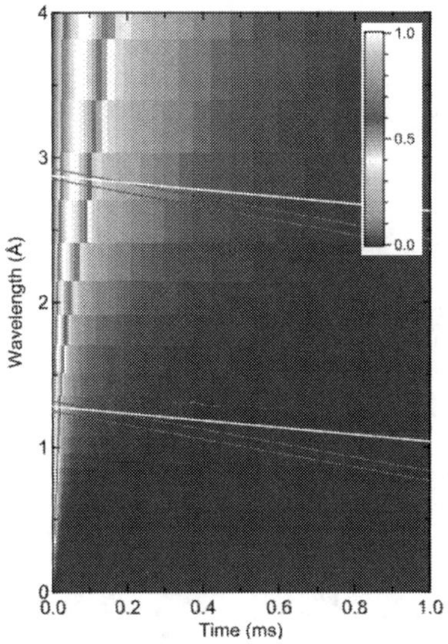

Figure 5. Time-wavelength acceptance diagram of the pulse-shaping chopper (red lines) and the Fermi chopper (white lines) for 10 meV and 50 meV neutrons. The rotation speed of the pulse-shaping chopper is 300 Hz, while that of the Fermi chopper is 200 Hz (10 meV) and 300 Hz (50 meV), respectively. Color map shows the time structure of the moderator at BL01, whose peak intensity is normalized to unity.

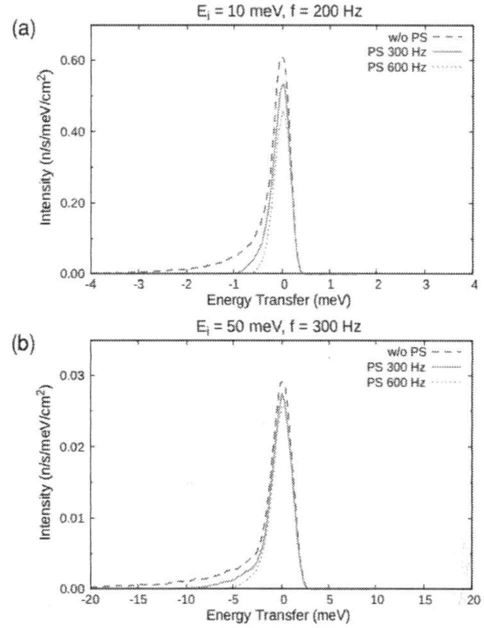

Figure 6. Energy-transfer spectra without and with the pulse shaping chopper obtained by Monte Carlo simulations. The sample is the same vanadium as used in the calculations of Fig. 3. (a) E_i = 10 meV and f = 200 Hz. (b) E_i = 50 meV and f = 300 Hz. Broken, solid, and dotted lines denote spectra without the pulse-shaping chopper, with the pulse-shaping chopper rotating at 300 Hz and at 600 Hz, respectively.

Conversely, the required conditions for Δt_{ps} should be mitigated for the tail cutter. The pulse tail is particularly prominent at $E_i \lesssim 50$ meV where the pulse width of the coupled moderator is enhanced. The tail hampers the quantitative analysis of the observed spectra irrespective of nominal energy resolutions in, for example, quantitative estimation of line width or observation of spin gap. Therefore, suppressing the pulse tail by the pulse-shaping chopper will be useful even if the nominal energy resolution is not improved.

We plot a time-wavelength acceptance diagram in Fig. 5, following Refs. [23, 41], to examine whether the pulse-shaping chopper can cut the pulse tail effectively. Red and white lines indicate the time-wavelength regions of neutrons at the moderator that pass through the pulse-shaping chopper and the Fermi chopper, respectively. We plot the cases for 10 meV (2.9 Å) and 50 meV (1.3 Å) neutrons. The pulse-shaping chopper has a 300-mm radius and an 80 mm-wide slit as discussed above, and its rotation speed is 300 Hz. The color map in Fig. 5 shows the pulse structure of the moderator at BL01. Here, Δt_m is 90 μs and 18 μs at E_i = 10 meV and 50 meV, respectively. Figure 5 indicates that the pulse-shaping chopper can substantially cut the pulse tail for 10 meV neutrons, though the pulse-shaping chopper is less effective for 50 meV or higher-energy neutrons due to the sharp moderator pulse width. We, thus, conducted a Monte Carlo simulation study using McStas. Figure 6 shows simulated energy-transfer spectra for E_i = 10 meV and 50 meV at the detector. The rotation speeds of the Fermi chopper f were chosen to be 200 Hz and 300 Hz for E_i = 10 meV and 50 meV, respectively, so that the energy resolutions $\Delta E/E_i$ became about 5%. The short guide section at ~8 m was removed to place the pulse-shaping chopper (see Fig. 4). Without the pulse-shaping chopper, the spectra show significant tails in the energy gain ($E < 0$) sides. The rotation of the pulse-shaping chopper at 300 Hz results in a finite decrease in the tails. In particular, the impact of the pulse-shaping chopper is evident at E_i = 10 meV [Fig. 6(a)]. The peak intensity slightly decreases by pulse-shaping, but it is acceptable. If we can rotate the pulse-shaping chopper at 600 Hz, the tail is significantly suppressed even at E_i = 50 meV [Fig. 6(b)]. Such a high rotation speed is not feasible with the currently available disk chopper, but a counter-rotating double-disk chopper can effectively achieve this speed. The tail cut by the pulse-shaping chopper becomes more effective at lower E_is where the moderator pulse width becomes broader, while it is less effective at E_i higher than 50 meV. However, the pulse tail becomes less prominent at high energies because of the decrease in the moderator pulse width (Fig. 3). Therefore, we conclude that the pulse-shaping chopper in 4SEASONS is effective.

3.4 Beam transport

The upgrades discussed above focused mainly on improving the resolution or pulse shape. However, these upgrades

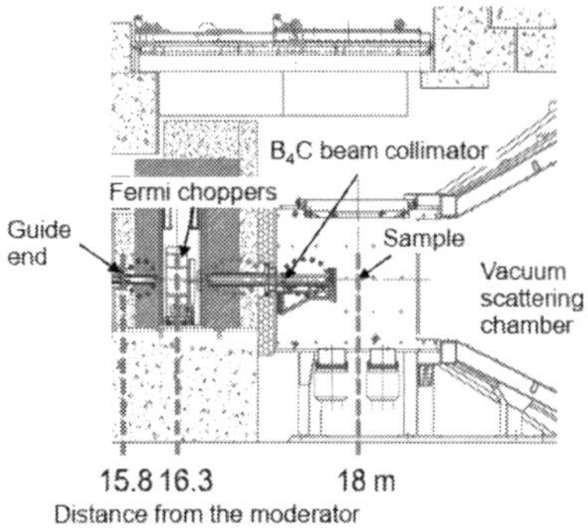

Figure 7. Cutaway view of 4SEASONS around the sample, Fermi chopper, and the end of the neutron guide. The sections marked by blue and red dotted circles are beam collimators made of B_4C (see text for details).

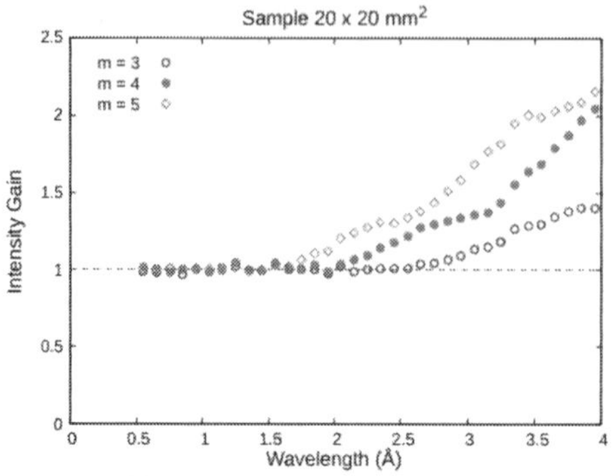

Figure 8. Gain in flux at a 20 mm × 20 mm sample by replacing the 50 mm × 50 mm × 580 mm beam collimator after the Fermi chopper with a supermirror neutron guide obtained by Monte Carlo simulation. Open circles, closed circles, and open diamonds show the gains when the *m* values of the supermirror are 3, 4, and 5, respectively.

generally decrease the neutron flux on the sample in compensation for the better pulse shape. Here, we discuss a possible upgrade of the beam transport to increase the neutron flux.

The current neutron guide is designed to transport high neutron flux at the sample [42]. It has an elliptically converging shape and uses high-critical-angle supermirrors whose *m* values are 3.2–4 (*m* is the critical angle relative to that of the nickel mirror). Given the relatively short flight path to the sample, modifying the design of the existing guide will not be effective in increasing the neutron flux.

On the other hand, 4SEASONS still has a room for extending the neutron guide. Figure 7 shows a cutaway view of 4SEASONS at the end of the neutron guide and the sample position. The neutron guide ended 0.5 m before the Fermi chopper, resulting in a substantially large space between the Fermi chopper and the sample. This space is left for future upgrade, and beam collimators made of B_4C are currently installed. Thus, the neutron guide can be extended by about 1 m. Currently, the total length of the neutron guide, excluding the length of gaps for the choppers, is about 12.7 m. The total flight path between the moderator and the sample *without* the neutron guide is $18\,\mathrm{m} - 12.7\,\mathrm{m} = 5.3\,\mathrm{m}$. If the flight path with the neutron guide is extended by 1 m, the flight path without the neutron guide becomes 4.3 m. Suppose the neutron guide transports neutrons by 100% for simplicity, then we need to consider only the flight path without the neutron guide. Thus, a rough estimate of the flux gain becomes $(5.3\,\mathrm{m}/4.3\,\mathrm{m})^2 = 1.5$. The 50% increase in flux is significant and should be examined in detail.

Figure 8 shows a more practical study of the gain by extending the neutron guide. We assumed that the 50 mm × 50 mm × 580 mm beam collimator placed after the Fermi chopper (indicated by a red dotted circle in Fig. 7) was replaced with a supermirror-coated neutron guide having the same dimension. Thus, the gain in intensity at a 20 mm × 20 mm sample was calculated by Monte Carlo simulation using McStas. By adding only this short neutron guide, we obtained a positive result. If we use the $m = 4$ supermirror, which is the same as that used in the downstream part of the present neutron guide, we can obtain a finite gain at a wavelength longer than 2 Å (~20 meV), and the gain reaches about two at 4 Å (~5 meV). Recently an *m* value higher than 5 is commercially available; this will be beneficial for shorter wavelengths. If the sample size is larger, then the gain at short wavelengths will be larger, while the gain at long wavelengths will be enhanced by utilizing a focusing shape. Thus, there is a need for a more detailed analytical and numerical study to optimize the specifications of the extended guide.

3.5 Flight paths

In the previous sections, we have discussed practical upgrades without changing the key flight paths of L_1, L_2, and L_3. This is because the modification of these flight paths requires complete redesigning of the instrument, which is out of the scope of the present paper. Nevertheless, understanding how changes in flight paths affect the performance of the instrument will help in preparing for future major upgrades or development of a new spectrometer. To finish this paper, we briefly discuss this issue in terms of the energy resolution.

The energy resolution of the direct-geometry chopper spectrometer improves with the increase in L_1 and L_2 and the decrease in L_3. However, the contribution of L_1 saturates as L_1 increases [see Eq. (1)], and L_1 longer than ~15 m is not advantageous to the energy resolution at short-pulse facilities, including J-PARC [17, 43]. Figure 9(a) shows calculated energy resolutions as a function of the energy transfer for $L_1 = 18$ m, 20 m, and 30 m when $E_i = 50$ meV and $f = 300$ Hz. Even if L_1 is extended by 12 m, the energy resolution will improve by only 0.5% at elastic scattering. Similarly, shortening L_3 shows little improvement in energy resolution, as shown in Fig. 9(c). If L_3 is reduced to 1 m, which is the practical minimum value

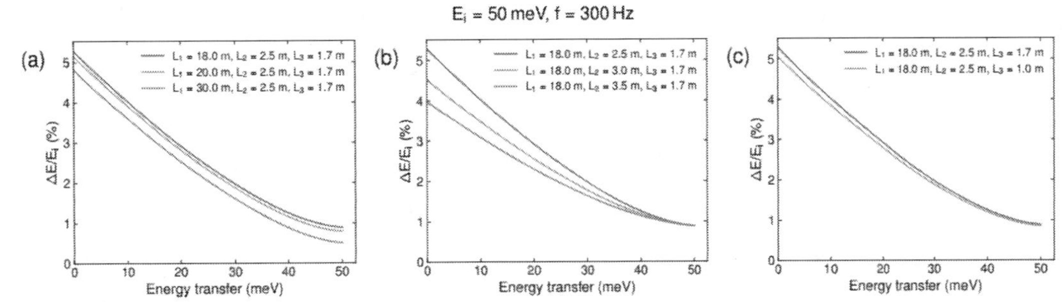

Figure 9. Calculated energy resolutions as a function of energy transfer for several sets of flight paths. E_i, f, and the sample diameter are 50 meV, 300 Hz, and 20 mm, respectively. (a) The case varying L_1 as 18 m, 20 m, and 30 m with L_2 and L_3 being fixed at 2.5 m and 1.7 m, respectively. (b) The case varying L_2 as 2.5 m, 3.0 m, and 3.5 m with L_1 and L_3 being fixed at 18 m and 1.7 m, respectively. (c) The case varying L_3 as 1.7 m and 1.0 m with L_1 and L_2 being fixed at 18 m and 2.5 m, respectively.

considering the required space for sample environment devices, the energy resolution will improve by only 0.2% at elastic scattering. Conversely, lengthening L_2 is the most effective way to improve energy resolution. As shown in Fig. 9(b), increasing L_2 by only 0.5 m improves the energy resolution by 0.8% at elastic scattering. In reality, the flight paths are limited by spatial constraints as long as 4SEASONS continues to reside at BL01. However, L_2 can be extended by ~0.5 m and should be considered when reconstructing the vacuum scattering chamber in the future.

4 Conclusion

In this paper, we discuss the possible upgrades of the key components of 4SEASONS to examine their feasibility in improving the performance and flexibility of the instrument. We demonstrated that 4SEASONS can improve the energy resolution by a factor of two, remove frame overlap of adjacent incident energies, significantly improve asymmetry in the pulse shape, and increase the flux by a factor of ~1.5, without major technical difficulties. The findings of this study will contribute to the development of future upgrades for the 4SEASONS spectrometer. More detailed quantitative studies, including numerical simulations and prioritization based on scientific needs, budget, and human resources, are required to further elucidate the proposed upgrades.

We thank the members of the MLF Spectroscopy Group for valuable discussions. This work was supported by JSPS KAKENHI Grant Numbers 19K12648 and 22K12663.

References

[1] R. Kajimoto, M. Nakamura, Y. Inamura, F. Mizuno, K. Nakajima, S. Ohira-Kawamura, T. Yokoo, T. Nakatani, R. Maruyama, K. Soyama et al., J. Phys. Soc. Jpn. **80**, SB025 (2011)

[2] H. Seto, S. Itoh, T. Yokoo, H. Endo, K. Nakajima, K. Shibata, R. Kajimoto, S. Ohira-Kawamura, M. Nakamura, Y. Kawakita et al., Biochim. Biophys. Acta, Gen. Subj. **1861**, 3651 (2017)

[3] K. Nakajima, Y. Kawakita, S. Itoh, J. Abe, K. Aizawa, H. Aoki, H. Endo, M. Fujita, K. Funakoshi, W. Gong et al., Quantum Beam Sci. **1**, 9 (2017)

[4] R. Kajimoto, T. Yokoo, M. Nakamura, Y. Kawakita, M. Matsuura, H. Endo, H. Seto, S. Itoh, K. Nakajima, S. Ohira-Kawamura, Physica B **562**, 148 (2019)

[5] M. Nakamura, R. Kajimoto, Y. Inamura, F. Mizuno, M. Fujita, T. Yokoo, M. Arai, J. Phys. Soc. Jpn. **78**, 093002 (2009)

[6] R. Kajimoto, M. Nakamura, Y. Inamura, K. Ikeuchi, S. Ji, K. Nakajima, S. Ohira-Kawamura, W. Kambara, M. Sawabe, A. Kamiya et al., J. Phys. Soc. Jpn. **82**, SA032 (2013)

[7] M. Nakamura, Y. Kawakita, W. Kambara, K. Aoyama, R. Kajimoto, K. Nakajima, S. Ohira-Kawamura, K. Ikeuchi, T. Kikuchi, Y. Inamura et al., JPS Conf. Proc. **8**, 036011 (2015)

[8] R. Kajimoto, M. Nakamura, Y. Inamura, K. Kamazawa, K. Ikeuchi, K. Iida, M. Ishikado, K. Nakajima, S. Ohira-Kawamura, T. Nakatani et al., in *Proc. ICANS-XXI* (2015), JAEA-Conf 2015-002/KEK Proceedings 2015-7, p. 319

[9] M. Nakamura, W. Kambara, K. Iida, R. Kajimoto, K. Kamazawa, K. Ikeuchi, M. Ishikado, K. Aoyama, Physica B **551**, 480 (2018)

[10] R. Kajimoto, M. Nakamura, Y. Inamura, K. Kamazawa, K. Ikeuchi, K. Iida, M. Ishikado, N. Murai, H. Kira, T. Nakatani et al., J. Phys.: Conf. Ser. **1021**, 012030 (2018)

[11] R. Kajimoto, M. Ishikado, H. Kira, K. Kaneko, M. Nakamura, K. Kamazawa, Y. Inamura, K. Ikeuchi, K. Iida, N. Murai et al., Physica B **556**, 26 (2019)

[12] H. Takada, K. Haga, M. Teshigawara, T. Aso, S. Meigo, H. Kogawa, T. Naoe, T. Wakui, M. Ooi, M. Harada et al., Quantum Beam Sci. **1**, 8 (2017)

[13] F. Mezei, J. Neutron Res. **6**, 3 (1997)

[14] F. Mezei, M. Russina, S. Schorr, Physica B **276-278**, 128 (2000)

[15] M. Russina, F. Mezei, Nucl. Instrum. Methods Phys. Res., Sect. A **604**, 624 (2009)

[16] M. Nakamura, R. Kajimoto, JPS Conf. Proc. **1**, 014018 (2014)

[17] S. Itoh, M. Arai, T. Otomo, K. Ohoyama, T. Osakabe, K. Kuwahara, J. Suzuki, M. Matsuda, KEK Report 2001-3 (2001)

[18] S. Itoh, T. Yokoo, S. Satoh, S. Yano, D. Kawana, J. Suzuki, T.J. Sato, Nucl. Instrum. Methods Phys. Res., Sect. A **631**, 90 (2011)

[19] M. Arai, R. Kajimoto, M. Nakamura, Y. Inamura, K. Nakajima, K. Shibata, N. Takahashi, J. Suzuki, S. Takata, T. Yamada et al., J. Phys. Soc. Jpn. **82**, SA024 (2013)

[20] R. Kajimoto, M. Nakamura, K. Iida, K. Kamazawa, K. Ikeuchi, Y. Inamura, M. Ishikado, J. Neutron Res. **22**, 99 (2020)

[21] R.E. Lechner, in *Proc. ICANS-XI* (1991), KEK Report 90-25, p. 717

[22] C.G. Windsor, *Pulsed Neutron Scattering* (Taylor & Francis Ltd., London, 1981)

[23] J. Voigt, N. Violini, T. Brückel, Nucl. Instrum. Methods Phys. Res., Sect. A **741**, 26 (2014)

[24] `http://j-parc.jp/researcher/MatLife/en/instrumentation/ns3.html`

[25] D. Mildner, R. Sinclair, Ann. Nucl. Energy **6**, 225 (1979)

[26] S. Itoh, T. Yokoo, T. Masuda, S. Asai, H. Saito, D. Kawana, R. Sugiura, T. Asami, Y. Ihata, Physica B **568**, 76 (2019)

[27] J.Y. Lin, A. Banerjee, F. Islam, M.D. Le, D.L. Abernathy, Physica B **562**, 26 (2019)

[28] K. Lefmann, K. Nielsen, Neutron News **10**, 20 (1999)

[29] P. Willendrup, E. Farhi, K. Lefmann, Physica B **350**, E735 (2004)

[30] P. Willendrup, E. Farhi, E. Knudsen, U. Filges, K. Lefmann, J. Neutron Res. **17**, 35 (2014)

[31] P.K. Willendrup, K. Lefmann, J. Neutron Res. **22**, 1 (2020)

[32] P.K. Willendrup, K. Lefmann, J. Neutron Res. **23**, 7 (2021)

[33] K. Nakajima, S. Ohira-Kawamura, T. Kikuchi, M. Nakamura, R. Kajimoto, Y. Inamura, N. Takahashi, K. Aizawa, K. Suzuya, K. Shibata et al., J. Phys. Soc. Jpn. **80**, SB028 (2011)

[34] G. E. Granroth, J. Phys. Soc. Jpn. **80**, SB016 (2011)

[35] P. Deen, A. Vickery, K. Andersen, R. Hall-Wilton, EPJ Web Conf. **83**, 03002 (2015)

[36] G. Sala, J.Y.Y. Lin, V.B. Graves, G. Ehlers, J. Appl. Crystallogr. **51**, 282 (2018)

[37] P.P. Deen, S. Longeville, W. Lohstroh, F. Moreira, G. Fabrèges, L. Loaiza, D. Noferini, Rev. Sci. Instrum. **92**, 105104 (2021)

[38] R. Bewley, J. Taylor, S. Bennington., Nucl. Instrum. Methods Phys. Res., Sect. A **637**, 128 (2011)

[39] M. Russina, G. Guenther, V. Grzimek, R. Gainov, M.C. Schlegel, L. Drescher, T. Kaulich, W. Graf, B. Urban, A. Daske et al., Physica B **551**, 506 (2018)

[40] R.A. Ewings, J.R. Stewart, T.G. Perring, R.I. Bewley, M.D. Le, D. Raspino, D.E. Pooley, G. Škoro, S.P. Waller, D. Zacek et al., Rev. Sci. Instrum. **90**, 035110 (2019)

[41] K.H. Andersen, J. Neutron Res. **10**, 179 (2002)

[42] R. Kajimoto, K. Nakajima, M. Nakamura, K. Soyama, T. Yokoo, K. Oikawa, M. Arai, Nucl. Instrum. Methods Phys. Res., Sect. A **600**, 185 (2009)

[43] K. Ohoyama, S. Itoh, T. Otomo, T. Osakabe, J. Suzuki, M. Matsuda, K. Kuwahara, M. Arai, in *Proc. ICANS-XV* (2001), JAERI-Conf 2001-002/KEK Proceedings 2000-22, p. 418

MIASANS at the longitudinal neutron resonant spin-echo spectrometer RESEDA

Jonathan C. Leiner,[1,3,*] *Christian* Franz,[2] *Johanna K.* Jochum,[1,3] and *Christian* Pfleiderer[1,4,5]

[1]Physik Department, Technische Universität München, D-85748 Garching, Germany
[2]Jülich Centre for Neutron Science JCNS, Forschungszentrum Jülich GmbH Outstation at MLZ FRM - II, D-85747 Garching, Germany
[3]Heinz Maier - Leibnitz Zentrum (MLZ), Technische Universität München, D-85748 Garching, Germany
[4]Center for Quantum Engineering (ZQE), Technische Universität München, D-85748 Garching, Germany
[5]Munich Center for Quantum Science and Technology (MCQST), Technische Universität München, D-85748 Garching, Germany

Abstract. The RESEDA (Resonant Spin-Echo for Diverse Applications) instrument has been optimized for neutron scattering measurements of quasi-elastic and inelastic processes over a wide parameter range. One spectrometer arm of RESEDA is configured for the MIEZE (Modulation of Intensity with Zero Effort) technique, where the measured signal is an oscillation in neutron intensity over time prepared by two precisely tuned radio-frequency (RF) flippers. With MIEZE, all spin-manipulations are performed before the beam reaches the sample, and thus the signal from sample scattering is not disrupted by any depolarizing conditions there (i.e. magnetic materials and fields). The MIEZE spectrometer is being further optimized for the requirements of small angle neutron scattering (MIASANS), a versatile combination of the spatial and dynamical resolving power of both techniques. We present the current status of (i) the newly installed superconducting solenoids as part of the RF flippers to significantly extend the dynamic range (ii) the development and installation of a new detector on a translation stage within a new larger SANS-type vacuum vessel for flexibility with angular coverage and resolution, and (iii) the efforts to reduce background.

1 Introduction

1.1 Motivation

Small angle neutron scattering (SANS) represents an exceptionally important probe for a wide range of scientific problems. SANS is a powerful tool for studying the structure of complex fluids and polymers [1, 2]. SANS is also routinely practical for studying the soft matter materials and structures such as biological macromolecules, e.g. proteins [3] and DNA [4]. In addition, SANS is particularly useful in resolving mesoscale magnetic structures [5] such as those of magnetic nanoparticles [6, 7] and vortices in superconductors [8]. The comprehensive utilities of SANS are difficult to fully enumerate, as shown by the breadth and depth contained in various review articles [9, 10].

In recent years, the interplay of processes on microscopic scales with texture formations and the slow dynamics on mesoscopic and macroscopic length and time scales has attracted great interest in many different areas of condensed matter physics, motivating investigations over a very large dynamic range [11]. Materials studied comprise soft matter and biological systems, complex bulk compounds, as well as hard condensed matter materials, such as complex metallic alloys, magnets, and superconductors. For one such material, the ferromagnetic superconductor UGe_2, momentum transfers Q down to 0.008 $Å^{-1}$ and an energy resolution of 1 μeV was necessary to fully characterize the spin fluctuations [12]. Such systems take the form of bulk samples, heterostructures, and nanoscale configurations.

Despite the versatility and heavy demand for SANS measurements, there is essentially no dedicated instrumentation for SANS studies (i.e. pinhole collimation of the incident neutron beam) of the dynamical properties. On the one hand, triple axis or time-of-flight spectrometers are operated at the limit of their accessible parameter-range without optimized background reduction. The upper limit of energy (Fourier time) resolution for such techniques is around 600 μeV (1 picosecond) [13]. For backscattering spectrometers the upper limit is 0.3 μeV (~2 ns) [11]. On the other hand, conventional neutron spin-echo spectrometers are optimized for large samples and large beam-divergences on the expense of the requirements of SANS. Optimized in this way they can reach Fourier times of 1000 ns with Q down to 0.001 $Å^{-1}$ [14] and as high as 4 $Å^{-1}$ [15]. However, being based on polarized neutrons, conventional spin-echo spectrometers suffer from large incoherent background in studies of hydrogen containing materials, while studies of depolarizing samples (such as magnetic materials) or use of depolarizing sample environment are prohibitively difficult in all but ideal cases [16].

*e-mail: jon.leiner@tum.de

Magnetic materials is a classification that envelops exceptional diversity, they represent a key material category where spectroscopic studies are a necessary tool for comprehensive exploration and characterization. For example, neutron spectroscopy of the magnetic fluctuations near quantum critical points is a means to extract crucial information about their behavior over broad spatial and dynamical ranges [17]. In magnetically frustrated spin systems, often spin-glass transitions can develop resulting in interesting disordered ground states, some of which contain the fingerprints of the Berry phase and other contributions from wavefunction topology. In general, the topological spin textures arising out of careful balances between the magnetic interactions continue to open up new potential cases for applications. Exploring and characterizing the behavior of these types of systems requires the resolution of small momentum and energy transfers, since in general mesoscale textures and microscopic dynamics are involved. This dual requirement naturally demands a detector with both the highest possible spatial (2D) and temporal resolutions, and ideally the capability to adjust the sample-detector distance for flexibility between the angular coverage and angular resolution.

A different way to illustrate the need for neutron spin-echo spectroscopy in small angle scattering configuration represent the requests for experimental beam time at the beamline RESEDA at FRM II. With the successful implementation of the longitudinal neutron resonance spin-echo and longitudinal MIEZE technique set-up, an increasing number of experimental proposals has been submitted for studies at momentum transfers below 0.01 $Å^{-1}$ - 1 $Å^{-1}$, i.e., in the regime of small angle scattering. Notably, since 2015 this concerned over 60% of the proposals submitted. Moreover, these proposals specifically required the MIEZE technique as they were dealing either with hydrogen-containing substances, ferromagnetic or superconducting materials, or strongly depolarizing sample environments. Examples include (undeuterated) water, Fe, UGe_2, and MnSi measurements as summarized in Ref. [18].

1.2 The MIEZE technique

Neutron spin-echo spectroscopy (NSE) [19–21] allows measurements at both high resolution *and* intensity as compared to triple-axis spectroscopy or time-of-flight spectroscopy, for which high energy resolution can only be reached at the expense of considerable losses of intensity. NSE gets around this tradeoff by producing neutron energy encoding via Larmor precession. In contrast to NSE, in neutron resonant spin-echo (NRSE) [22] the large magnetic solenoids used to encode the neutron energies in their Larmor precessions are replaced by pairs of resonant spin flippers which are driven by a oscillating radio-frequency magnetic fields [23, 24]. Note these oscillating RF fields must be surrounded by the appropriate static field B_0 tuned to give the Larmor resonance condition. This is essentially the Nuclear Magnetic Resonance (NMR) methodology for neutrons instead of atomic nuclei. Resonant spin flippers inherently reduce the differences in magnetic field integral that divergent neutron trajectories experience, and so the complicated field correction coils used in NSE [25] are not necessary to reach the same instrument performance. More details about the current status of the NRSE technique and its relationship to other variants of spin-echo spectrometers can be found in Refs. [18, 26].

Modulation of Intensity with Zero Effort (MIEZE) is a neutron resonant spin-echo technique variation where potential information loss via depolarization of the Larmor precessing spins prepared by the resonant flippers is preempted by placing a spin analyzer just before the sample position, which results in an intensity modulated neutron beam downstream of the analyzer [27–33]. No further beam manipulation components are placed between the sample and the detector in contrast to conventional NSE and NRSE. The end result is that the same information which was encoded in the Larmor precession (i.e. in the interference between the two spin states of a neutron) is instead encoded in the intensity modulation of the neutron beam. More specifically, before the analyzer the two RF flippers couple together an individual neutron's spin and energy degrees of freedom via resonant spin flips. By selecting just one spin state, the analyzer removes the spin degree of freedom from the neutron. What remains then is a neutron with two separated energy states, prepared in such a way that the separation length goes to zero at the detector position (without any sample interactions). For many neutrons combined, the cumulative consequence of this energy state overlap is an intensity beating with a large amplitude facilitating its observation with a detector.

MIEZE requires a position sensitive detector which can accurately measure the time-of-flight of the neutrons [34] after they have scattered from the sample. The detector must be placed at a specific position given by the MIEZE condition, where an oscillating intensity will be observed as determined by the frequencies of the RF flippers f_1 and f_2:

$$\frac{L_1}{L_2} = \frac{f_2 - f_1}{f_1} \tag{1}$$

Here, L_1 is the distance between the two RF-flippers, whereas L_2 is the distance between the second RF-flipper and the detector.

As has been demonstrated a number of times [12, 16, 35–38], this setup imparts the capability to measure with magnetic samples, magnetic fields in the sample environment, and samples with a high incoherent scattering cross section. These capabilities are acquired without the need for the instrumental complexities faced in ferromagnetic spin-echo [16, 39]. Indeed, proof-of-concept measurements under large magnetic fields up to 17 Tesla [40] at RESEDA underscore the inherent advantages of the MIEZE technique as compared to the ferromagnetic spin-echo set-up, where the stray field of the magnet would interfere significantly with the instrument components. As a result, MIEZE has the highest energy resolution (and Q range) for magnetic scattering [11].

As spin-echo spectroscopy encodes the sample scattering processes in the spin phase of the Larmor precession induced in the neutrons, it is inherently sensitive to path

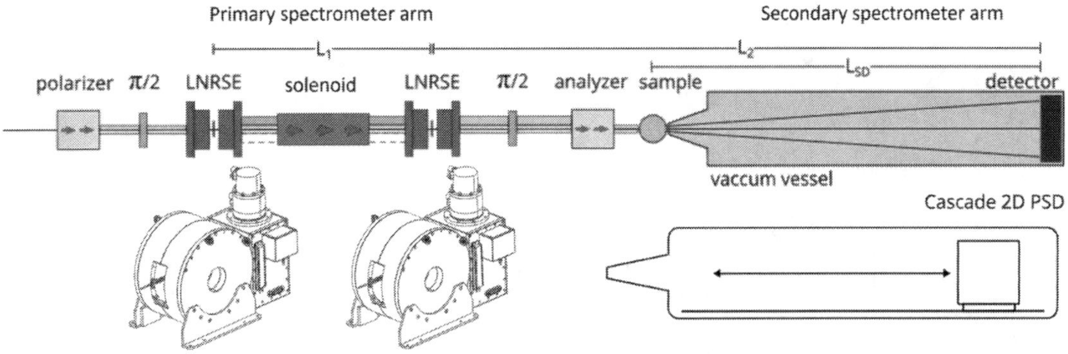

Figure 1. Schematic depiction of the RESEDA-MIASANS (Modulation of Intensity Applied to Small Angle Neutron Scattering) set up with sketches of the main new components. The RESEDA beamline is comprised of the primary spectrometer arm with the two resonant spin flippers (left), the sample environment indicated in the center, and secondary spectrometer with the vacuum vessel containing the position sensitive CASCADE detector (right). Shown below the diagram of the beamline components are sketches of the new B_0 superconducting solenoids for the RF spin flippers and new detector tank with space for a larger CASCADE detector and on-demand detector translation capability over the full L_{SD} range.

length differences of different neutron trajectories [41]. A major advantage of the longitudinal NRSE and MIEZE set-up concerns the self-correction of path length differences of parallel neutron trajectories, and a strongly reduced sensitivity to differences of divergent beam trajectories [42]. However, the signal in MIEZE studies degrades for large scattering angles due to path length differences, for which there are currently no feasible compensation methods for the longitudinal configuration of RESEDA. (However, note that in the transverse MIEZE configuration there are multiple viable ways to do such compensation [33, 43]) The strength of this signal reduction has been experimentally quantified for different momentum transfers [18] and agrees well with the calculated expectations [44]. Fortunately though, all commissioning and scientific studies clearly show that the low angle regime is favorable for the MIEZE technique, as flight path differences of neutron trajectories from the sample to the detector [45] play only a minor role [46, 47] and the full resolution can be exploited without restrictions on the sample geometry. For this essential reason, the MIEZE technique is especially compatible with SANS type measurements. This special compatibility has been known for over two decades and demonstrations [48–50] and experiments [51] have been carried out over the years under the MISANS (MI-SANS) designation.

1.3 The RESEDA instrument

RESEDA (REsonance Spin-Echo for Diverse Applications) is a high resolution resonance spin-echo spectrometer located at the cold neutron guide NL5-S in the Neutron Guide Hall West of the FRM II reactor in Garching, Germany [52, 53]. In its present configuration the instrument provides access to a large time and scattering vector range for quasi-elastic measurements [18, 28, 54].

A critical milestone in the development of RESEDA is represented by the development of electronics and coils for the generation of high frequency oscillatory magnetic fields. Using a combination of signal generators, amplifiers, impedance matching transformers, and a carefully designed cascade of tunable capacitors and customized coils, it is possible to achieve a very broad bandwidth from 35 kHz (lower limit due to the Bloch-Siegert shift [55]) to unprecedented 3.6 MHz [56], leading to the extremely high resolution of the spectrometer.

At present in the MIEZE set-up a transmission bender analyzer is used in front of the sample for the spin analysis. This transmission bender meets its high specifications, i.e. high polarisation (> 95%), without affecting the beam properties (such as divergence). However, for small angle scattering geometries, a sufficient degree of collimation is necessary because a residual background arises from the roughness of the wafers in the transmission bender [57]. The background may be suppressed by means of a 20' (horizontal and/or vertical) collimation behind the bender. It is important to note that the beam divergence for the small angle scattering geometry is determined by the pinhole collimation on the primary spectrometer arm. In turn, no neutrons are lost with the addition of a 20' collimator.

2 MIASANS Objectives

In its previous configuration, RESEDA was optimized to cover a wide parameter range, but there is still opportunity to significantly expand this range in a cost effective manner. The aim of the RESEDA-MIASANS (longitudinal Modulation of Intensity Applied and optimized for Small Angle Neutron Scattering) project has been to further develop the resonant neutron spin-echo spectrometer RESEDA in the area of small angle scattering with higher energy resolution. Two major component upgrades provide these objectives: (i) upgrading the resonant flippers with compact, cryogen free, superconducting magnet coils for generating the static B_0 field (ii) equipping the secondary spectrometer arm used for MIEZE studies with

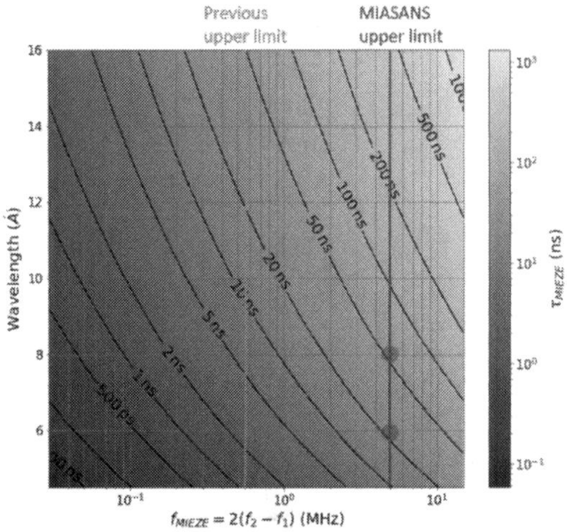

Figure 2. The resolution, i.e. τ_{MIEZE}, of a MIEZE instrument given as a function of neutron wavelength and the frequency difference ($f_2 - f_1$) between the two resonant flippers according to Eq. 2. τ_{MIEZE} for selected wavelengths (6 Å is the most common wavelength setting for experiments at RESEDA) are indicated as described in the main text. The newly accessible regimes of τ_{MIEZE} made possible by the MIASANS project (assuming spin-flips at 8 MHz; B_0 = 300 mT) are labeled accordingly.

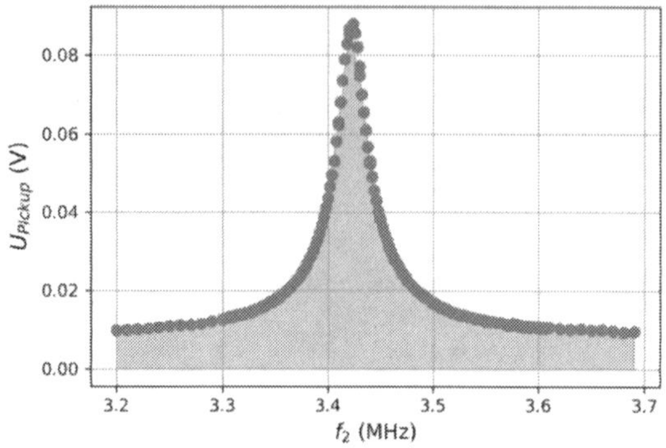

Figure 3. A representative resonance scan (as described in Ref. [56]) of the second MIEZE RF flipper coil showing a peak at 3.42 MHz using the newly installed superconducting coils which provide the static field B_0 around the resonance coils for longitudinal resonant spin-echo. As shown in Fig. 9 of Ref. [56], these new magnets will allow for resonant flips at a record high frequency f = 3.6 MHz at RESEDA.

a detector tank for background reduction and translation stage for increased versatility in positioning the detector and (iii) reducing the background from the primary spectrometer arm.

A schematic depiction of the RESEDA instrument with the new MIASANS components is shown in Fig. 1. Note that RESEDA is a modular instrument and development of this tank will not affect the possibility to perform longitundinal NRSE measurements. In addition, it is crucial to further optimize the neutron beam generated on the primary spectrometer arm. As introduced in detail in Sec. 1.2 above, in a MIEZE spectrometer all spin manipulations are carried out on the primary spectrometer arm. For high scattering angles and relaxed resolutions, the neutron flux on the sample is maximised by means of m = 1.2 supermirror guides. However, for small angle scattering and high resolution this would severely limit the accessible momentum-range and momentum resolution. Even with a pinhole collimation, small angle scattering from the aluminum in the RF-flippers and the air surrounding them in the space outside the neutron guides creates a significant amount of unwanted background. An objective in the MIASANS project is to quantify this background and implement the most effective ways to mitigate it.

3 Primary spectrometer arm

3.1 Superconducting static field coils

Here we briefly describe the significantly increased resolution, i.e. τ_{MIEZE}, capabilities from higher frequency spin flips generated in the two resonant RF flippers on the primary arm of RESEDA. Fig. 2 illustrates the increase in the upper limits of τ_{MIEZE} as the MIEZE frequency is raised. Analogous to the so-called spin-echo time, τ_{MIEZE} depends on the neutron wavelength λ, the frequency difference between the two RF flippers ($f_2 - f_1$), and the sample-detector distance L_{SD}:

$$\tau_{MIEZE} = \frac{m^2}{h^2} \cdot \lambda^3 \cdot 2\pi(f_2 - f_1) \cdot L_{SD} \qquad (2)$$

(m = neutron mass, h = Planck's constant). The substantial improvement of accessible τ_{MIEZE} is made possible with the newly procured superconducting solenoids which provide static B_0 fields up to 300 mT with the required field homogeneity over the active resonant flipping areas.

To reach the maximum resolution and full dynamic range [29] of the existing MIEZE set-up at RESEDA, RF spin flips at frequencies up to 4 MHz are required [56]. However, the maximum flipping frequency was previously limited to roughly 1 MHz due to capacitive stray coupling between the the RF coils and conventional Helmholtz coils used to generate the static B_0 field. This was attributed to capacitive coupling causing a parasitic damping of the RF oscillations, and as such this situation could be resolved by reducing/eliminating such couplings.

Indeed, this solution was demonstrated using a small high-T_C superconducting magnet of the sample environment group at FRM II to generate the static B_0 field, and with that the resonant flip at a frequency of 3.6 MHz using a neutron wavelength of 6 Å was confirmed as shown in Fig. 9 of Ref. [56]. For this test a static field B_0 = 122 mT was required.

Therefore, two new 300mT high temperature superconductor (HTS) magnet systems were designed together with the company HTS-110 (based in New Zealand) to the specifications achieved in the aforementioned test-set up. Following successful performance characterization, these new customized superconducting magnets from HTS-110

https://doi.org/10.1051/epjconf/202227202008

were integrated on the RESEDA beamline. The first functional tests of these magnets show resonant frequencies up to 3.42 MHz, as seen in Fig. 3. Future test with neutrons on this new RESEDA configuration are expected to reproduce the data shown in Fig 9 of Ref. [56]

The newly installed superconducting B_0 coils have been shown to provide field strengths up to 320 mT. This static field strength will permit resonant spin-flips up to 8 MHz on the primary arm of RESEDA, which corresponds to maximum $f_{MIEZE} = 2(f_2 - f_1)$ of 4.9 MHz. This enables τ_{MIEZE} for practical experiments to reach 20 and 50 nanoseconds (with L_{SD} = 3340 mm), for neutron wavelengths of 6 and 8 Å respectively as can be seen in Fig. 2. Using wavelengths above 8 Å to reach higher τ_{MIEZE} is possible in principle, but impractical due to the substantial loss of neutron flux at those wavelengths. With these improvements in the field integral (~0.5 Tm) and τ_{MIEZE} at RESEDA, the instrument makes a considerable step closer to the performance of world leading NSE spectrometers such as J-NSE Phoenix [58] and IN15 [14].

Thus in the future it is planned to construct and install compact resonant spin-flipping solenoids which can go up to 8 MHz instead of the 3.6 MHz [56]. At such high frequencies the density of phase rings on the detector will increase. In order to extract meaningful results from measurements, the phase difference contained within one detector pixel should be less than 20° [45, 50]. For the case of a MIEZE frequency f_{MIEZE} = 4.9 MHz (and time resolution of 20 MHz using the 4 point method [59]), then on the outermost edge of the detector where the phase rings are densest there will be a phase difference of 115° within one pixel. Such an extreme phase difference in one pixel would result in severe reduction in the measured signal (contrast). In order to try and avoid this in such extreme cases, the area of the individual detector pixels needs to be further reduced, which may be feasible in the next iterations of the ^{10}B GEM foil production for CASCADE detectors [60, 61].

3.2 Background reduction

A substantial source of background scattering at the RESEDA instrument has been identified to originate from the air in front of the sample location. This scattering from air may be reduced by customized evacuated adapters matched to the different sample environments available, namely the closed cycle cryostats, furnaces, and 5 T and 12 T superconducting magnets. Using a system of standard KF flanges in combination with sapphire windows, a considerable reduction of the air within the neutron flight paths has been achieved without compromising the full range of required distances and geometries for SANS. However, there are still several portions of the primary arm where small volumes of air cut through the neutron paths, such as gaps in between the different evacuated neutron guides to accomodate the resonant flippers and other vital components. Further gains in background reduction could be achieved by minimizing these portions of air.

In addition to the air, another source of SANS background originates from the material of the longitudinal RF coils, which consist of pure aluminium with a thickness of 0.2 mm. It is noteworthy that this is significantly less material in the path of the neutron beam in comparison to the transverse NRSE configuration, where the B_0 coil has a thickness of several mm (in the previous TNRSE design at RESEDA, this thickness was 6 mm plus a layer of anodized aluminium). Replacing the anodized aluminium with pure aluminium alone reduced the background scattering produced by the coils by a factor of 100 [62]. In the current configuration of RESEDA, no transmission losses through the RF coils have been observed in either McStas simulations or experiments.

In an attempt to distinguish the contributions from these different sources of background arising within RESEDA's primary arm with pinhole slit apertures configured, we utilized the RESEDA instrument file setup in the Monte Carlo ray tracing program McStas [63, 64].

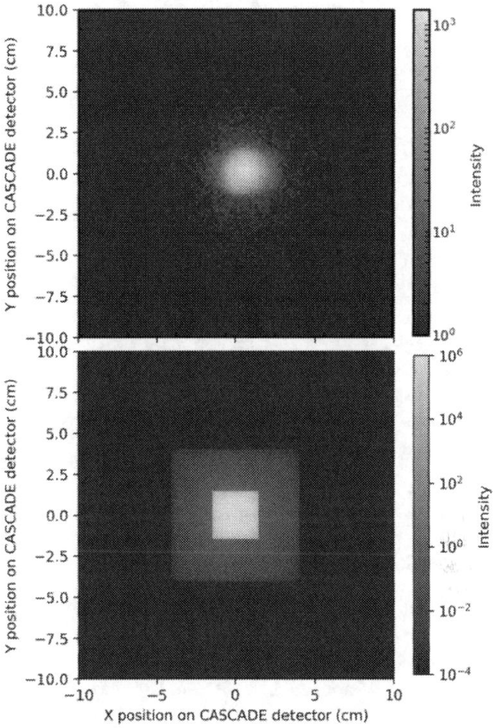

Figure 4. Typical detector image from the 2D position sensitive CASCADE detector [60, 61] at RESEDA, illustrating the small angle scattering scattering background halo surrounding the direct beam. (top) Experimentally measured detector image, with standard 20' horizontal and vertical collimation. (bottom) McStas simulation of the same configuration, without any air scattering being simulated, showing that the background from SANS scattering at the RF flipper components accounts for some of this halo. This can be qualitatively compared with the SANS measurements from pure aluminum RF coils shown in Fig 2.5 of Ref. [62]. Note that because diffraction and air scattering are not accounted for in the McStas simulation, the square shape from the guides and apertures is traced through to the detector image.

With the information from these simulations, the theoretical configurations for optimized SANS background reduction may be discerned.

As shown in Fig. 4, a halo of background around the direct beam can be observed in McStas simulations, where it is possible to easily insert and remove the background sources. Based on such simulations, it can be deduced that the majority of this background is incoherent scattering originating from air and aluminum components in the neutron flight path on the primary arm of RESEDA. Thus, steps will be taken to reduce these remaining portions of air on the primary arm. Importantly, the McStas simulations allowed us to test the hypothesis that the reflections carried through the m = 1.2 neutron guides may enhance the background halo around the direct beam. Simulation results with an m = 0 guide were negligibly different in comparsion with the m = 1.2 guide.

4 Secondary spectrometer arm

4.1 Vacuum Vessel

The secondary spectrometer arm currently available at RESEDA for the MIEZE set-up, which consists of an evacuated flight tube (only 30 cm in diameter) with the CASCADE detector attached to the outside of the far end, requires massive improvements for optimization of SANS measurements. The main reasons such an upgrade is necessary are (i) the reduction in the background scattering in SANS geometry and (ii) the facilitation of maximum flexibility between the largest angular coverage with relaxed time resolutions and reduced angular coverage for the highest τ_{MIEZE}. For the purpose of this flexibility, it is mandatory to mount the detector inside a vacuum tank large enough to accommodate a translation rail system akin a conventional SANS detector tank. In addition, mounted in front of the detector will be three customized beamstops, which can be horizontally adjusted on-demand anywhere along the active detector area at the height of the direct beam. These automated beamstops will save valuable experiment time over the manual manner in which this direct beam attenuation was done before.

The new 101 cm diameter vacuum vessel, manufactured by the Added Value Solutions (AVS) firm, is configured to accommodate a position sensitive CASCADE detector [60, 61] covering an area of 30 x 30 cm^2 expected to be commercially available in the coming years. This will enable potential increases of the count rate by a factor of two as compared to the current 20 x 20 cm^2 detectors discussed in the following section.

4.2 CASCADE detector

The existing 20 x 20 cm^2 CASCADE detector [60, 61] at RESEDA has proven very effective [35] especially since the foil waviness has been fully characterized [28]. Such waviness can be corrected in the standard data reduction procedure (MIEZEPY [47]). Spin-echo signals up to 100 ns have been recorded in the direct beam, albeit with very low intensity.

As noted previously in Sec. 3.1, phase effects across the pixels will eventually become one of the resolution bottlenecks of the (flat) CASCADE detector. The 1.56 mm pixel size of the CASCADE detector is (much) smaller than the average pixel size of a typical SANS detector (8 mm), and therefore any effects due to the pixelated nature of the detector have not yet been observed in experiments. For example, given a MIEZE frequency of 2.5 MHz (time resolution of 10 MHz using the 4 point method [59]), and using 6 Å neutrons with L_{SD} = 3340 mm, one estimates a phase variation in a single pixel of ~15°. This value is still below the threshold of 20° given in [45, 50]. Even though the neutron counts in each individual small pixel will be quite low, with the proper phase correction procedure implemented with the MIEZEPY software [47], the cumulative intensity of a custom group (bin) of pixels can be determined.

The setting accuracy and stability of the power supplies yields a sufficient phase stability for the frequencies used so far at RESEDA (i.e. the phase does not drift significantly over time when the settings are maintained). Furthermore, all frequency generators and RF circuits receive their signal from the same high precision Rubidium atomic clock time-standard. However, in this context, other environmental contributions to the phase stability need to be considered, particularly effects from temperature variations. Based on the thermal expansion of alumninum, a 1 K temperature variation with the aforementioned parameters induces a ΔL_{SD} and a corresponding phase shift below the critical 20° threshold. This however does further motivate a reduction in detector pixel size in order to better account for temperature variations at the highest MIEZE frequencies. For these reasons, once the time resolution of the CASCADE detector is pushed beyond 10 MHz, it will become necessary to improve upon the spatial resolution for use at the upper resolution limits of the MIEZE technique. Fortunately, recent innovations in the GEM foil production processes [65] will allow CASCADE-type detectors with a larger active area and smaller pixel size to be manufactured, making such improvements in spatial resolution feasible [45, 50].

CDT CASCADE Detector Technologies has completed the design and construction of a detector housing large enough to fit future ^{10}B GEM detectors. The housing is equipped with a circular detector window of 44 cm diameter which surrounds a 30 x 30 cm^2 square active detector area. The detector window thickness is minimized to avoid secondary scattering effects as much as possible. Furthermore, the new detector is adapted to minimize local power dissipation such that active cooling of the detector electronics is unnecessary. Initially this new 20 x 20 cm^2 CASCADE detector will be placed in this new housing, for use at RESEDA until the aforementioned 30 x 30 cm^2 detector with increased spatial resolution becomes available.

5 Summary

We have designed a MIEZE set-up as part of the spectrometer RESEDA at FRM II that is optimized for the requirements of small angle neutron scattering (MIASANS). This project builds upon major progress with the longitudinal neutron resonance spin-echo method, achieved in recent years at RESEDA [18, 28, 29, 42, 54, 56, 59]. We have so far demonstrated resonant spin flips up to 3.6 MHz, and established that our CASCADE detector is sufficiently sensitive to the highest frequency intensity modulations currently available. As a crucial milestone, the superconducting magnet systems from HTS-110 were delivered and successfully put into operation at the RESEDA beamline.

The results of the Monte Carlo simulations indicate substantial contributions from scattering by air and aluminum components on the primary arm of RESEDA in front of the sample environment. As the amount of aluminum in the neutron beam is already minimized, further steps will be taken to minimize the volume of air in the neutron flight paths.

Another important milestone was the design of the detector tank, which is currently in production by AVS. In comparison to standard SANS beamlines, the new RESEDA detector tank will be mounted onto a reconstructed support structure for the MIEZE spectrometer arm, permitting continued studies at large scattering angles with relaxed resolutions in addition to SANS measurements with high resolutions. As an additional feature needed to attenuate the direct beam, three customized movable beam stops will be installed.

Finally, a newly enhanced CASCADE detector covering an area of 20 x 20 cm^2 and its vacuum housing has been manufactured by CDT. The vacuum housing is large enough for a detector with an active area of 30 x 30 cm^2 expected to be commercially available in the not-to-distant future. This will result in increases of the count rate by a factor of two under optimal conditions as compared to the current set-up of the RESEDA instrument.

We emphatically thank P. Böni, C. Fuchs, K. Fellner, P. Bender, P. Wild, L. Vogl, M. Kleinhans, N. Huber, A. Wendl, the E51 group in the Physics Department of the Technical University of Munich, and the Forschungs-Neutronenquelle Heinz Maier-Leibnitz (FRM II) for support, assistance with setup and testing of the equipment described here, and fruitful discussions. We also wish to thank T. Huang from HTS-110, P. Noguera Crespo and F. Cacho-Nerin from Added Value Solutions (AVS), and M. Klein and C. Schmidt from CDT technologies GmBH for their support in manufacturing the MIASANS equipment described here. This project has been funded by the by the German Bundesministerium für Bildung und Forschung (BMBF, Federal Ministry for Education and Research) through the Project No. 05K19W05 (Resonante Longitudinale MIASANS Spin-Echo Spektroskopie an RESEDA). We also acknowledge support of the German excellence initiative under EXC-2111 (MCQST, project-id 390814868).

References

[1] A.R. Brás, R. Pasquino, T. Koukoulas, G. Tsolou, O. Holderer, A. Radulescu, J. Allgaier, V.G. Mavrantzas, W. Pyckhout-Hintzen, A. Wischnewski et al., Soft Matter **7**, 11169 (2011)

[2] H. Yao, B.D. Olsen, Soft Matter **17**, 5303 (2021)

[3] S.J. Perkins, Biochemical Journal **254**, 313 (1988)

[4] C.K. Kennaway, J.E. Taylor, C.F. Song, W. Potrzebowski, W. Nicholson, J.H. White, A. Swiderska, A. Obarska-Kosinska, P. Callow, L.P. Cooper et al., Genes & development **26**, 92 (2012)

[5] J. Kindervater, I. Stasinopoulos, A. Bauer, F.X. Haslbeck, F. Rucker, A. Chacon, S. Mühlbauer, C. Franz, M. Garst, D. Grundler et al., Phys. Rev. X **9**, 041059 (2019)

[6] D. Honecker, M. Bersweiler, S. Erokhin, D. Berkov, K. Chesnel, D.A. Venero, A. Qdemat, S. Disch, J.K. Jochum, A. Michels et al., Nanoscale Adv. **4**, 1026 (2022)

[7] P. Bender, D. Honecker, M. Bersweiler, R. Costo, T. Kahmann, F. Ludwig, J. Leiner, J.K. Jochum, Journal of Applied Crystallography **55**, 586 (2022)

[8] M.R. Eskildsen, E.M. Forgan, H. Kawano-Furukawa, Reports on Progress in Physics **74**, 124504 (2011)

[9] S. Mühlbauer, D. Honecker, E.A. Périgo, F. Bergner, S. Disch, A. Heinemann, S. Erokhin, D. Berkov, C. Leighton, M.R. Eskildsen et al., Rev. Mod. Phys. **91**, 015004 (2019)

[10] C.M. Jeffries, J. Ilavsky, A. Martel, S. Hinrichs, A. Meyer, J.S. Pedersen, A.V. Sokolova, D.I. Svergun, Nature Reviews Methods Primers **1**, 1 (2021)

[11] J.S. Gardner, G. Ehlers, A. Faraone, V. García Sakai, Nature Reviews Physics **2**, 103 (2020)

[12] F. Haslbeck, S. Säubert, M. Seifert, C. Franz, M. Schulz, A. Heinemann, T. Keller, P. Das, J.D. Thompson, E.D. Bauer et al., Phys. Rev. B **99**, 014429 (2019)

[13] R. Bewley, J. Taylor, S. Bennington, Nuclear Instruments and Methods in Physics Research Section A: Accelerators, Spectrometers, Detectors and Associated Equipment **637**, 128 (2011)

[14] B. Farago, P. Falus, I. Hoffmann, M. Gradzielski, F. Thomas, C. Gomez, Neutron News **26**, 15 (2015)

[15] M. Ohl, M. Monkenbusch, N. Arend, T. Kozielewski, G. Vehres, C. Tiemann, M. Butzek, H. Soltner, U. Giesen, R. Achten et al., Nuclear Instruments and Methods in Physics Research Section A: Accelerators, Spectrometers, Detectors and Associated Equipment **696**, 85 (2012)

[16] B. Farago, F. Mezei, Physica B+C **136**, 100 (1986), neutron Scattering

[17] F. Mezei, Phys. Rev. Lett. **49**, 1096 (1982)

[18] C. Franz, S. Säubert, A. Wendl, F.X. Haslbeck, O. Soltwedel, J.K. Jochum, L. Spitz, J. Kindervater, A. Bauer, P. Böni et al., Journal of the Physical Society of Japan **88**, 081002 (2019)

[19] F. Mezei, Zeitschrift für Physik A Hadrons and nuclei **255**, 146 (1972)

[20] F. Mezei, in *Neutron Spin Echo. Lecture Notes in Physics* (Springer, Berlin, Heidelberg, 1980), Vol. 128

[21] T. Keller, R. Golub, R. Gähler, in *Scattering*, edited by R. Pike, P. Sabatier (Academic Press, 2002), p. 1264

[22] M. Köppe, M. Bleuel, R. Gähler, R. Golub, P. Hank, T. Keller, S. Longeville, U. Rauch, J. Wuttke, Physica B: Condensed Matter **266**, 75 (1999)

[23] R. Gähler, R. Golub, T. Keller, Physica B: Condensed Matter **180**, 899 (1992)

[24] R. Dadisman, D. Wasilko, H. Kaiser, S.J. Kuhn, Z. Buck, J. Schaeperkoetter, L. Crow, R. Riedel, L. Robertson, C. Jiang et al., Review of Scientific Instruments **91**, 015117 (2020)

[25] B. Farago, P. Falus, I. Hoffmann, M. Gradzielski, F. Thomas, C. Gomez, Neutron News **26**, 15 (2015)

[26] R. Golub, R. Gähler, physics letters A **123**, 43 (1987)

[27] W. Besenböck, R. Gähler, P. Hank, R. Kahn, M. Köppe, C.H. De Novion, W. Petry, J. Wuttke, Journal of Neutron Research **7**, 65 (1998)

[28] C. Franz, O. Soltwedel, C. Fuchs, S. Säubert, F. Haslbeck, A. Wendl, J.K. Jochum, P. Böni, C. Pfleiderer, Nuclear Instruments and Methods in Physics Research Section A: Accelerators, Spectrometers, Detectors and Associated Equipment **939**, 22 (2019)

[29] J.K. Jochum, A. Wendl, T. Keller, C. Franz, Measurement Science and Technology **31**, 035902 (2019)

[30] N. Geerits, S.R. Parnell, M.A. Thijs, A.A. van Well, C. Franz, A.L. Washington, D. Raspino, R.M. Dalgliesh, J. Plomp, Review of Scientific Instruments **90**, 125101 (2019)

[31] T. Oda, M. Hino, H. Endo, H. Seto, Y. Kawabata, Physical Review Applied **14**, 054032 (2020)

[32] B. Liu, Z. Wang, Y. Wang, C. Huang, G. Sun, Journal of Instrumentation **15**, P04007 (2020)

[33] R. Dadisman, G. Ehlers, F. Li, Review of Scientific Instruments **93**, 013301 (2022)

[34] F. Funama, M. Hino, T. Oda, H. Endo, T. Hosobata, Y. Yamagata, S. Tasaki, Y. Kawabata, *A study of focusing TOF-MIEZE spectrometer with small-angle neutron scattering*, in *Proceedings of the 3rd J-PARC Symposium (J-PARC2019)* (2021), Vol. 33, p. 011088, `https://journals.jps.jp/doi/abs/10.7566/JPSCP.33.011088`

[35] W. Häussler, P. Böni, M. Klein, C.J. Schmidt, U. Schmidt, F. Groitl, J. Kindervater, Review of Scientific Instruments **82**, 045101 (2011)

[36] S. Säubert, J. Kindervater, F. Haslbeck, C. Franz, M. Skoulatos, P. Böni, Phys. Rev. B **99**, 184423 (2019)

[37] A. Wendl, Neutron spin echo spectroscopy on geometrically frustrated magnets, Master thesis, Technical University of Munich (2018)

[38] N. Martin, I. Mirebeau, C. Franz, G. Chaboussant, L.N. Fomicheva, A.V. Tsvyashchenko, Phys. Rev. B **99**, 100402 (2019)

[39] J.P. Boucher, F. Mezei, L.P. Regnault, J.P. Renard, Phys. Rev. Lett. **55**, 1778 (1985)

[40] J. Kindervater, N. Martin, W. Häussler, M. Krautloher, C. Fuchs, S. Mühlbauer, J. Lim, E. Blackburn, P. Böni, C. Pfleiderer, *Neutron spin echo spectroscopy under 17 T magnetic field at RESEDA*, in *EPJ Web of Conferences* (EDP Sciences, 2015), Vol. 83, p. 03008

[41] T. Weber, G. Brandl, R. Georgii, W. Häußler, S. Weichselbaumer, P. Böni, Nuclear Instruments and Methods in Physics Research Section A: Accelerators, Spectrometers, Detectors and Associated Equipment **713**, 71 (2013)

[42] M. Krautloher, J. Kindervater, T. Keller, W. Häussler, Review of Scientific Instruments **87**, 125110 (2016)

[43] F. Li, Journal of Applied Crystallography **55** (2022)

[44] L. Spitz, Reduction factor calculator (2019), `https://webapps.frm2.tum.de/reseda_redfac/`

[45] N. Martin, Nuclear Instruments and Methods in Physics Research Section A: Accelerators, Spectrometers, Detectors and Associated Equipment **882**, 11 (2018)

[46] W. Häussler, U. Schmidt, Phys. Chem. Chem. Phys. **7**, 1245 (2005)

[47] A. Schober, A. Wendl, F.X. Haslbeck, J.K. Jochum, L. Spitz, C. Franz, Journal of Physics Communications **3**, 103001 (2019)

[48] M. Bleuel, K. Littrell, R. Gähler, J. Lal, Physica B: Condensed Matter **356**, 213 (2005)

[49] M. Bleuel, R. Gahler, E. Lang, J. Lal, Nuclear Instruments and Methods in Physics Research Section A: Accelerators, Spectrometers, Detectors and Associated Equipment **600**, 220 (2009)

[50] G. Brandl, R. Georgii, W. Häussler, S. Mühlbauer, P. Böni, Nuclear Instruments and Methods in Physics Research Section A: Accelerators, Spectrometers, Detectors and Associated Equipment **654**, 394 (2011)

[51] S.J. Kuhn, N. Geerits, C. Franz, J. Plomp, R.M. Dalgliesh, S.R. Parnell, Journal of Applied Crystallography **54**, 751 (2021), number: 3 Publisher: International Union of Crystallography

[52] W. Häussler, B. Gohla-Neudecker, R. Schwikowski, D. Streibl, P. Böni, Physica B: Condensed Matter **397**, 112 (2007)

[53] W. Häussler, D. Streibl, P. Böni, Measurement Science and Technology **19**, 034015 (2008)

[54] C. Franz, O. Soltwedel, S. Säubert, A. Wendl, W. Gottwald, F. Haslbeck, L. Spitz, C. Pfleiderer, Conference Proceedings (2019)

[55] F. Bloch, A. Siegert, Phys. Rev. **57**, 522 (1940)

[56] J. Jochum, A. Hecht, O. Soltwedel, C. Fuchs, J. Frank, E. Faulhaber, J. Leiner, C. Pfleiderer, C. Franz, Measurement Science and Technology **32**, 045902 (2021)

[57] J.K. Jochum, J.F. Cooper, L.M. Vogl, P. Link, O. Soltwedel, P. Böni, C. Pfleiderer, C. Franz, Quantum Beam Science **6**, 26 (2022)

[58] S. Pasini, O. Holderer, T. Kozielewski, D. Richter, M. Monkenbusch, Review of Scientific Instruments **90**, 043107 (2019)

[59] J.K. Jochum, L. Spitz, C. Franz, A. Wendl, J.C. Leiner, C. Pfleiderer, O. Soltwedel, Journal of Applied Crystallography **55**, 14 (2022)

[60] M. Köhli, M. Klein, F. Allmendinger, A.K. Perrevoort, T. Schröder, N. Martin, C.J. Schmidt, U. Schmidt, Journal of Physics: Conference Series **746**, 012003 (2016)

[61] M.O. Köhli, Ph.D. thesis, University of Heidelberg (2019)

[62] J. Kindervater, Ph.D. thesis, Technische Universität München (2015)

[63] P.K. Willendrup, K. Lefmann, Journal of Neutron Research **22**, 1 (2020)

[64] P.K. Willendrup, K. Lefmann, Journal of Neutron Research **23**, 7 (2021)

[65] A. Shah, A. Sharma, A. Kumar, J. Merlin, M. Naimuddin, O. behalf of the CMS-Muon et al., Nuclear Instruments and Methods in Physics Research Section A: Accelerators, Spectrometers, Detectors and Associated Equipment **936**, 459 (2019)

Two Pressure Cells for Quasielastic and Inelastic Neutron Scatterings

Bao Yuan[1], *Richard* Mole[2], *Chin-Wei* Wang[3], *Amy* Shumack[2], *Rachel* White[2], *Bing* Li[4], *Xin* Tong[1*], *and Dehong* Yu[2*]

[1]Spallation Neutron Source Science Center, Dongguan 523803; Institute of High Energy Physics, Chinese Academy of Sciences (CAS), Beijing 100049, China
[2]Australian Nuclear Science and Technology Organisation, Lucas Heights, NSW 2234, Australia
[3]National Synchrotron Radiation Research Center, Hsinchu 30077, Taiwan
[4]Shenyang National Laboratory for Materials Science, Institute of Metal Research, Chinese Academy of Sciences, 72 Wenhua Road, Shenyang 110016, China.

Abstract. Two clamp pressure cells for QENS and INS have been developed. One is a hybrid CuBe/NiCrAl cell which is for relative high pressure up to 2.5 GPa and another one is made from high strength aluminium alloy (mesolite NA723) with pressure up to 0.5 GPa. The sample volume is 0.3 mL and 1 mL, respectively. The pressure cells have been thoroughly calibrated and tested. In addition, the contribution to phonon density of states from the pressure cells has been evaluated. Measurements of the phonon density of states for two perfluorocarbon polymer liquids FOMBLIN oil and Fluorinert have indicated that they are suitable to serve as the low background pressure transmission media for high pressure INS experiments. The applications of the pressure cells for INS and QENS are demonstrated by studies of pressure-induced phase transition of plastic crystals.

1 Introduction

The application of high pressure is one of the powerful techniques, among high or low temperature, high magnetic and electrical fields, used to induce unique changes in the structure, dynamics, and the associated properties of matter. Combined with neutron scattering, high-pressure research covers a wide range of areas of science to tackle many challenging problems. This includes, but is not limited to, emergent phenomena in quantum materials; synthesis of novel materials; mimic of planetary condition for geoscientific problems and tackling problems related to biosystems such as improving food processing or enhancing protein stability.

Neutron diffraction under high pressure has become a routine technique to study material structure evolution versus pressure, however, it is still a challenging task to perform quasielastic (QENS) and inelastic neutron scattering (INS) under high pressure for material dynamics investigation. This is mainly attributed to the intrinsic several orders of magnitude lower cross section for the QENS and INS processes than elastic scattering. Consequently, a large amount of sample and minimum background contribution from the sample cell is essential for a successful high pressure QENS and INS.

There are, in general, three types of high-pressure cells widely used for neutron scattering. The well-known Paris-Edinburgh anvils cell have been routinely used for neutron diffraction up to 10 GPa pressure with limited sample volume [1]. The McWhan-type cell can accommodate a slightly larger sample volume, but with lower pressure up to around 4 GPa [2]. The piston-cylinder type cell overcomes some limitations of the McWhan cell and increases the sample volume further [3]. The piston-cylinder type of cells has seen increasing popularity in INS and QENS studies in recent years [3-4]. In terms of materials requirement, a handful of materials have been tested and used for these pressure cells. This includes CuBe, NiCrAl, TiZr alloy, NiCr alloy, and high strength Al, etc. A detailed comparison of the background contribution from these materials for high-pressure inelastic neutron scattering has been reported [5].

Two piston-cylinder type pressure cells for QENS and INS have been developed by Chinese Spallation Neutron Source (CSNS) and fully tested on the Pelican time of flight cold neutron spectrometer at Australian Nuclear Science and Technology Organisation (ANSTO) [6-7]. One is a hybrid CuBe/NiCrAl cell which is for relative high pressure up to 2.5 GPa and another one is made from high strength aluminium alloy (mesolite NA723) with pressure up to 0.5 GPa. We call the two cells as Hybrid-cell and HAS-cell respectively. The outline of the paper is as the following. The design aspect of these pressure cells will be described first. The calibration of the cells will be given followed by the evaluations on the contributions to QENS and INS from the pressure cells, and pressure transmission media. In

* Corresponding author: tongx@ihep.ac.cn;dyu@ansto.gov.au

the end, the application of the pressure cells will be demonstrated by pressure-induced phase transition in plastic crystals.

2 Design and Load Test

The design and selection of materials for the pressure cells are largely based on existing knowledge as reported previously [8-9]. The sketch, finite element analysis (FEA) and load test will be outlined here.

2.1. CuBe/NiCrAl – Hybrid Cell

The sketch, dimensions, and FEA maps are shown in Fig. 1 for the Hybrid-cell. This hybrid design allows the realisation of a maximum pressure of 2.5 GPa with a relatively large sample volume of 0.3 mL. The maximum principal stress distribution with a load of 2.6 GPa is analysed through FEA as shown on the right panel of Fig.1. A maximum stress of 2.1 GPa and deformation of 0.014 mm are found for the inside cylinder, while the corresponding values for the outside cylinder are 1.22 GPa and 0.012 mm. The sample is filled into a Teflon tube which is sealed through a CuBe insert and a carefully shaped plug ring. The Teflon tube is pushed into the inside cavity of the pressure cell. The bottom of the cell is locked with a CuBe screw with a piece of Tungsten Carbide (WC) block against the sample tube plug. On top of the sample tube is a WC piston and another CuBe block through which force is applied to the sample tube with the WC push rod. The top CuBe screw is used to lock the position of the piston and maintain the pressure.

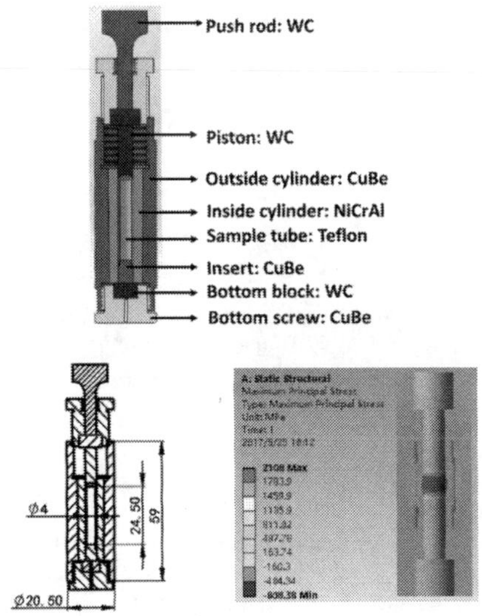

Fig. 1. The sketch, dimensions, and maximum stress distribution with FEA are shown for the Hybrid – Cell.

2.2 Mesolite NA723 – HSA Cell

As shown in Fig. 2, the same design has been applied to the HSA-Cell. Taking advantage of the high strength of the Mesolite NA723 Al alloy, this cell can deliver up to 0.5 GPa pressure with a much larger sample volume of 1 mL as compared to the Hybrid-Cell. The operation of this cell follows the same procedure as described above for the Hybrid-cell.

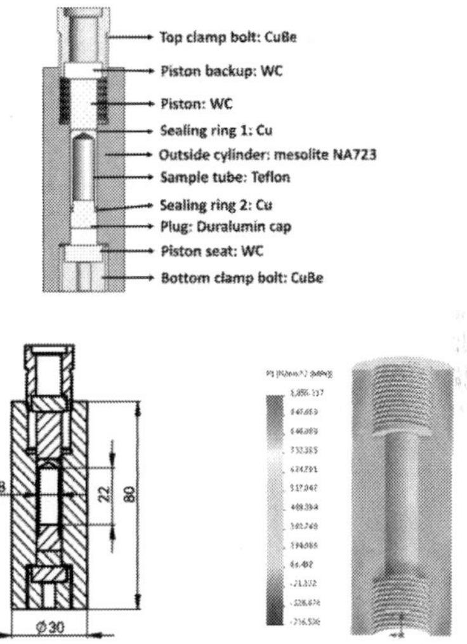

Fig. 2. The sketch, dimensions, and maximum stress distribution with FEA are shown for the HSA – Cell.

2.3 Load Test

By monitoring the resistance changes of a bismuth sensor as a function of force applied to the cell, preliminary load tests have been performed on both cells [8]. For the Hybrid-Cell, the maximum pressure of 2.55 GPa has been achieved. For safety purpose, it is recommended to keep the maximum operating pressure at 1.5 GPa. For the Mesolite NA723 Al cell, maximum safe pressure of 500 MPa has been achieved.

3 Pressure Cell Calibration

The pressure cells have been calibrated with neutron diffraction of KBr using the Wombat neutron diffractometer at ANSTO [10]. The Equation of State (EOS) for KBr [11] was used to calculate the pressure achieved corresponding to the force applied to the cells as shown in Fig. 3 and Fig. 4 below. A small amount of Fluorinert liquid (1/3 of the total sample in weight) was added to KBr powder as the pressure transmission medium. The estimated errors of about 8% are mainly from the uncertainty in the force applied.

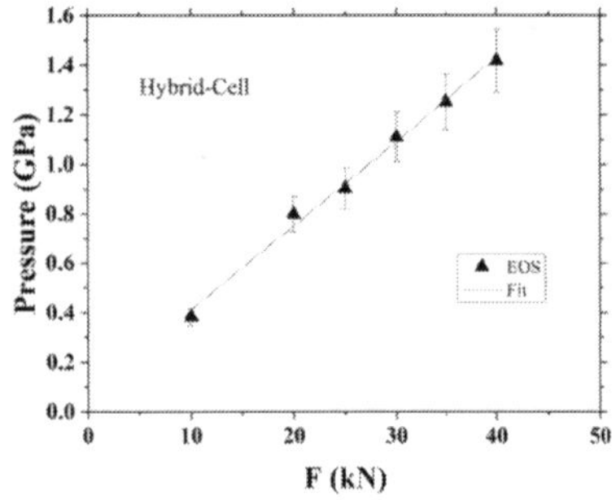

Fig. 3. The pressure calibration curve for the Hybrid-Cell.

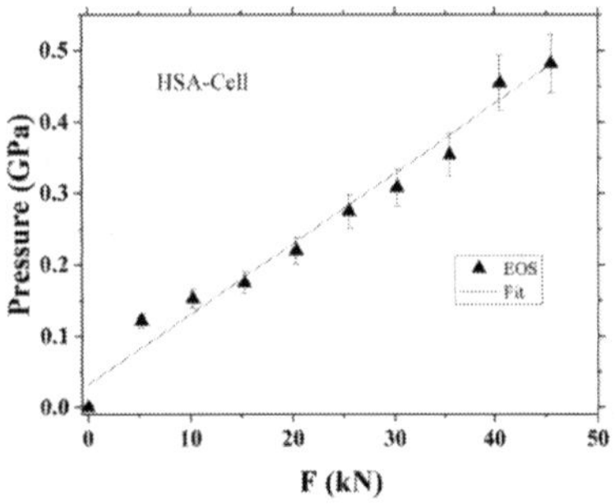

Fig. 4. The pressure calibration curve for the HSA-Cell.

4 QENS and INS contribution of the Pressure Cells

A proper evaluation of these pressure cells for quasielastic (QENS) and inelastic neutron scatterings (INS) have been performed on Pelican instrument. In addition to the empty pressure cells, the QENS and INS have been measured on two non-H contained liquids of Fomblin and Fluorinert which are frequently used as low background pressure transmission medium. Vanadium results are shown for reference. The QENS results are integrated over all Q values. The incident neutron energy is 3.7 meV corresponding to 0.13 meV energy resolution at the elastic peak. The QENS and generalised phonon density of state (GDOS), as presented in Fig. 5 and Fig. 6, respectively, are all normalised to vanadium signal. These measurements are performed at 300 K. For both Fomblin and Fluorinert, the results correspond to 0.3 g of sample weight. There are clearly QENS from these liquids. For the two empty cells, the lower QENS signal is mainly from the Teflon tube inside. For the GDOS, the contributions from the two pressure cells are much stronger (about 2 orders of magnitude) than the two pressure transmission media. These results indicate clearly that care must be taken to properly subtract backgrounds from both cells and pressure transmission medium, especially for weak scattering samples. Certainly, these pressure cells are most suitable for hydrogen-contained samples. For proper background subtraction, the following precautions need to be taken: Time for empty can measurements should be at least the same as that for sample measurements to minimise error bars due to background subtraction; The pressure cell should be masked by neutron absorption materials (Cd, for example) to ensure the same amount of cell exposed to neutrons for both sample and empty can measurements.

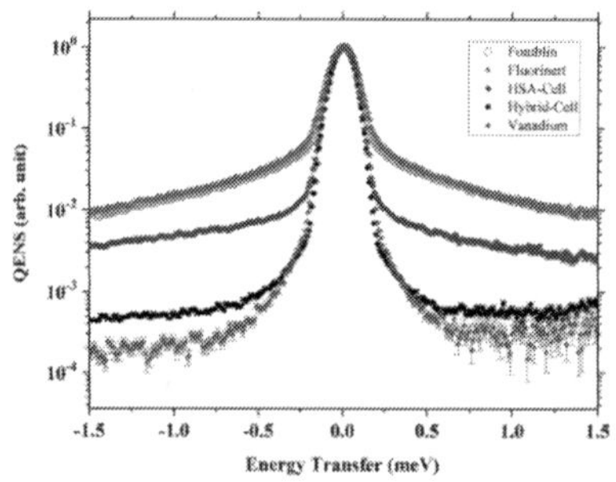

Fig. 5. QENS contributions from the pressure cells and several other materials as described in the text.

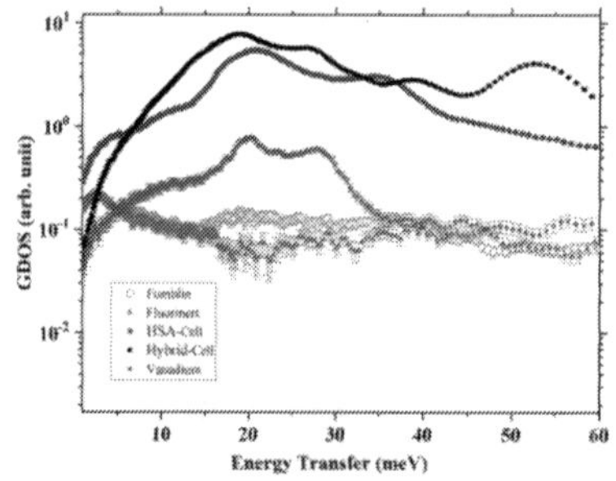

Fig. 6. GDOS from the pressure cells and several other materials as described in the text.

5 Applications

The application of the pressure cell (HSA-cell) has been demonstrated with pressure induced phase transition of NH_4I. As shown in figures 7 and 8 below, both QENS and the imaginary part of the dynamic susceptibility (χ'') have reflected the phase transition from the α-phase to β-phase upon applying 300 MPa pressure. This measurement confirms the promising application of NH_4I in solid state cooling technology through barocaloric effect [12].

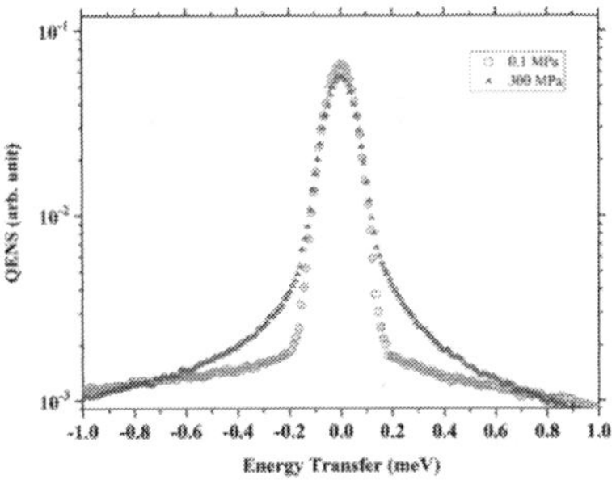

Fig. 7. QENS indicates the phase transition from the α-phase to β-phase upon applying 300 MPa pressure for NH_4I.

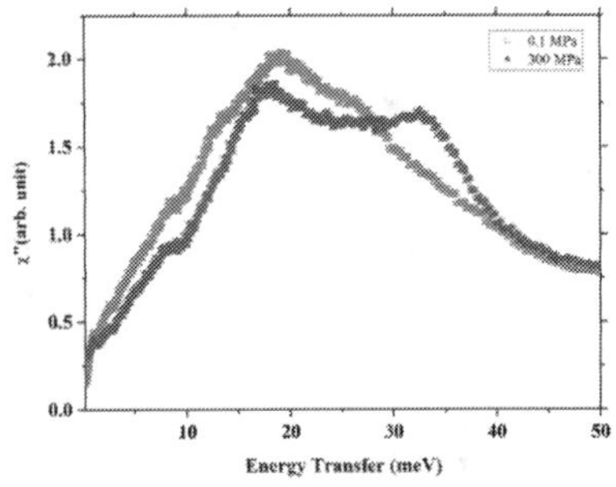

Fig. 8. The imaginary part of the dynamic susceptibility indicates the phase transition from the α-phase to β-phase upon applying 300 MPa pressure for NH_4I.

6 Summary

Two pressure cells have been developed and thoroughly tested. The contributions to quasielastic neutron scattering and phonon density of states from the pressure cells have been evaluated. A proper correction of the cell contribution must be taken especially for weak scatters. While these cells can be certainly used for diffraction experiments requiring large amount of sample for structure determination through refinement, they may be best suitable for hydrogen-contained samples for QENS and INS experiments, which has been demonstrated with pressure induced phase transition of NH_4I.

BY and XT would like to acknowledge the support from the National Key R&D Program by the Ministry of Science and Technology (MOST) of China (Grant no. 2016YFA0401503). The authors are grateful for the generous support of neutron beam time through commission proposals at Australian Centre for Neutron Scattering.

References

1. S. Klotz, J. M. Besson, G. Hamel, R. J. Nelmes, J. S. Loveday, and W.G. Marshall, High Press. Res. 14, 249 (1996)
2. D. B. McWhan, D. Bloch, and G. Parisot, Rev. Sci. Instrum.45,643 (1974)
3. W. Wang, D. A. Sokolov, A. D. Huxley, and K. V. Kamenev, Rev. Sci. Instrum **82**, 073903 (2011)
4. Q. Chen, S. Holdsworth, J. Embs, V. Pomjakushin, B. Frick, A. Braun, High Pressure Research **32** 471 (2012)
5. M.G. Kibble, V. Laliena, C.M. Goodway, E. Lelièvre-Berna, K.V. Kamenev, S. Klotze and O. Kirichek, J. Neutron Research, **21**, 105 (2019)
6. D. Yu, R. A. Mole, T. Noakes, S. Kennedy and R. Robinson, J. Phys. Soc. Jpn. **82**, SA027 (2013).
7. D. Yu, R. A. Mole and G. J. Kearley, EPJ Web of Conferences **83**, 03019 (2015)
8. H. Taniguchi, S. Takeda, R. Satoh, A.Taniguchi, H. Komatsu, and K. Satoh, Rev. Sci. Instrum **81**, 033903 (2010)
9. N. Aso, T. Fujiwara, Y. Uwatoko, H. Miyano and H. Yoshizawa, J. Phys. Soc. Jpn. **76,** SA228 (2007)
10. A. J. Studer, M. E. Hagen, T. J. Noakes, Phys. B Condens. Matter **385–386**, 1013 (2006)
11. A. Dewaele, A.B. Belonoshko, G. Garbarino, F. Occelli, P. Bouvier, M. Hanfland, and M. Mezouar, Phys. Rev. B **85**, 214105 (2012)
12. Qi. Ren, J. Qi, D. Yu, Z. Zhang, R. Song, W. Song, B. Yuan, T. Wang, W. Ren, Z. Zhang, X. Tong and B. Li, Nature Communications **13**, 2293 (2022)

Conceptual design of a radial collimator for MIRACLES, the time-of-flight backscattering spectrometer at the European Spallation Source

Roberto Martínez,[1*] *Marton* Marko,[2] *Alexander* Conde,[1] *Aitor* Zugazaga,[1] *Idoia* Mazkiaran,[1] *Octavio* G. del Moral,[1] *Giles* Harper,[1] *José E.M.* Pereira,[1] *Heloisa* N. Bordallo,[3,4] and *Félix* J. Villacorta[1]

[1]ESS-Bilbao. Parque Científico y Tecnológico Bizkaia Nave 201, 48170 Zamudio, Spain
[2]Neutron Spectroscopy Department, Wigner Research Centre for Physics, Konkoly Th. M. 29-33, 1122 Budapest, Hungary
[3]Niels Bohr Institute-University of Copenhagen. Universitetsparken 5 -2100 Copenhagen, Denmark
[4]ESS ERIC, P.O. Box 176, SE-22100 Lund, Sweden

Abstract. A parametrized conceptual design for a radial collimator in a neutron backscattering instrument is presented, with application to the characteristic geometry of the MIRACLES spectrometer, that will be constructed at the European Spallation Source (ESS). The analytic development of this design has considered both the forward scattering (sample-analyzer) and backscattering (analyzer-detectors) pathways. All the characteristic dimensions (internal and external radii, slit angle) and figures of merit (such as transmission and estimated background reduction) of the device are calculated as a function of the focal points. Finally, the estimated performance of the final concept has been validated by Monte Carlo simulations.

1 Introduction

An essential aspect in the design of any neutron spectrometer is the optimization of the signal-to-background ratio, that is, the minimization of neutrons scattered from sample environment equipment or from any in-vessel component that could arrive at the detectors and degrade the quality of the spectra. This is especially critical in indirect-geometry spectrometers in neutron time-of-flight instruments, that feature a significant flux at the sample due to their large-bandwidth [1,2]. These neutrons can scatter (Bragg, incoherent…) with devices around the sample environment, and can contribute negatively to the correct performance of the experiments.

Radial collimators are the elements whose aim is to block radiation sourced from these background-generating regions [3–5]. From the first radial collimators, developed at the ILL (Institut Laue-Langevin, Grenoble, France) [6], a number of neutron instruments included a radial oscillating collimator (ROC) – with low oscillation periods of the order of the ~10 s – as a valuable component to further improve the background ratio [7].

This work reports on the conceptual design of a radial oscillating collimator for MIRACLES, the time-of-flight backscattering spectrometer at the European Spallation Source [8,9]. Most published concepts on radial collimators are related to diffractometers [10,11] or direct-geometry spectrometers [7,12]. This work has extended the design considerations towards backscattering spectrometers. Along the process, the main variables (inner and outer radii, slit size…) have been parametrized to offer a solution that can be adapted to changes in the design of the secondary spectrometer. The final parameters for MIRACLES are based on the specificity of the geometry of the secondary spectrometer; in particular, the detector bank layout has defined the slit size and the inner radius of the radial collimator. Finally, the estimated performance has been validated by Monte Carlo simulations.

2 Material and methods

The analytical calculations have been carried out for both scattering and backscattering flightpaths. Intuitively, the radial oscillating collimator provides a better functionality by placing it close to the detectors, to suppress neutrons stemming from spurious scattering in the backscattering pathway (Fig. 1a). This collimator is oscillating, aiming to average the shadowing effect provided by the blades to the nearby detectors. Thus, the oscillation range is set to be of the order of the angular coverage of a detector tube. In order to reach this averaging effect, the oscillation period of the collimator should be both slower than the ESS pulse period (71.4 *ms*) but faster than the measurement time (~30 – 60 *min*). However, all calculations in this paper have been carried out for a collimator in a static state, without loss of generality of the final results for an oscillating case.

Moreover, if the collimator simply accounts for one only pathway, either the sample-analyzer or the

* Corresponding author: rmartinez@essbilbao.org

backscattering analyzer-detector pathway, some undesirable shadowing issues due to the mechanical frame can reduce the angular acceptance of the instrument (for MIRACLES the angular coverage is 9.5 to 165° in the horizontal, and ±22° in the vertical). Thus, for a backscattering spectrometer, neutrons must go through the collimator channels twice, in the scattering and the backscattering pathways (Fig. 1a).

The radial collimator considers the following parameters (see also Fig. 1b):

- X-axis: scattering axis (direction of the scattered beam).
- Y-axis: perpendicular to X-axis, in the same scattering plane (equatorial).
- 2θ: scattering angle (angle incoming beam – scattered beam).
- r_m: inner radius of the collimator.
- R_M: outer radius of the collimator (channel/blade length: R_M - r_m).
- ϕ: angle between blades (slit size).
- $2d_l$: width of a blade (a constant thickness blade [5] is assumed).
- $2F_Y$: focal width (sight of a slit at the sample position).
- d_s: sample size (cylinder diameter).
- R_d: radius of the detector array.

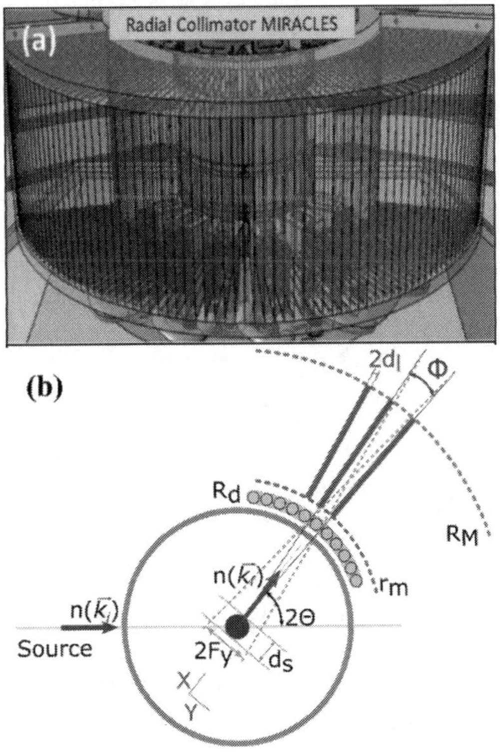

Fig. 1. (a) 3D preliminary model of the MIRACLES radial collimator; (b) Geometry of the radial collimator calculations.

The main parameters needed to design a radial collimator (signal transmission, *T*, background reduction, *R*) will be parametrized as a function of the collimator characteristics and the sample size (ϕ, r_m, R_M, d_s). The estimated performance has been validated by Monte Carlo neutron optics calculations using the McStas [13] available components (in particular, *Exact_radial_coll* is used to simulate the physical radial collimator. Finally, analytical calculations of the neutron absorbing layer used to coat the collimator blades were corroborated by simple MCNP calculations.

3 Results and discussion

3.1 General model

Let us focus our attention on one slit of the ROC. The concept behind the design efforts resides in the characterization of any possible radial collimator by defining two focal lines that describe the extreme neutron paths transmitted by a slit of the collimator (orange dash lines in Fig. 2). These lines define in turn two focal lengths: a focal abscissa F_x that marks the crossing between the two focal lines along the X-axis, and a focal ordinate F_Y which defines the distance between the crossing points of the focal line with the Y-axis, located at the sample position (Fig. 2). With this layout, the range $(-F_Y, +F_Y)$ of length $2F_Y$ defines the focal width. This focal width includes not only the sample (cylinder of diameter d_s) but also the area around it (sample environment). Any neutron scattered by any point y_s within this range in the Y-axis is transmitted through the collimator slit if the scattering angle is limited by the angle between the focal lines, β. Additionally, the intersection points between focal lines, F_x, is related to the inner and outer radii for each possible radial collimator (r_m and R_M respectively).

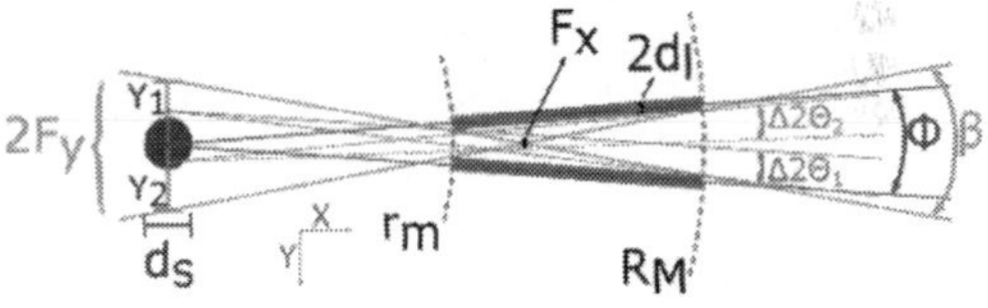

Fig. 2. Sight of the sample area by a single slit (blue radial blades) showing the main parameters defining the radial collimator: focal lines (orange), focal lengths F_x and F_Y, slit aperture angles Φ and β, inner (r_m) and outer radius (R_M), blade length (or collimation length, R_M - r_m) and blade width ($2d_l$).

Thus, let y_s be scattering points within the $(-F_Y, +F_Y)$ range that scatters neutrons through a slit of the collimator (see Fig. 2). Assuming a slit periodicity of ϕ, and approaching to small angles, the actual angular acceptance of a slit for the radial collimator (transmittance divergence) is $\Delta 2\theta$, that ranges from $\Delta 2\theta = 0$ (for the focal lines, starting at points $\pm F_Y$) to $\Delta 2\theta = \phi$ (for a central point sample, at y = 0). Two different points, y_1 and y_2, representing scattering points of the sample environment and the sample, respectively (Fig. 2), deliver neutrons inside a collimator channel (slit) with divergence $\Delta 2\theta_1$ and $\Delta 2\theta_2$ (represented by green and pink dash lines, respectively). This divergence $\Delta 2\theta$ are delimited by angles $2\theta_+$ and $2\theta_-$:

$$2\theta_+ = 2\theta_0 + \frac{\phi}{2} - \frac{(y+d_l)}{r_m} \quad (1)$$

$$2\theta_- = 2\theta_0 - \frac{\phi}{2} - \frac{(y-d_l)}{R_M} \quad (2)$$

$$\Delta 2\theta(y) = 2\theta_+ - 2\theta_- = \phi - y\left(\frac{1}{r_m} - \frac{1}{R_M}\right) - d_l\left(\frac{1}{r_m} + \frac{1}{R_M}\right) \quad (3)$$

This general expression, that displays the finite sample size and the finite blade effects, results in $\Delta 2\theta = \phi$ in the ideal case of a point sample (y = 0) and zero blade thickness ($d_l = 0$).

As previously mentioned, focal lines (scattering trajectories starting at points $\pm F_Y$ display a transmitted divergence is $\Delta 2\theta = 0$:

$$\Delta 2\theta(F_Y) = \phi - F_Y\left(\frac{1}{r_m} - \frac{1}{R_M}\right) - d_l\left(\frac{1}{r_m} + \frac{1}{R_M}\right) = 0 \quad (4)$$

From here, two characteristic radii r_c and r_c' can be defined as:

$$r_c' \cong \left(\frac{1}{r_m} + \frac{1}{R_M}\right)^{-1} \quad (5)$$

$$r_c \cong \left(\frac{1}{r_m} - \frac{1}{R_M}\right)^{-1} \quad (6)$$

Then, from eq. (4):

$$F_Y = \left(\phi - \frac{d_l}{r_c'}\right) r_c \quad (7)$$

Since d_l is typically small (~0.1 mm, thus $d_l \ll r_m$, R_M), there is a direct relationship between the slit size and the observable focal width around the sample, in agreement with previous results [4,5,10]:

$$F_Y \cong \phi r_c \quad (8)$$

The focal abscissa F_X is also related to the parameters of the collimator. The angle delimited by the focal lines, β, can be linked to the inner and outer radius:

$$\tan\frac{\beta}{2} = \frac{r_m \tan\left(\phi/_2\right)}{F_X - r_m} = \frac{R_M \tan\left(\phi/_2\right)}{R_M - F_X} \quad (9)$$

This second term in (9) results in an expression for F_X:

$$F_X = 2r_c' \quad (10)$$

Somehow, the location of the focal lines and their characteristic focal point F_X and focal width F_Y can reveal at first glance the angular acceptance, and the slit dimensions (angle ϕ and radii r_m *and* R_M, see Fig. 3 for variable F_X and fixed $2F_Y$ =30 mm).

3.2 Transmission

The transmitted neutrons are estimated using the transmission quality factor T, defined as the ratio between the number of neutrons scattered from the sample at the exit of a collimator slit and the number of neutrons at its entrance ($T = \frac{I_t}{I_i}$).

A relationship (q ratio) can be established between the sample size, d_s, and the focal width, $2F_Y$:

$$q = \frac{d_S}{2F_Y} \quad (11)$$

where $0 < q < 1$. In the case where $q=1$ the sample region would completely lie inside the focal width of the slit and $d_s=2F_Y$.

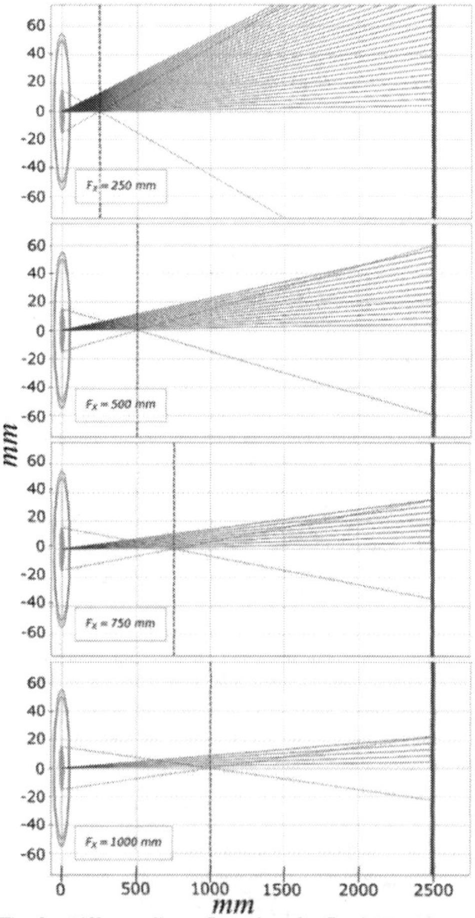

Fig. 3. Different slit configurations for fixed $2F_Y$ =30 *mm* and variable F_X from 250 to 1000 *mm*. The solutions for r_m and R_M radii are the intersections of the radial blue lines (having an angle ϕ) with the red-dashed focal lines.

Considering neutrons scattered by the sample, the total intensity, I_i, at the entrance of the slit is related to the angle distribution (aperture), ϕ:

$$I_i = \int_{-\frac{d_s}{2}}^{\frac{d_s}{2}} \phi(y)dy = 2\int_{0}^{\frac{d_s}{2}} \phi(y)dy = \\ = 2\int_{0}^{qF_Y} \phi(y)dy = \\ = 2qF_Y{}^2 r_c^{-1} \quad (12)$$

The total intensity at the exit (transmitted) from one channel is the integral of the total transmittance divergence stated in Eq. (3) along the sample along the Y-axis:

$$I_t = \int_{-\frac{d_s}{2}}^{\frac{d_s}{2}} \Delta 2\theta(y)dy = 2\int_{0}^{\frac{d_s}{2}} \Delta 2\theta(y)dy \\ = 2\int_{0}^{qF_Y} \Delta 2\theta(y)dy \quad (13)$$

Assuming that $d_s \ll r_m, R_M$

$$I_t = 2\int_{0}^{qF_Y} \phi(y)dy - 2\int_{0}^{qF_Y} y r_c^{-1}dy \\ = I_i - 2\int_{0}^{qF_Y} y r_c^{-1}dy \\ = I_i - (qF_Y)^2 r_c^{-1} \quad (14)$$

The transmission parameter of a slit belonging to the radial collimator is then:

$$T = \frac{I_t}{I_i} = 1 - \frac{(qF_Y)^2 r_c^{-1}}{2qF_Y{}^2 r_c^{-1}} = 1 - \frac{q}{2} \quad (15)$$

Thus, in the common case that $d_s \ll r_m$, R_M, the transmission depends on the ratio q between the sample size and the focus size. Retrieving the expression for the ROC parameters ($2F_Y$, ϕ, r_m, R_M) and the sample size, d_s:

$$T = 1 - \frac{d_s}{4F_Y} = 1 - \frac{d_s}{4\phi r_c} \quad (16)$$

3.3 Background reduction

Background neutrons come mainly from spurious and not controlled scattering with in-vessel components. This randomness renders a challenging observation of the background reduction functionality and performance of the component. However, a qualitative estimation can be attempted, using sample environment equipment as a primary source of background.

On one note, the sample environment can be simplified to a hollow cylindrical model with radius R_{SE}, (a circumference in the sample plane with length $2\pi R_{SE}$).

Furthermore, the number of neutrons scattered by the sample environment and transmitted through the slit is proportional to the arc lengths in the direct line of sight of the sample by the slit, Δl, which is always $4F_Y$ (see Figure 4, being $\Delta l = 2d_s$ when $d_s=2F_Y$, i.e., when q=1, thus T = ½).

Thus, for a hollow cylindrical sample environment equipment, the background reduction can be defined as the ratio between the number of neutrons scattered by the sample environment and transmitted through the slits (background after implementation of the ROC, B_{coll}) with respect to the total number of neutrons scattered by the sample environment (background without using the collimator, B_0):

$$R = \frac{B_{coll}}{B_0}T = \frac{\Delta l}{2\pi R_{SE}}T \approx \frac{2F_Y}{\pi R_{SE}}T \quad (17)$$

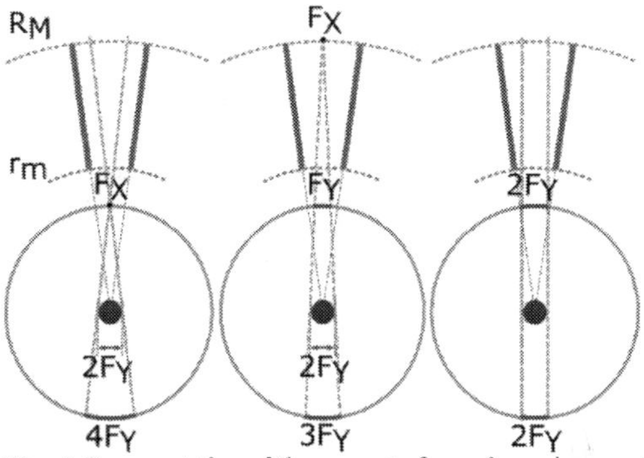

Fig. 4. Representation of the amount of sample environment region (orange circle) "observed" by the slit (green arcs) for three extreme cases of blade configuration corresponding to focal lenghts F_Y lying either before r_m (left), between R_M and r_m (center), or beyond R_M (right). For each case the arcs observed by the slit can be effectively approximated to ~$4F_Y$.

3.4 Backscattering pathway

The background is minimized largely in the sample-analyzer pathway. Neutrons scattered from a sample passing through a collimation channel (slit) will reach a segment (see Fig. 5a) of the analyzer given by:

$$A_A = 2R_A \sin\frac{\phi}{2} \quad (18)$$

Recalling the description of Figure 2 for focal lines, the total projection of neutrons passing through a slit to the analyzer has a length, P_A, that covers an angle β ($\tan\beta = \frac{r_c}{r_{c'}}\tan\phi$, see Fig. 5a), where:

$$\frac{F_Y}{F_X} = \tan{}^{\beta}/_{2} = \frac{{}^{P_A}/_{2}}{R_A - F_X} \quad (19)$$

$$P_A = 2F_Y\left(\frac{R_A}{F_X} - 1\right) \quad (20)$$

The backscattering trajectory, after reflection from the analyzer, follows a trajectory distribution centered at the same slit. There is a possibility some neutrons arrive not at the corresponding ^{3}He tube detector, but in the adjacent ones (see inset of Figure 5b), since the collimator is not in contact with the detector array ($r_m > R_d$). The arc of the detector array circumference where the neutrons from a determined slit impact (it can be called "detector impact arc", a_m) is defined by the slit aperture for the minor radius:

$$a_m = 2r_m \sin\frac{\phi}{2} \quad (21)$$

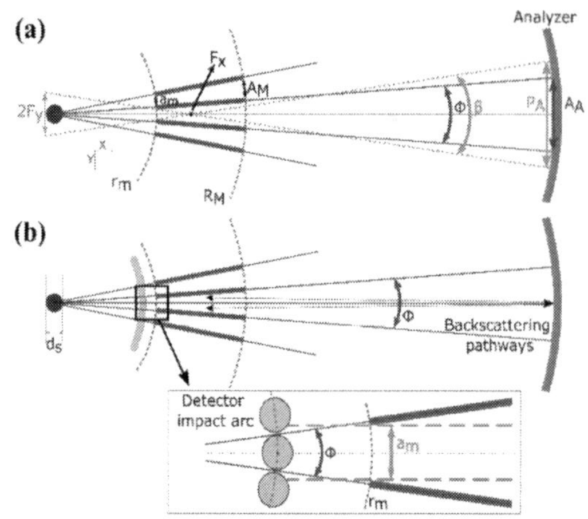

Fig. 5. (a) Schematic representation of forward scattered neutrons passing through a single slit of the collimator and arriving to the analyzer. (b) Schematic representation of backscattered neutrons arriving to the detector array (green dashed lines) define the *detector impact arc* a_m.

3.5 Application to MIRACLES

The previous general method has been conceived to generate all the possible radial collimator configurations. Nevertheless, constraints imposed by realistic design of the MIRACLES secondary spectrometer eventually determine the final parameters. For instance, the slit angle ϕ is determined by the angular sector covered by the detector tubes: since the detector banks for MIRACLES are cylindrical arrays with radius R_d = 230 *mm*, and tubes have 12.8 *mm* diameter, the repetition angle is $\phi = 3.24°$.

Additionally, the inner radius, r_m, should be placed as close as possible to the detectors, to optimize the dimensions of the collimator, and minimize the arrival of neutrons to adjacent detector tubes. Considering the detector frame and other spatial considerations, the minimum value is r_m = 300 mm. Finally, MIRACLES is designed to measure sample sizes of 30 *x* 30 mm^2 (thus, d_s = 30 *mm*), and by default, the radius of any sample environment equipment used in MIRACLES shall be smaller than the sample chamber radius, thus $R_{se} \leq 210$ *mm*).

A key factor is that the focal width, $2F_r$, can be larger than the sample size, d_s; however, to minimize the inclusion of sample environment equipment such $2F_r$ parameter should be as close as possible to the sample bore, the small vessel that hosts the sample inside the SE equipment. In an orange cryostat, this bore size is around 50 - 70 *mm*, and some small magnets show a sample space of 80 *mm*. Thus, selecting a $2F_r$ of about this size minimizes the transmission of neutrons scattered by the SE equipment, beyond the sample bore walls.

Table 1 shows results of different radial collimator options as a function of selected focal widths, $2F_r$, and the inner radii, r_m. The transmission T has been compared in some examples with the transmission obtained after Monte Carlo calculations (using McStas, see also Figs. 6a and 6b), providing a good match between them (with an underestimation of the first with respect to the latter within 10%). For the background reduction figure of merit, R, the sample environment radius has been fixed to R_{se} = 210 *mm*.

Assuming an analyzer radius of R_A = 2500 *mm*, a detector array radius of R_d = 230 *mm*, and a radius of the ½-inch detector tubes of r_d = 6.35 *mm*, the view (spot) of a slit on the analyzer is:

$$A_A = 2R_A \sin\frac{\phi}{2} = 141\, mm \qquad (22)$$

Finally, Fig. 6c shows the dependence of the collimation length (also called blade length, $R_M - r_m$) on the focal width, $2F_r$, and the inner radius, r_m. It is intuitive to

Table 1. A list of potential options for the MIRACLES radial collimator and estimated performance in terms of transmission (comparing analytical and Monte Carlo calculations), background reduction, and detector impact.

r_m *(mm)*	R_M *(mm)*	$2F_r$ *(mm)*	*T*	*T (MC)*	*R*	a_m *(mm)*	$a_m/2r_d$ *(# tubes)*
300	582	70.0	0.79	-	0.083	17	1.3
300	599	68.0	0.78	0.84	0.080	17	1.3
300	690	60.0	0.75	0.82	0.068	17	1.3
300	935	50.0	0.70	0.77	0.053	17	1.3
330	707	70.0	0.79	-	0.083	18.6	1.5
350	805	70.0	0.79	-	0.083	19.8	1.6
350	693	80.0	0.81	0.86	0.098	19.8	1.6
400	920	80.0	0.81	0.85	0.099	22.6	1.8

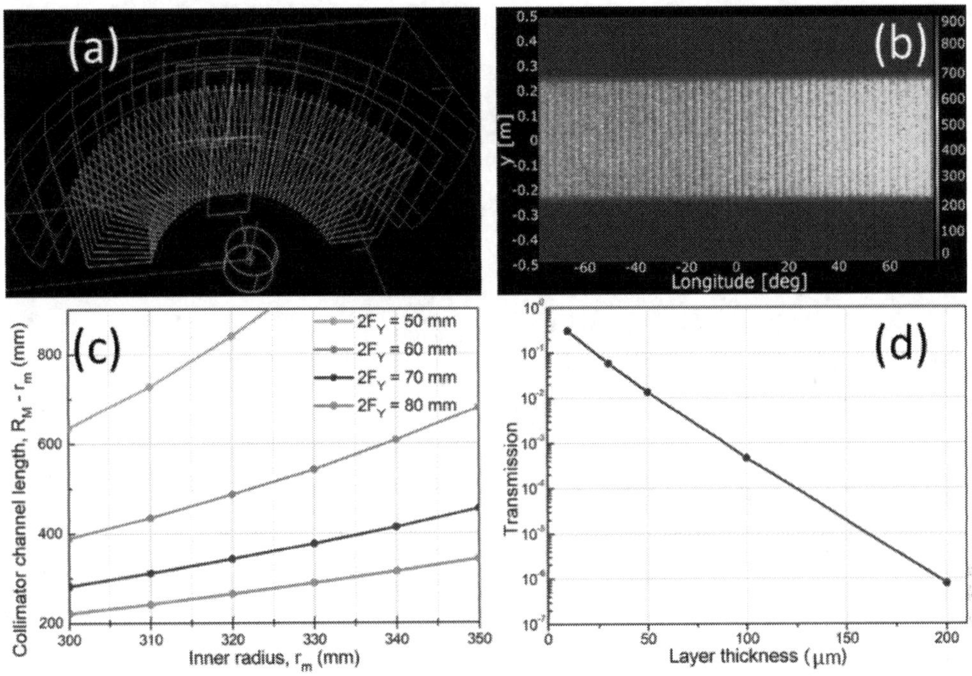

Fig. 6. (a) McStas visualization of the radial collimator for MIRACLES (b) Signal collected by a McStas position sensitive monitor at the exit of the collimator in the forward scattering (c) Collimation length (that can serve or relative comparison of ROC dimensions and cost) as a function of the inner radius and focal length; (d) Thermal neutron absorption factor (1/attenuation factor) of a Gd_2O_3 layer as a function of the layer thickness, obtained by MCNP.

deduce that this collimation length is directly related with the final cost of the ROC. Results show that, on one hand, in order to cover the same focal width (determined by the sample bore), the closest the collimator is to the detector bank ($r_m \rightarrow R_d$), the shorter the collimation length required. On the other hand, it is observed that to reach smaller focal widths (for smaller sample bore dimensions), the collimation lengths need to be longer, for a given value of r_m.

From all the potential results, the best option consists of reaching the closest position possible of the collimator with respect to the detectors bank (r_m = 300 *mm*) and, assuming a sample bore size of the sample environment equipment (e.g., MIRACLES cryostat) of 70 *mm*, fixing a focal width that reduces 1 *mm* at each side of the sample bore, (i.e., $2F_Y$ = 68 *mm*). With these parameters an external radius of R_M = 599 *mm*, is obtained, with a transmission estimated of about ~80%, and a detector impact arc of $a_d/2r_m$=1.3 detectors.

3.6 Thickness of the blade shielding

The collimating blades, typically made of thin aluminum, PETP or Kapton sheets of thickness ~100 *mm*, less than 1% of the channel width, are coated with a neutron capturing layer, usually Gd_2O_3 paint. An estimation of the minimum thickness of this coating, *t*, has been carried out, considering that the blade transmission can be calculated as $\tau = e^{-N\sigma t}$, where $N = N_A\rho/M$ is the molecular density of Gd_2O_3 (N_A is the Avogadro number, M = 362.5 *g/mol* is the molecular weight and ρ is the density: 7.407 g/cm^3) and $\sigma = 5 \times 10^{-20}$ $cm^2/atom$ is the cross section of natural Gd at thermal neutrons with $E = 25$ *meV*.

Aiming at an attenuation factor of $\sim 10^{5}$, ($\tau = 10^{-5}$) and assuming that density is reduced by ½ to simulate that the paint is Gd_2O_3 powder mixed in epoxy, the minimum layer thickness required is ~360 *μm* (180 *μm* on each side of the blade). This estimation is corroborated by simple Monte Carlo calculations (see Fig. 6d). The maximum grazing incidence angle, ξ, that a neutron scattered by sample environment equipment with radius R_{SE} = 210 *mm* can have crossing a single blade (that is, higher angles will convey neutrons that go through more than one blade) is given by the expression:

$$\tan\xi \simeq \frac{R_M}{(R_M - R_{SE})}\tan 2\phi \qquad (23)$$

From here, ξ = 9.6°, and thus, the minimum thickness of the gadolinium layer is $t \simeq 30$ *μm*.

4 Conclusions

This work reports on the conceptual design for a radial collimator to be installed in a neutron backscattering instrument. A general design concept is developed, by which the focal lines and focal points that define every collimation channel serve in turn to parametrize the dimensional features of the collimator (inner and outer radius, slit/channel angle) and figures of merit that provide information about the performance of the collimator (transmission, background reduction, detector impact) in the forward and backscattering neutron pathways.

The general expressions obtained were eventually applied to MIRACLES, the time-of-flight backscattering spectrometer at the European Spallation Source. For MIRACLES, the focal width has been determined by the sample bore of the cryostat dedicated to the instrument; the slit aperture has been defined by the solid angle covered by the tube detectors; and the inner radius has been determined by the position of the detectors, as well as by mechanical design constraints. Finally, an estimation of the neutron capturing layer that coats the collimation blades has been estimated considering the grazing incidence angles at which neutrons scattered by the sample environment reach the collimation channels.

This research is partially funded by the Next Generation EU–Recovery and Resilience Facility (RRF). We would like to express our gratitude to Gorka Mujika (ESS-Bilbao); his continued encouragement along the whole research process, from the initial conception of the idea to the draft writing, has been determinant for the successful accomplishment of this work.

References

1. E. Mamontov, K. W. Herwig, Rev. Sci. Instrum. **82**, 085109 (2011)
2. K. H. Andersen, D. Martín y Marero, M. J. Barlow, Appl. Phys. A Mater. Sci. Process. **74**, (2002)
3. M. B. Stone, J. L. Niedziela, M. A. Overbay, D. L. Abernathy, EPJ Web Conf. **83**, 03014 (2015)
4. M. B. Stone, J. L. Niedziela, M. J. Loguillo, M. A. Overbay, D. L. Abernathy, Rev. Sci. Instrum. **85**, 085101 (2014)
5. J. R. D. Copley J. C. Cook, Nucl. Inst. Methods Phys. Res. A **345**, 313 (1994)
6. J. Bouillot, J. Torregrossa, Rev. Phys. Appliquée **19**, 799 (1984)
7. M. Nakamura, W. Kambara, K. Iida, R. Kajimoto, K. Kamazawa, K. Ikeuchi, M. Ishikado, K. Aoyama, Phys. B Condens. Matter **551**, 480 (2018)
8. K. H. Andersen *et al.* Nucl. Instruments Methods Phys. Res. Sect. A Accel. Spectrometers, Detect. Assoc. Equip. **957**, 163402 (2020)
9. N. Tsapatsaris, R. E. Lechner, M. Markó, H. N. Bordallo, Rev. Sci. Instrum. **87**, 085118 (2016)
10. M. Yonemura, K. Mori, T. Kamiyama, T. Fukunaga, S. Torii, M. Nagao, Y. Ishikawa, Y. Onodera, D. S. Adipranoto, H. Arai, Y. Uchimoto, Z. Ogumi, in *J. Phys. Conf. Ser.* (Institute of Physics Publishing, 2014)
11. I. P. Orench, J. F. Clergeau, S. Martínez, M. Olmos, O. Fabelo, J. Campo, J. Phys. Conf. Ser. **549**, 012003 (2014)
12. D. L. Abernathy, M. B. Stone, M. J. Loguillo, M. S. Lucas, O. Delaire, X. Tang, J. Y. Y. Lin, B. Fultz, Rev. Sci. Instrum. **83**, 015114 (2012)
13. K. Lefmann, K. Nielsen, Neutron News **10**, 20 (1999)

Performances of neutron cold multi-chopper instruments with different source natures among reactor, short and long pulse sources

Masatoshi Arai*

European Spallation Source ERIC, P.O. Box 176, 221 00 Lund, Sweden

Abstract. Cold multi-chopper instrument is a popular instrument commonly used for variety of scientific fields, built in reactors and sharp pulse sources and will be built in a long pulse source. In this article I compare the performance of those instruments, IN5 at ILL, AMATERAS at J-PARC and CSPEC at ESS, and show a merit/demerit come from the different characters of the source itself, and finally give a suggestion for a further improvement.

1 Introduction

Cold multi-chopper instrument is very popular and very useful, which serves a very important platform to study dynamical aspect of condensed matter in various scientific fileds. Its optimization scheme at reactor source was established in early 1990's [1], and since then relevant instruments have been constructed and operated at various kind of neutron sources. The design of the instrument is performed at a balance between two aspects, namely those are flux and resolution as a regular guideline. However, different characters of the source itself gives another dimension to be considered. Recent analysis on an expected performance of ESS instruments to ones at other sources with a different character stimulated us to revisit a way of optimization and use cases [2]. In this report, I present the consideration to optimize the performance of the relevant instruments at reactor (IN5), short-pulse (AMATERAS) and long-pulse (CSPEC), and show not only merit/demerit of source but also tunability emerged from the different source nature.

Here, we mostly focus on an argument for a typical use case (hereafter referred as "Typical Case") for E_i = $3\ meV$ (5Å) with resolution $\Delta\varepsilon / E_i = 3\%$ and angular divergence of $\Delta\theta = \pm 1.0°$, but extend to tunability by varying the energy resolution in the end.

2 Flux calculation

Figure 1 shows the time averaged brightness, B_{ta} [n/cm²/s/str/eV], of ESS (5MW), ILL (57MW) and J-PARC's coupled moderator (1MW) [2]. This is the important base information to go further, from which we can calculate flux at sample, once we know parameters of instrument such as energy range, energy resolution, angular divergence at sample. However, here we assume the brilliance transformation efficiency of guide system is perfect, $T_s = 1.0$, and we assume that there is no beam loss in it. Since it highly depends on the guide design, therefore, we neglect it for the time being, but it will be discussed in the later section.

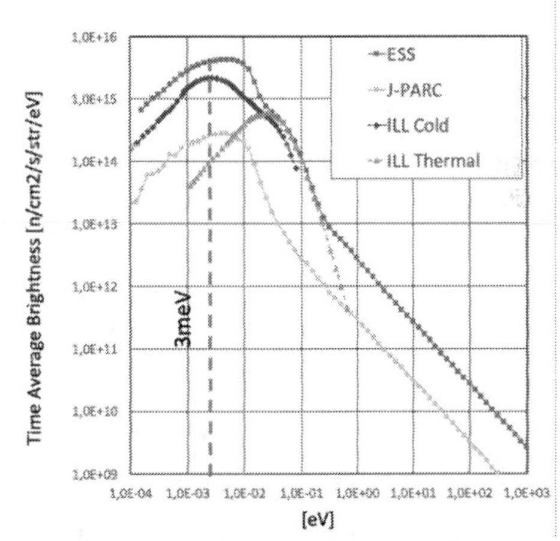

Fig. 1. Time averaged brightness of ESS (5 MW), ILL (57 MW) and J-PARC's coupled moderator (1 MW) [2]. In the report we will discuss for Ei = 3meV as Typical Case.

Fig. 2 shows a layout of a multi-chopper instrument. There are several time widths to be considered depending on sources. Those are,

- Δt_m: chopper opening time of the monochromatic chopper (MC)
- Δt_{PSC}: chopper opening time of the pulse shaping chopper (PSC)
- Δt_m: pulse width at moderator

* Corresponding author: masatoshi.arai@ess.eu

Here, we assume an instrument at short pulse does not use PSC, although there is its option on AMATERAS. Also, all of widths are defined as a full width at half maximum of a peak structure.

There are also three key distances, d_1, d_2, and d_3. Those are:

- d_1: the primary distance from the source to MC for short pulse, and from PSC to MC for long pulse or reactor instrument.
- d_2: the secondary distance from sample to detector.
- d_3: the distance between MC and sample. Normally a shorter d_3 is better in terms of the resolution.

In case for a pulsed neutron source, nevertheless short or long pulse source, the flux at sample is approximately defined as,

$$F(E_i) = B_{TA}(E_i) \cdot (\Delta t_{ch}/\Delta t_m) \cdot \Delta E_i \cdot (2\Delta\theta_s)^2 \cdot T_B \quad (1)$$

where the monochromatic chopper cuts out the time width Δt_{ch} from the original width of Δt_m at moderator for a given energy E_i, giving the term $\Delta t_{ch}/\Delta t_m$ with the condition of $\Delta t_{ch} < \Delta t_{PSC} < \Delta t_m$. Practically it is $\Delta t_{ch} \ll \Delta t_m$ for long pulse and reactor sources. The angular divergence, $2\Delta\theta_s$, at sample should be less than that of the guide system and incident energy resolution ΔE_i is also much less than the available energy range of $B_{TA}(E_i)$.

On the other hand, in case for reactor source, PSC creates pulses at frequency of f, then the resultant pulse width can be defined as $\Delta t_m = 1/f$. Therefore, the flux is described as,

$$F(E_i) = B_{TA}(E_i) \cdot (\Delta t_{ch} \cdot f) \cdot \Delta E_i \cdot (2\Delta\theta_s)^2 \cdot T_B \quad (2)$$

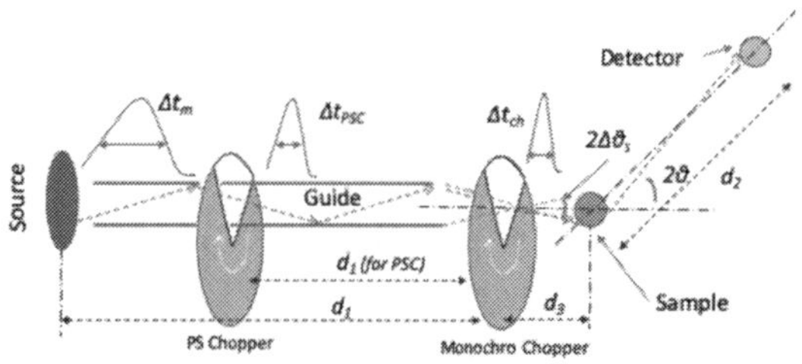

Fig. 2. Layout of a multi-chopper instrument. In case for a long pulse or a reactor source, a pulse shaping chopper (PSC) sharpens the pulse. And monochromatic chopper defines Ei in any sources.

The incident energy resolution ΔE_i after MC contains two components. One comes from MC's time width, Δt_{ch}, and another from that at the source or PSC, $\Delta t_{m/PSC}$.

$$\left(\frac{\Delta E_i}{E_i}\right)^2 = \left(2\frac{h}{m \cdot \lambda \cdot d_1}\right)^2 \left(\Delta t_{ch}{}^2 + \Delta t_{\frac{m}{PSC}}{}^2\right)$$
$$= \left(2\frac{0.3956}{\lambda \cdot d_1}\right)^2 \left(\Delta t_{ch}{}^2 + \Delta t_{m/PSC}{}^2\right) \quad (3)$$

Here, we define $\Delta t_{m/PSC}$ so as Δt_m is for short pulse source and Δt_{PSC} for long and continuous sources. The conversion factor 0.3956 is applied for distance in 'cm', time in 'μs' and wavelength λ in '$\AA$'.

The time of flight of neutrons is described as,

$$t[\mu s] = \frac{L}{v} = \frac{m \cdot \lambda \cdot L}{h} = \frac{\lambda[\AA] \cdot L[cm]}{0.3956} \quad (4)$$

Therefore, from Eq. (3),

$$\Delta E_i = \frac{2 \times 0.3956}{\lambda \cdot d_1} E_i \cdot \sqrt{\left(\Delta t_{ch}{}^2 + \Delta t_{m/PSC}{}^2\right)} \quad (5)$$

For a convention, we rewrite Eq. (5) as

$$\Delta E_i = D \cdot \sqrt{x^2 + y^2} \quad (6)$$

$$D = \frac{2 \times 0.3956}{\lambda \cdot d_1} E_i, \ \Delta t_{ch} = x, \ \Delta t_{m/PSC} = y \quad (7)$$

3 Operation mode of instruments

Figure 3 shows a TOF-distance diagram of operation of multi-chopper instruments at different kind of sources (time of flight: TOF). At reactor, a set of PSC and MC extracts multi pulsed beams at the same incident energy with a frequency of f, called repetition rate multiplicity (RRM) [3]. At long pulse source, the chopper system is the same or similar as at reactor, but it extracts slightly different incident energies by RRM at a long flight distance from moderator to detector. The energy variation depends on the flight distance. At short pulse, only MC is needed to extract different incident energies, and creates well separated incident energies in most cases with a shorter d_3.

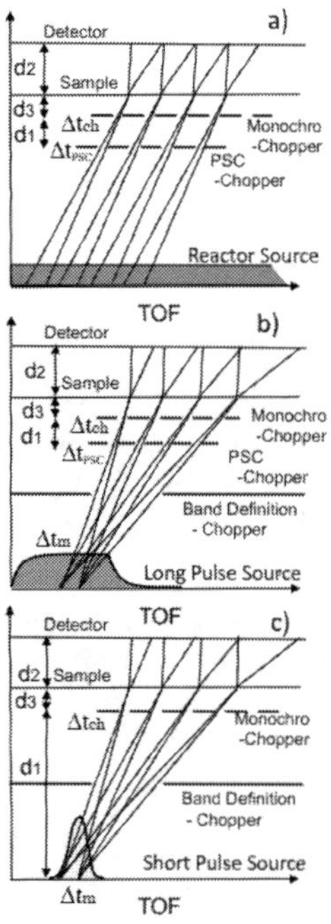

Fig. 3. Operation mode of multi-chopper instruments at different kind of sources. Reactor; a set of PSC and MC create monochromatic multi pulsed beam at a frequency of f. Long pulse source; it extracts RRM pulses with slightly different incident energies with a long flight distance. Short pulse; Only MC extracts well separated different incident energies.

This ability is preferably called "multi-Ei" a synonym of RRM developed on a Fermi-chopper instrument [4]. In the comparison in the next section, we mainly compare the RRM of IN5 at 142 Hz and single-mode (one of single incidence within RRM) of CSPEC and AMATERAS, where the source is operated at repetition rate of 14 Hz and 25 Hz respectively.

4 Resolution of a chopper instrument

The energy resolution of chopper instrument is described as Eq. (8) [5],

$$\frac{\Delta\varepsilon}{E_i} = \sqrt{\left\{\left[\frac{2\Delta t_{ch}}{t_{ch}}\left(1+\frac{d_1+d_3}{d_2}\left(1-\frac{\varepsilon}{E_i}\right)^{\frac{3}{2}}\right)\right]^2+\left[\frac{2\Delta t_{m/PSC}}{t_{ch}}\left(1+\frac{d_3}{d_2}\left(1-\frac{\varepsilon}{E_i}\right)^{\frac{3}{2}}\right)\right]^2\right\}} \quad (8)$$

However, in order to simplify the argument, here, we concentrate on the energy resolution at the elastic position ($\varepsilon = 0$), where most neutron intensity is counted, and we can analyze the major aspect of optimization. Therefore, the resolution formula at the elastic position can be written as follows,

$$R = \frac{\Delta\varepsilon}{E_i} = \frac{2\times 0.3956}{d_1\cdot\lambda}\times \sqrt{\left\{\left[\Delta t_{ch}\left(\frac{d_1+d_2+d_3}{d_2}\right)\right]^2+\left[\Delta t_{m/PSC}\left(\frac{d_2+d_3}{d_2}\right)\right]^2\right\}} \quad (9)$$

The Parameters in Eq. (9) are depicted in Fig. 2.
Eq. (9) is simplified as,

$$c^2 = \{(ax)^2 + (by)^2\} \quad (10)$$

where,

$$x = \Delta t_{ch},\ y = \Delta t_{m/PSC}$$

and

$$c = \frac{R\cdot d_1\cdot\lambda}{2\times 0.3956},\ a = \frac{d_1+d_2+d_3}{d_2},\ b = \frac{d_2+d_3}{d_2} \quad (11)$$

This is converted and is a formula of an ellipse,

$$g = \left(\frac{a}{c}\right)^2 x^2 + \left(\frac{b}{c}\right)^2 y^2 - 1 = Ax^2 + By^2 - 1 = 0 \quad (12)$$

where,

$$A = \left(\frac{a}{c}\right)^2 = \frac{\left(\frac{d_1+d_2+d_3}{d_2}\right)^2}{\left(\frac{R\cdot d_1\cdot\lambda}{2\times 0.3956}\right)^2},\ B = \left(\frac{b}{c}\right)^2 = \frac{\left(\frac{d_2+d_3}{d_2}\right)^2}{\left(\frac{R\cdot d_1\cdot\lambda}{2\times 0.3956}\right)^2} \quad (13)$$

5 Optimization Concepts

Here, we assume the geometrical parameters, such as d_1, d_2, and d_3 have been given for the time being, and optimization is a subject to determine the timing aspects. In case for a short pulse source Δt_m is given from the moderator characteristics for each energy, and cannot control from the instrument [2]. Therefore, Δt_m is uniquely determined for the required energy resolution in Eq. (9). On the other hand, in a long pulse source or a reactor source, two chopper windows, Δt_{ch} and Δt_{PSC}, are tunable, and the optimization of the chopper opening time should be done by finding the maximum flux at sample by keeping a required resolution R through the constraint of Eq. (12).

5.1 Lechner's Optimization Concept

Lechner assumed the flux at sample is proportional to $\propto \Delta t_{ch}\cdot\Delta t_{PSC} = x\cdot y$ [1], [6]. This corresponds to the maximization of the area of,

$$S = x\cdot y \text{ for } x, y \geq 0 \quad (14)$$

under the condition of Eq. (12) as shown in Fig. 4. The solution is given by the Lagrange multiplier method,

$$L(x,y) = S - \lambda\cdot g = x\cdot y - \lambda\cdot(Ax^2 + By^2 - 1) \quad (15)$$

by taking partial derivatives to find local minima, $\frac{\partial L}{\partial x} = 0$, $\frac{\partial L}{\partial y} = 0$, and we can have the following solution.

$$x = \frac{1}{\sqrt{2A}},\quad y = \frac{1}{\sqrt{2B}} \quad (16)$$

And the optimum opening time has a relation as,

$$\frac{x}{y} = \frac{\Delta t_{ch}}{\Delta t_{PSC}} = \sqrt{\frac{B}{A}} = \frac{d_2+d_3}{d_1+d_2+d_3} \quad (17)$$

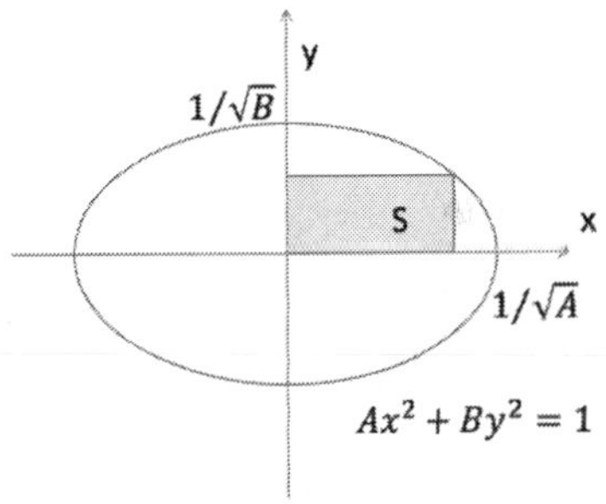

Fig. 4. Ellipse of Eq. (12). Maximizing "S" of Eq. (14) is subject.

5.2 Another Optimization concept

As we expressed in Eq. (1), Eq. (2) and Eq. (6) under a given beam divergence, $2\Delta\theta_s$ the flux at sample is proportional to,

$$flux \propto \frac{\Delta t_{ch}}{\Delta t_m}\cdot\Delta E_i = \frac{\Delta t_{ch}}{\Delta t_m}\cdot D\cdot\sqrt{\Delta t_{ch}^{\ 2} + \Delta t_{PSC}^{\ 2}} \propto f = x\sqrt{x^2+y^2} \quad (18)$$

where we left only relevant terms, x and y, in the rightmost equation, since Δt_m and D is given from the source character, and d_i and E_i are also given. Now we maximize Eq. (18) under a constraint of Eq. (12) by Lagrange's multiplier method;

$$L(x,y) = f - \lambda\cdot g = x\sqrt{x^2+y^2} - \lambda(Ax^2 + By^2 - 1) \quad (19)$$

The solution is following,

$$x = \sqrt{1/2(A-B)},\ y = \sqrt{(A-2B)/(2B(A-B))} \quad (20)$$

namely,

$$x/y = \sqrt{B/(A-2B)} \quad (21)$$

This contrasts with Lechner's result, $x/y = \sqrt{B/A}$, Eq.(17). A correction term "2B" comes into the denominator in Eq. (21).

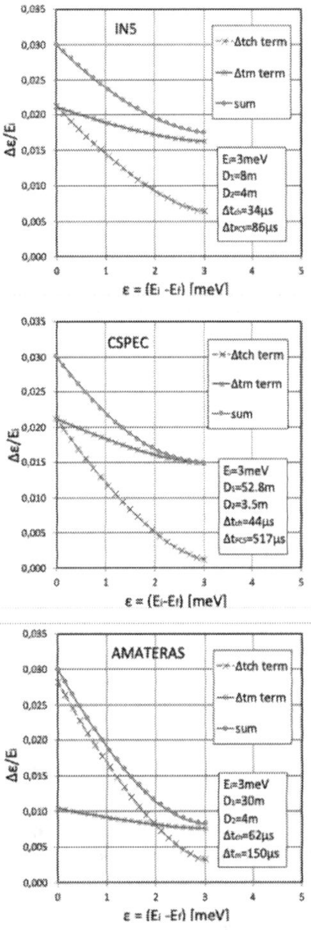

Fig. 5. Lechner's optimization; Energy resolution diagram for CSPEC, AMATERAS and IN5, where Δε/Ei is kept at 3% at the elastic position. In each panel, the first term with (Δtch), the second term with (Δtm/PSC) in Eq. (8) and the sum of them are depicted.

Lechner's results gives that a matching condition for the partial resolution regarding Δt_{ch} and Δt_{psc} and the first and the second term of Eq. (9) have the same value at the elastic position as shown in Fig. 5 for CSPEC and IN5. Fig. 5 was calculated by Eq. (8) in the applicable energy transfer range up to $\varepsilon = E_i$.

On the other hand, the resolution diagram for the present optimization is shown in Fig.6. CSPEC does not have significant change in the diagram because d_1 is relatively very longer than d_2 and d_3. However, the diagram of IN5 shows a minor but definite difference by the correction term in Eq. (21) as shown at the elastic position in Fig.6.

AMATERAS has decisive value, x = Δt_{ch}, determined by a required resolution with a given y = Δt_m as shown in Eq. (22), and there is no difference whatever the optimization method, and the resolution diagram is only shown in Fig.5.

$$x = \sqrt{\left(\frac{c}{a}\right)^2 - \left(\frac{c}{b}\right)^2 y^2} \quad (22)$$

Therefore, the present optimization can give a minor correction over Lechner's optimization only for a short instrument.

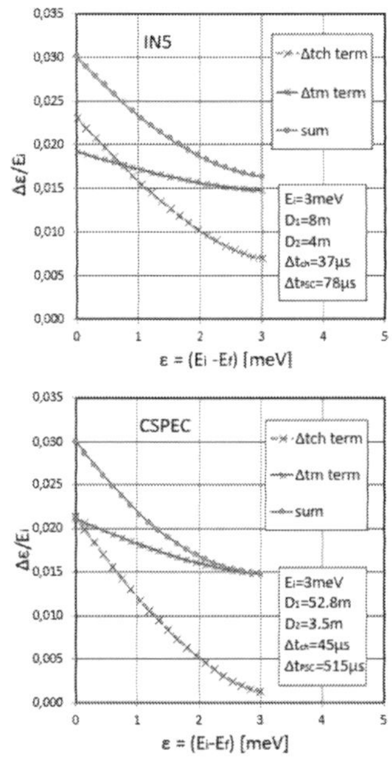

Fig. 6 Resolution diagram for the present optimization. Now the partial resolution regarding Δt_{ch} and Δt_m does not necessarily match at the elastic energy as seen in the panel of IN5, although that of CSPEC's has negligibly small change.

6 Estimated Flux of IN5, AMATERAS and CSPEC

Here, we estimate flux of IN5, AMATERAS and CSPEC on Typical Case.

In Table 1, we summarized the operation parameters in comparison with the Lechner's concept [1]. The difference is almost negligible for CSPEC but very small gain is found for IN5 which has a short d_1 in comparison to CSPEC.

At the third bottom row in Table 1, the estimated absolute fluxes are listed. The IN5's flux of 6.44 X 10^5 is very closed to one listed in Ref. [10], and CSPEC's flux 2.64 X 10^6 is also very close to one in Ref. [8]. These results show the validity of the present calculation independently estimated from the brightness in Fig. 1 and give a base to make further estimates.

At the second bottom row, we can see that the performance of the single mode of AMATERAS [7] is almost comparable to CSPEC [8], and it is about four times higher than IN5 at 142 Hz operation [9], [10]. If we can reasonably utilize the RRM operation of CSPEC, we can have the performance shown at the bottom row, where the CSPEC's operation mode is assumed to be with 20% single mode and 80% of 6 - RRM mode giving a factor 5.0. But here we remark that AMATERAS regularly has 4 multi-Ei measurement, although the incident energy separation is well apart from each other as specified in Table 1 [7]. Therefore, the practical comparison between CSPEC and AMATERAS cannot simply estimated by the factor 5.0. RRM gives very different aspects for two instruments; CSPEC can increase statistics by adding RRM data in a certain range of incident energies, instead AMATERAS can observe a wide Q-E spectrum simultaneously.

Table 1. Parameters and performance in comparison with Lechner's concept [1], where the time averaged brightnesses for 1MW of J-PARC, 5MW of ESS and 57MW of ILL are taken as shown in Fig. 1.

Source	ILL (Lechner)	ILL (Present)	J-PARC	ESS (Lechner)	ESS (Present)	ESS
Time Averaged Brightness at 3meV, 5A, at the moderator surface [n/cm^2/s/str/eV]	2,09E+15	2,09E+15	2,0E+14	3,1E+15	3,1E+15	3,1E+15
Instrument operation mode	IN5 (142Hz)	IN5 (142Hz)	AMATERAS (single)	CSPEC (single)	CSPEC (single)	CSPEC (RRM) (single 20% + 6 RRM 80%)
d1 [m]	8	8	30	52,8	52,8	
d2 [m]	4	4	4	3,5	3,5	
d3 [m]	1,2	1,2	1,5	1,5	1,5	
Band width / frame [Å]			5,3	1,88	1,88	1,88
RRM or Multi-Ei	142 Hz	142 Hz	1,7 3,1 7,7 42,0 meV			2,2 2,4 2,6 3,0 3,4 3,9 meV
Monochro Chopper time Δtch [μs]	34	37	62	44,8	44,9	
Moderator or PSC time width Δtm [μs]	86	78	150	517,4	515,5	
2Δθ [deg]	2,0	2,0	2,0	2,0	2,0	
Δε[μeV] at elastic	90,0	90,0	90,0	90,0	90,0	
ΔEi[μeV]	52,5	49,1	24,6	44,7	44,5	
Flux (Estimated)	6,44E+05	6,55E+05	2,48E+06	2,64E+06	2,64E+06	1,32E+07
Ratio (1/AMATERAS)	0,26	0,26	1,0	1,07	1,07	5,34
Ratio /(CSPEC (Multi))	0,05	0,05	0,19	0,20	0,20	1,0

7 Optimization of d1

Because of constraint of a realistic dimension of the secondary spectrometer, d_2 could be 4 m and d_3 1.5 m, which achieves most of required resolution within a reasonable construction cost. However, d_1 is a kind of an optimizable length although we again need to consider the cost for a long guide. Here, we try to find an optimal length of d_1. Figure 7 shows the expected flux as a function of d_1 for Typical Case. Very interestingly the behavior of the flux curves have an opposite behavior in d_1 between AMATERAS and CSPEC or IN5. This is because Δt_m of AMATERAS is given, and shortening of d_1 makes Δt_{ch} a little shorter, and the argument in square-root of Eq. (5) has small reduction. But the division by d_1 in Eq. (5) eventually makes ΔE_i larger, and it makes a higher flux through Eq. (1). Contrarily to this, increasing d_1 for either CSPEC or IN5 makes increases both Δt_m and Δt_{mc}, but it cancels out with d_1 in the denominator in Eq. (5), and eventually ΔE_i stays almost constant independent from d_1 at a large d_1. Therefore, through Eq. (1) or Eq. (2), an increase of d_1 only makes gradual increase in the flux, and it shows a completely opposite d_1 dependences against one at a short pulse source.

For a long guide, however, we expect a beam loss. So, here, we show a very simple analysis of the beam loss through guide, where, we assumed a reflection performance parameter of m = Q/Q_c(Ni) = 2 with the critical refection factor of 0.94 with a 10 cm wide straight guide, which gives the required angular divergence of $\Delta\theta = \pm 1°$ at 5 Å [11]. The estimated flux of CSPEC multiplied by the beam loss is plotted in Fig. 7. It shows the flux above d_1 = 30 m does not have significant increase with d_1. However, since this estimate highly depends on the guide design and a detailed estimate should be necessary by using a computer simulation, and the present results should be taken just as a reference.

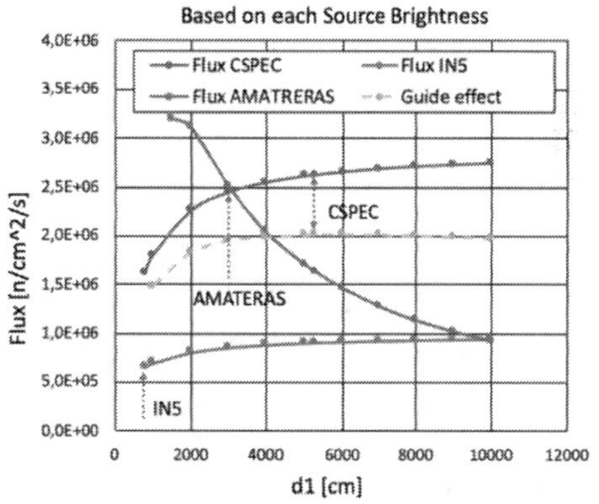

Fig. 7. Expected flux as a function of d_1 for a given resolution of 3% for Ei = 3meV for single mode.

From those results, it seems that CSPEC's d_1 = 52.8 m (PSC to sample) is a good choice, where flux increase tends to saturate. CSPEC has a 100 m guide-section before PSC from the moderator [8], which could make a beam loss of ca 20%, but can eliminate high energy neutrons and the spectrum is spread enough at PSC to have 6 RRM pulses. On the other hand, in case for IN5, d_1 seems to be too short and increasing d_1 to, say, 30 m may gain 20 - 30 % more flux, if the spatial constraint is not an issue. We have noticed a similar d_1 of 30 m was adapted to an upgrade of NEAT instrument giving a high performance [12].

AMATERAS had better to reduce d_1 as much as possible, however, this has a constraint given in Eq. (22). The widest Δt_m at practically lowest energy at 1 meV is 200 μs, therefore, the choice of d_1 = 30 m is also reasonable. However, for Typical Case of *Ei* = 3meV, a shorter distance of d_1 = 20 m gains 20 % more in flux.

8 Tunability in resolution and flux

So far, we have sticked on Typical Case. However, as we have noticed in Ref. [13], long pulse source inherently has a tunability in its resolution and flux. It can be also true even for spectrometers. Therefore, Fig. 8 shows the flux variation as a function of the energy

https://doi.org/10.1051/epjconf/202227202011

resolution at Ei = 3meV. Since Δt_m of AMATERAS is fixed to be 150 μs at 3meV, the available resolution range is very limited between 1,5% and 6%, below or above which Eq. (22) or Eq. (1) is not valid anymore because of negative Δt_m or $\Delta t_m > \Delta t_m$ respectively. On the other hand CSPEC and IN5 can choose almost freely the resolution in trade of flux because of a tunability in both Δt_m and Δt_m. However, the maximum available resolution 15% of CSPEC is also limited by Δt_m= 2.96 ms of ESS's long pulse width. Because of this tunability, CSPEC and IN5 can have a very high flux by relaxing resolution, if it is demanded, which is inherently not available on AMATERAS.

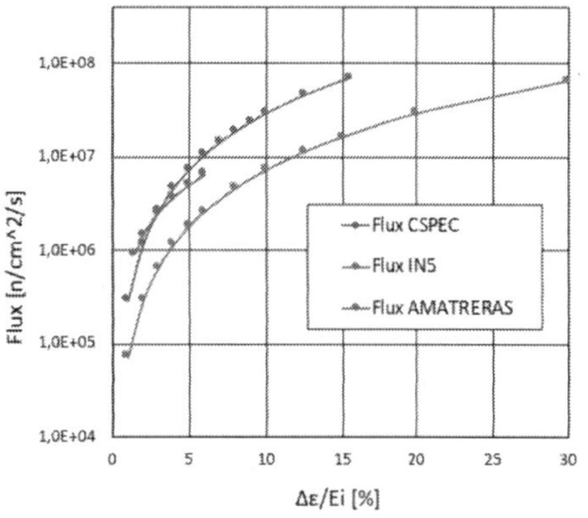

Fig. 8. Flux variation as a function of energy resolution.

9 Conclusion

In this report, I have developed a more precise optimization to determine Δt_m and Δt_m of multi-chopper instrument. However, the result only gives a very minor improvement to a long d_1 instrument over the well-known Lechner's optimization. It can only show an improvement to a short instrument, and an elongation of d_1 of IN5 could gain another 20 to 30 % flux, if a spatial and cost constraint is not an issue. The analysis also showed that the choice of the current d_1 of CSPEC and AMATERAS is very reasonable. The analysis showed how this class of instrument is suitable at a sharp pulse source for a high-resolution mode. However, CSPEC and IN5 has a great tunability in resolution and flux, giving various options from high resolution to high flux configuration covering a wide range of use cases, which is inherently difficult at a short pulsed source. Application of RRM of CSPEC could enhance its performance, although some detailed assessment could be necessary in terms of resolution effect by adding data of different wavelengths. On the other hand, AMATERAS has another class of multiplicity with multi-Ei, which gives a quick survey in a well separated Q-E space.

I acknowledge Dr. Ruep. E. Lechner very much, who made an outstanding guideline of design of this class of instrument, lead the community for a long time, and helped us to design CSPEC at ESS and AMATERAS at J-PARC.

10 References

1. R. E. Lechner, ICANS – XI International Collaboration on Advanced Neutron Sources, pp. 716–732 (1990)
2. M. Arai, L. Zanini, K. Andersen et al., Journal of Neutron Research, **22**, 71 (2020)
3. M. Russina and F. Mezei, Nuclear Instruments and Methods in Physics Research, **A604**, 624 (2009)
4. M. Nakamura, R. Kajimoto et al., Journal of the Physical Society of Japan, **78**(9), 093002 (2009)
5. M. Arai, Experimental Methods in the Physical Science, Volume 44, Chapter 3 pp. 245-320, Neutron Scattering-Fundamentals, Edited by Felix Fernandez-Alonso and David L. Price, (Elsevier Inc., 2014)
6. R. Lechner, Physica B, **276-278**,67 (2000)
7. K. Nakajima, S. Ohira-Kawamura et al, J. Phys. Soc. Jpn., **80**, SB028-1~ SB028-6, (2011)
8. K. Andersen et al., Nuclear Inst. and Methods in Physics Research, **A 957**, 163402 (2020)
9. J. Ollivier and H. Mutka, J. Phys. Soc. Jpn., **80**, SB003 (2011)
10. "Disk chopper time-of-flight spectrometer IN5": https://www.ill.eu/users/instruments/instruments-list/in5/description/instrument-layout .
11. "Super mirror" Swiss Neutronics, https://www.swissneutronics.ch/index.php?id=24.
12. M. Russina et al., Physica B: Condensed Matter **551**, 506 (2018)
13. M. Arai, K. Andersen, et al., Journal of Neutron Research, **23** (4), 215 (2021)

Possible options for efficient wide-band polychromatic measurements using chopper spectrometers at pulsed sources

Kenji Nakajima[1, 2,*], *Tatsuya* Kikuchi[1, 3, 4], *Seiko* Ohira-Kawamura[1], and *Wataru* Kambara[1]

[1]Materials & Life Science Division, J-PARC Center, JAEA, Tokai, Ibaraki, Japan
[2]Materials Sciences Research Center, JAEA, Tokai, Ibaraki, Japan
[3]Research & Development HQ., Sumitomo Rubber Industries, Ltd., Kobe, Hyogo, Japan
[4]Institute of Materials Structure Science, KEK, Tsukuba, Ibaraki, Japan

Abstract. The polychromatic incident energy measurements using chopper spectrometers at pulsed sources enables us to measure dynamics with different energy resolutions, i.e., different time-windows with a single-measurement. By increasing the possible time-window for a single-measurement, the efficiency and usefulness of such measurements is enhanced. This is achieved by allowing the crossing of trajectories of different E_is from different source pulses on the time-of-flight diagram (the interveined trajectory). We will present an actual case of the interveined trajectory on AMATERAS, a cold-neutron chopper spectrometer installed at J-PARC, together with our experience. Some of our attempts to formulate the conditions for performing polychromatic E_i measurements employing interveined trajectories are discussed. These which can be applied to existing spectrometers, and considered as design criteria for future chopper spectrometers. Further related plan of AMATERAS to improve measuring efficiency is also mentioned.

1 Introduction

Direct access to quantitative measurements of the static and dynamical correlations of atoms, molecules and spins in energy and momentum spaces is one of prominent futures of the neutron scattering technique. Based on many of techniques and technologies invented so far, inelastic and quasi-elastic neutron scattering instruments have been continuously improved their performance to achieve higher and flexible resolution, higher intensity and higher efficiency, to open the door to a rich frontier of nature science.

To capture dynamics of phenomena in materials, a large coverage of different degrees of time windows is one of the keys. A legacy way to achieve such measurements is to measure the same dynamics with different neutron spectrometers having different energy resolutions. Of late, polychromatic incident energy (E_i) measurements carried out by utilizing repetition rate multiplication technique have considerably improved the efficiency of chopper spectrometers at pulsed sources [1, 2]. The polychromatic E_i technique, which has become common in chopper spectrometers in recent decades, has enabled wide-time window dynamic measurements by a single measurement, using a single spectrometer. This capability has stimulated the emergence of new types of data handling, such as the mode distribution analysis [3], which enables us to evaluate the number of modes and their characteristic time scale from quasi-elastic measurements, without making any assumptions. Then it is natural that our interest goes to seek the technique to obtain dynamical response function in wider range of time window with higher efficiency with a single measurement.

Some efforts have been made on AMATERAS, a cold-neutron chopper spectrometer installed at a pulsed source [4], to increase accessible time-windows with a single measurement condition. As it is known, by allowing the crossing of trajectories of different E_is from different source pulses on the time-of-flight (TOF) diagram, the usable width for the magnitude of E_is can be increased. We call this the interveined trajectory. The interveined trajectory is a powerful technique for spectrometers operating in the polychromatic E_i mode, and has been employed on AMATERAS along with other attempts, including to increase the number of usable E_is from the source by modifying the chopper system.

In this paper, we will present our attempts and experiences, and a plan of using AMATERAS to increase time-windows and enhance measuring efficiency in a single measurement. Measurements utilizing the interveined trajectory form the crux. Our experiences and considerations of the interveined trajectory will be described, which includes an actual case using AMATERAS. An introductory consideration, and our attempts to formulate the conditions are included. We also mention a future plan of AMATERAS to improve the efficiency of wide-band measurements by modifying the pulse shaping chopper. We hope that these will help to existing and future planned spectrometers with wide-band polychromatic E_i measurements.

* Corresponding author: kenji.nakajima@j-parc.jp

2 Bandwidth and interveined trajectory

For instruments installed at a conventional pulse neutron source, useful band width depends on the repetition rate of the source [5]. In the case of instruments utilizing a white neutron beam from the source (e.g., diffractometers, inverted geometry spectrometers and neutron imaging facilities), the band width in the wave length, $\Delta\lambda_{max}$, can be written as shown in Eq. (1).

$$\Delta\lambda_{max} = h\tau/m_n L_{tot} \quad (1)$$

where h is the Plunk constant, τ is the interval time between the source pulses, m_n is the mass of neutron, and

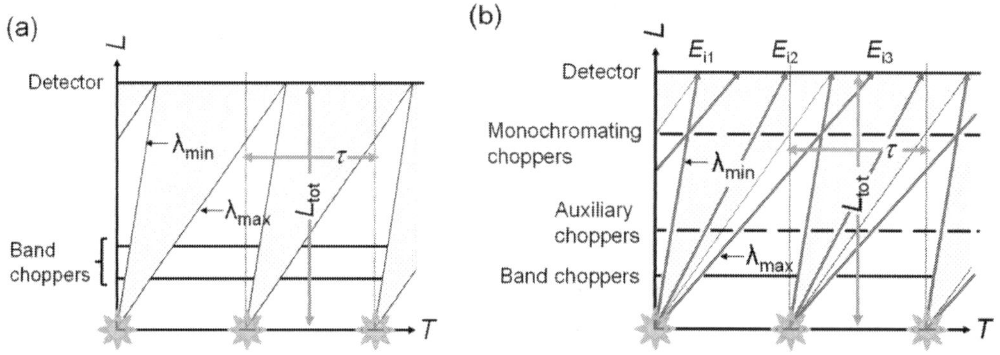

Fig. 1. A TOF diagram explaining the possible band width at a pulsed source. (a) In the conventional case. Available maximum time window at the detector position is *t*, the interval time of the source pulses. The possible wave length band, $\Delta\lambda = \lambda_{max} - \lambda_{min}$ is determined by *t* and the total flight distance, L_{tot}. (b) The interveined trajectory case. Three E_is are selected in the figure, namely E_{i1}, E_{i2}, and E_{i3}. Here, trajectories of E_{i1} and E_{i3} are crossing at the same monochromating chopper opening, i.e., the interveined trajectory. The effective wave length band, $\Delta\lambda = \lambda_{max} - \lambda_{min}$ is wider than it is in the conventional case. The auxiliary chopper in the figure is optional. In the figures, *L* and *T* stand for the flight path length and the time-of-flight, respectively.

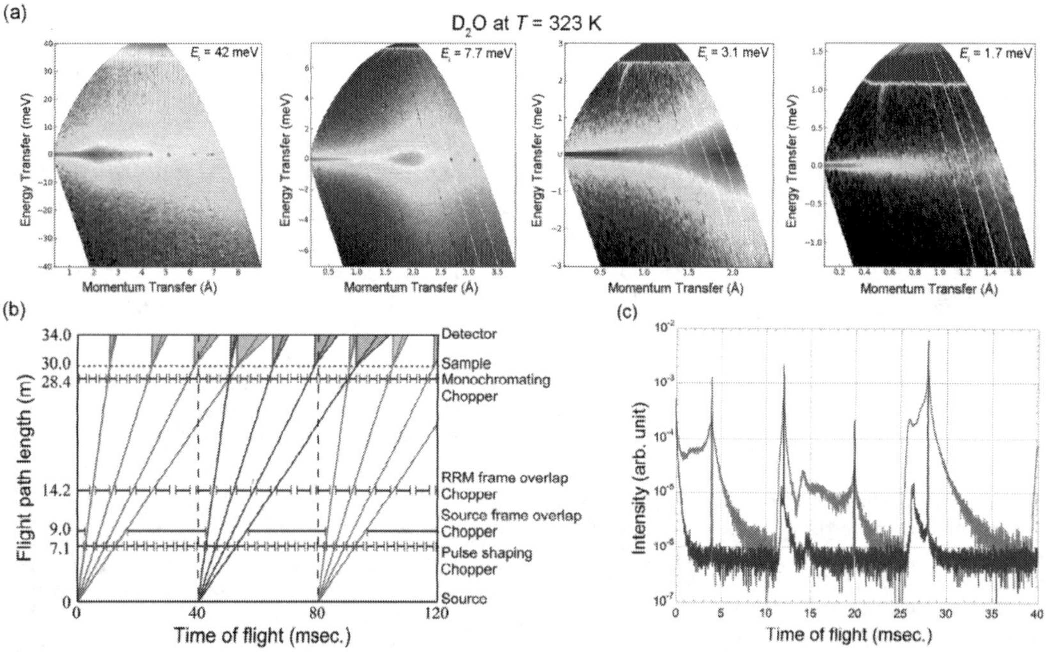

Fig. 2. An actual use of the interveined trajectory in AMATERAS at J-PARC, where we could realize the measurements with four E_is—42, 7.7, 3.1, and 1.7 meV—and energy resolutions in the 0.9–0.01 meV range, which corresponds to 0.02 ps–500 ps time windows. (a) Example of data obtained from D_2O measurements at 323 K. (b) Time-of-flight (TOF) diagram of this condition (E_is from the $T < 0$ are omitted in the figure). (c) TOF spectrum measured of D_2O data shown in (a) (red line) together with a vanadium cylinder at room temperature data (blue line). The data obtained at the equatorial position detector pixels were used. Note that AMATERAS has another band chopper at 13.7 m position, which is not used in this condition, and is therefore removed in panel (b). It should be noted that the value of the flight path length from the source to the detector shown in panel (b) is at the equatorial position of the detector bank [4]. Actual distance from the sample to the detector pixels spreads from 4 m to 4.4 m.

Table 1. AMATERAS chopper configuration

Role [7]	Position from the source	Maximum revolution	Number of disks	Slit size or opening angle
Pulse-shaping chopper	7.1 m	350 Hz	2 (counter-rotating)	Slit size: 30 mm
Monochromating chopper	28.4 m	350 Hz	2 (counter-rotating)	Slit size: 10 mm and 30 mm
RRM frame overlap chopper	14.2 m	350 Hz	1	Slit size: 30 mm
Source frame overlap chopper 1	9 m	25 Hz	2 (co-rotating)	Open angle: 0°–175° (variable)
Source frame overlap chopper 2	13.7 m	25 Hz	2 (co-rotating)	Open angle: 0°–175° (variable)

L_{tot} is the total flight path length. The arrangement is shown in Fig. 1 (a). If one needs the band width to be longer than that in Eq. (1), one can discard several source pulses by some means. Removing one of two pulses can double it, and removing two of three can triple it. This is usually done with instruments upon request. However, it sacrifices time averaged neutron flux.

It is known to those who are experienced with polychromatic E_i measurements that we can employ an alternative way to extend $\Delta\lambda$, in the case of chopper spectrometers. By allowing trajectories of different E_is from different source pulses to cross each other on the TOF diagram, we can have extension of the effective band width, that is interveined trajectory. An example is shown in Fig. 1 (b), in which allowing different E_is to transmit the same opening of the monochromating chopper, that is one of the ways to realize an interveined trajectory. Using an interveined trajectory, a chopper spectrometer can select the set of E_is in the range of the band width wider than in the conventional case.

An interveined trajectory is employed at AMATERAS in actual measurements. We have already reported the result of measurement on D_2O at 273 K in our earlier manuscript [4]. Here, we show another example, i.e., the result of measurement on D_2O at 323 K in Fig. 2 (a). Data taken with four different E_is, namely 42, 7.7, 3.1, and 1.7 meV, were obtained by a single measurement. Measurement with energy resolutions from 0.9 meV to 0.01 meV corresponds to 0.02 ps–500 ps time windows, were realized. Variation of four E_is in wave length (effective band width) is 5.6 Å. The value is wider than the maximum band width defined by Eq. (1), 4.7 Å for t = 40 ms and L_{tot} = 34 m. The time-of-flight (TOF) diagram for this condition is shown in Fig. 2(b). Detailed information on the AMATERAS disk-choppers can be found in the reports [4, 6], while simple information is shown in Table 1. The details on the roles of each chopper written can be found in Ref. 7. The pulse-shaping chopper, the monochromating chopper, the RRM frame overlap chopper, and the source frame overlap chopper 1 were operated at revolution rates 300 Hz, 300Hz, 150 Hz, and 25 Hz, respectively. We did not operate (i.e., stopped at the open position) the source frame overlap chopper 2 in this condition. We present D_2O spectrum data with vanadium data in Fig. 2 (c). Overlapping of signals (contamination) on the TOF spectrum is important for measurements on chopper spectrometers [8]. We will discuss the contamination issue later.

We obtained some experiences on the interveined trajectory measurements on AMATERAS. An auxiliary chopper shown in Fig. 1 (b) is optional. Although one can implement an interveined trajectory without this chopper (when the monochromating chopper runs at half of the frequency, we can remove the auxiliary chopper of Fig. 1 (b)), by using such choppers to remove selected pulses from the source, we can increase possible options of operation modes of the monochromating chopper. For this purpose, we sometime use the RRM frame overlap chopper. A fan chopper [9], a selective pulse suppression chopper [10] or a more flexible type of selective pulse suppression chopper [11] may provide more flexibility. On the other hand, slow disk-choppers with variable opening are indispensable. Since the width of the band definition in experiments with/without interveined trajectories are different, the ability to define different bandwidths is needed in slow disk-choppers. Ultimately, the redundancy of chopper systems is important. The flexibility allows us to perform new types of experiments, even those not considered before construction.

3 Some of attempts to formulate the interveined trajectory

This section documents our experience with the algebra used to find conditions suitable for the interveined trajectory in actual spectrometers. Our attempts to formulate the relationship between the pulse repetition rate of the source, the flight distances of the spectrometer, the chopper frequency, and other relevant parameters in two selected conditions for carrying out interveined trajectory measurements, are presented. These are just examples. One can extend these ideas upon one's requests for constructing new spectrometers or make a design of new types of measurements.

In this section, we assume the following *a priori* conditions.

i. The monochromating chopper is synchronized with the source pulse, and it opens at least once in the time period τ.
ii. Choppers with irregular opening timing (e.g., a fan chopper and a selective pulse suppression chopper), are not used.

3.1 Case I: Near-elastic (equal time interval case)

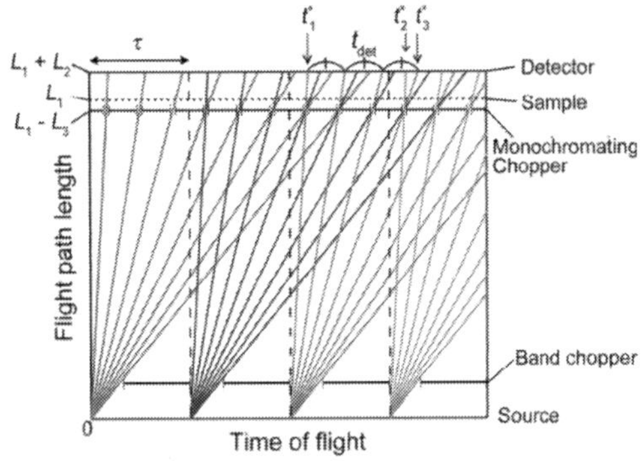

Fig. 3. Example of the case to obtain E_is arriving at the detector position at a regular interval. In this figure, the combination of parameters $n_f = 3$, $n_i = 3$, $n_p = 8$, and $n_d = 1$ is considered. Detail can be seen in the text. E_is from the time < 0 are omitted in the figure.

As one of the simplest examples, let us consider the case obtaining all E_is arriving at the detector position at regular intervals. If focusing on elastic and/or near elastic channels (e.g., elastic incoherent structure factor (EISF) measurements), we can consider this simple case, in which it is not necessary to consider the time window for the inelastic region. As case I is simple, it may not have many applications but is a good for an introduction.

An example of such a case is in Fig. 3. Here after, it is defined as follows.

L_1: Distance from the source to the sample,
L_2: Distance from the sample to the detector,
L_3: Distance from the monochromating chopper to the sample.

In addition, we introduce the following parameters.

n_p: Number of E_is from one source pulse,
n_f: Number of frames that E_is from one source pulse spread over,
n_i: Maximum number of E_is from one source can be accommodated in one frame,
t_{det}: Time interval between E_is from one source pulse at the detector position.

The relationship between these parameters is illustrated using the following ceiling functions.

$$n_f = \mathrm{ceil}\left(\frac{n_p t_{det}}{\tau}\right), \tag{2}$$

$$n_i = \mathrm{ceil}\left(\frac{\tau}{t_{det}}\right). \tag{3}$$

A priori conditions guarantee that $n_p \geq n_f$.

An example in Fig. 3 corresponds to $n_p = 8$, $n_f = 3$, and $n_i = 3$. We also introduce some definitions regarding the order of arrival of neutrons from different source pulses.

N_n: n is an integer index for the E_i to identify the source pulse from which it emerged. The E_i of N_1 means the E_i emerged from the most recent source pulse. The E_i of N_2 means the E_i emerged from the penultimate source pulse, and so on.

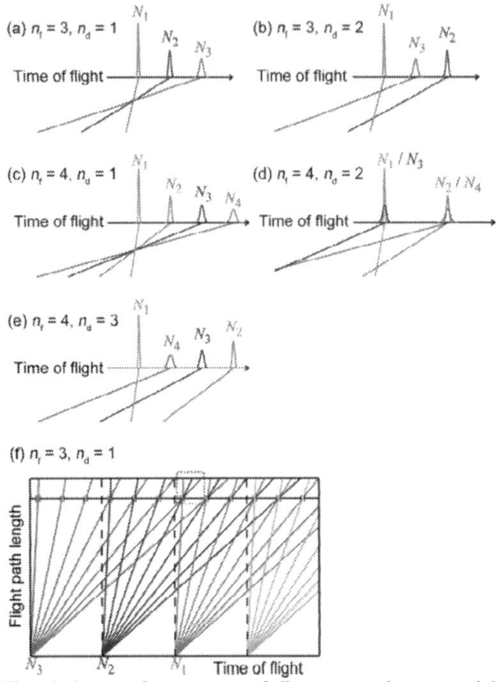

Fig. 4. Some of sequences of E_i as examples to explain definition of Nn and n_d. Possible sequences of E_is are described in the cases of n_f = 3 ((a) and (b)) and 4 ((c) ~ (e)). Corresponding n_ds are shown together. When n_d and n_f are not relatively prime numbers, neutrons from different source pulses arrive at same time at the detector position as can be seen in (d). (f) The full TOF diagram of the case (a) (n_f = 3, n_d = 1). E_is from the time < 0 are omitted in the figure. The panel (a) corresponds to the area surrounded by the red square in (f).

n_d: The index stands for the order of the arrival at the detector position of E_i of N_2 counted from the earliest E_i of N_1.

Examples are shown in Fig. 4. When $n_i \geq 3$, there are choices for a permutation-combination arrangement in the order of arriving E_is from different source pulses. Some of examples are shown in Fig. 4 for the cases of $n_f = 3$ and 4. Once n_f and n_d are determined, orders of N_n ($n \geq 3$) are automatically determined when *a priori* conditions are satisfied. When n_d and n_f are not relatively prime numbers, neutrons from different source pulses arrive simultaneously at the detector, as can be seen in Fig. 4 (d).

For E_is arriving at the detector at regular intervals, we have to accommodate E_is of N_2 ~ N_{nf} in the time window between two consecutive E_is of N_1s at the maximum. That means the time interval of all E_is is t_{det}/n_f. Therefore, the relationship is defined as follows.

$$\tau = \left(n_i - \frac{n_d}{n_f}\right) t_{det} \tag{4}$$

An empirical approach can help explain Eq. (4). One can see the timing of the highest E_is from two consecutive source pulses, t^*_1 and t^*_2 in Fig. 3. Then, $t^*_2 - t^*_1$ is τ within *a priori* conditions. Conversely, after t^*_2, (n_i+1)th E_i from the earlier source pulse arrives at t^*_3 and t^*_3 is equal to (n_i+1)×t_{det}. To go back from t^*_3 to t^*_2, the duration (interval time of E_is at the detector position)×n_d should be subtracted from t^*_3. Since the time interval of all E_is is t_{det}/n_f, τ is $((n_i+1)\times t_{det})-(n_d\times(t_{det}/n_f))$, that is Eq. (4).

There is choice in the location of the monochromating chopper. For example, in Fig. 1 (b), when we remove the monochromating chopper, only with the auxiliary chopper in the figure, we can obtain the same set of E_is, if we can ignore the resolution which becomes worse substantially with shorter distance from the source to the monochromating chopper. Under assumption of *a priori* conditions, one of the effective locations for the monochromating chopper in case I may be the position at which the interval between two consecutive opening timing of the monochromating chopper can be the longest. The longer interval time chopper location means more downstream position. Slower revolution rate of the chopper makes the resolution coarse. However, one may be able to obtain higher resolution by multiplying the revolution rate. That is the position on the TOF diagram at which number of the crossing of lines of E_is is maximum. This situation is described in Fig. 3. We define this location by L^*_3.

L^*_3: Distance from the monochromating chopper to the sample, where the interval between two consecutive opening timing of the chopper can be the longest.

Using L^*_3, we can show equations that are relevant to designing spectrometer dimensions and the lowest revolution rate of the chopper as follows.

$$\frac{L_2+L_3^*}{L_1-L_3^*}=\frac{t_{det}/n_f}{\tau}n_d=\frac{n_d}{n_f n_i-n_d}, \tag{5}$$

$$f_{chop}^*=n_i\nu. \tag{6}$$

Here, f_{chop}^* is the frequency of the monochromating chopper located at L_3^* and ν is the frequency of the source pulse. Eqs. (5) and (6) offer a choice of conditions for obtaining E_is at the same interval time at the detector position by utilizing the interveined trajectory with a simple chopper set—i.e., one monochromating chopper and band chopper, and no pulse removing choppers.

3.2 Case II: Inelastic case

Case I is an apt exercise for considering the interveined trajectory. However, it does not fit to inelastic measurements, which is the main task of chopper spectrometers. In inelastic measurements, the arrival time of neutrons scattered by inelastic processes is spread asymmetrically before and after the elastic channel. Moreover, time windows depend on the magnitude of E_i itself. Therefore, the regular interval time of case I is not suitable. In case II, we consider asymmetric arrival times and E_i dependent time spreading.

First, we define contamination of the inelastic spectra from other E_is. For simplify, we assume that (a) the contamination on inelastic spectrum from the specific E_i (E_{i0}) is caused only by the E_i (E_{i+}) arriving next to E_{i0}, and we ignore the contamination from the E_i (E_{i-}) arriving before the E_{i0}. We also assume that (b) the contamination starts from the time that E_{i+} arrives at the sample position. Cases that deviate from the above assumption are expected too. Regarding to assumption (a), contaminations from E_{i-} is less effective, since the inelastic signals are spreading in rather longer time ranges in the case of the neutron-energy-loss side. However, it depends on the intensity of the inelastic signals; if they have high intensity, we cannot ignore them [8]. About assumption (b), the time when the contamination from E_{i+} starts depends on the sample (scatterer) and temperature of it (the Bose factor). The actual contamination starts slightly later in some cases as we mention in Sec. 3.3. Although there are exceptions in our assumptions, our experience (and that of others working on inelastic spectrometers) suggests that these assumptions are reasonable. In addition, such simplification allows us to define the conditions using simple algebra, as we will show here.

We consider the inelastic spectra in only the neutron-energy-loss side. We define the largest energy transfer, ΔE, which is clean from the contamination as follow.

$$\Delta E\le rE_i. \tag{7}$$

Our task then is to find the condition to optimize the set of r for possible E_is to determine the most usable spectrometer parameters. By utilizing the idea used in case I, we can derive the following equation.

$$\tau=\left(n_i-\frac{n_d}{n_f}\right)t_{sam}+\Delta t \tag{8}$$

where t_{sam} is the time interval of E_is emitted from the same source pulse at the sample position, which is an analog of t_{det}. One should note that, under *a priori* conditions, what we can do is only adjusting Δt by changing the spectrometer parameters. When $\Delta t = 0$, all E_is, including those from different source pulses, arrive at the sample position at regular intervals. By changing Δt, the magnitude of r can be varied. After Δt was introduced, even in the case of n_d and n_f, which are not relative prime numbers, neutrons from different source pulses do not arrive at exactly same time at the detector position if Δt is not 0. However, these neutrons arrive closely and cannot again be used.

There can be various criteria to optimize the set of r. The lower E_i needs a wider time spread. Therefore, it is natural to focus on the r (r_L) for the lowest E_i (E_{iL}). Comparing r_L with r values of other higher E_is, leads to ideas for optimizing the set of r values. We compare r_L with r of the E_i (E_{iL-}) arrived at the sample position just before E_{iL}. We will find the condition to make rs for E_{iL} and E_{iL-} equal, and that is our criterion. We understand that this criterion may not be the best. We may have to compare with r values of all E_is. Other more sophisticated conditions, e.g., giving weights to some of specific E_is, which need wider windows to cover the signals that need to be measured may have to be employed. However, our simple criterion should work in most of experiments as a standard to determine the condition for the interveined trajectory measurements. Moreover, this criterion allows

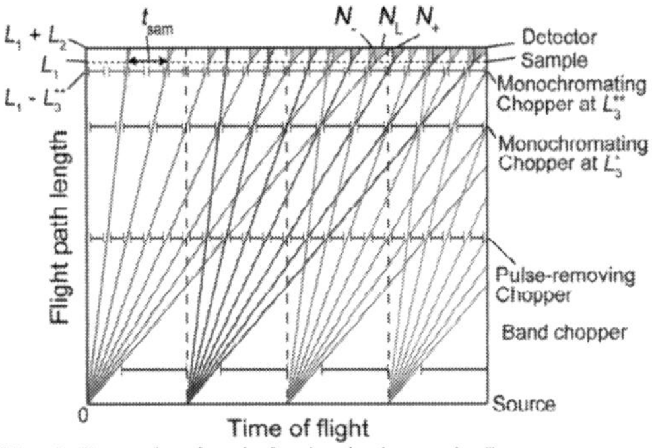

Fig. 5. Example of optimized polychromatic E_i measurements considering time spreading of neutrons' arrival time by inelastic scattering. Corresponding parameters are $n_f = 3$, $n_i = 3$, $n_d = 2$, and $s = 7$ (or $s' = 2.872$). Two possible monochromating chopper positions are shown at L_3^* and L_3^{**}. When the monochromating chopper at L_3^{**} is employed, a pulse-removing chopper should be introduced (see text). E_is from the time < 0 are omitted in the figure.

us to express the condition using relatively simple algebra. The condition depends on n_d and n_f. Here, we concentrate on the case of $n_d = n_f - 1$. That is the case used in the actual measurements on AMATERAS (Fig. 2) and, in this condition, it is easy to obtain the necessary formulas, as we will see later.

An example is shown in Fig. 5, in which n_f, n_i, and n_d are 3, 3, and 2, respectively. In this case, we introduce a pulse-removing chopper in addition to band choppers. One should note that, r values for other E_is are larger than that for E_{iL} and E_{iL-} ($= r_L$) accidentally.

We will express our criterion using equations. The flight time of E_{iL} from the source to the sample position, t_L is written as follows.

$$t_L = L_1\sqrt{\frac{m_n}{2}\cdot\frac{1}{E_{iL}}} \tag{9}$$

The possible minimum scattered (final) energy from E_{iL} is $(1-r_L)E_{iL}$ from the definition (Eq. (7)). The time from the source to the detector position, at which the maximum energy transfer in neutron-energy-loss side is detected, (t_{L_tot}) is given in Eq. (10).

$$t_{L\mathrm{tot}} = L_1\sqrt{\frac{m_n}{2}\cdot\frac{1}{E_{iL}}} + L_2\sqrt{\frac{m_n}{2}\cdot\frac{1}{(1-r_L)E_{iL}}},$$
$$= (1+\chi)t_L, \tag{10}$$

$$\chi = \frac{L_2}{L_1}\frac{1}{\sqrt{1-r_L}}. \tag{11}$$

The condition $n_d = n_f - 1$ gives the following constraints. The E_i (E_{iL+}) arriving at the sample position next to E_{iL} is from the source pulse next to where E_{iL} is emitted from. The flight time of E_{iL+} from the source to the sample is t_L-$(n_i - 1)t_{sam}$. When we count the time from the source pulse which E_{iL} emerged from, we should add τ. According to our criterion, this time coincides with t_{L_tot}.

$$(1+\chi)t_L = t_L - (n_i - 1)t_{sam} + \tau. \tag{12}$$

Then we consider E_{iL-} which is emitted from the source pulse $(n_f - 1) = n_d$ times next to that E_{iL} started from, under the condition of $n_d = n_f - 1$. The flight time of E_{iL-} from the source to the sample is $t_L - ((n_f - 1)(n_i - 1) + 1)t_{sam}$. When we count the time from the source pulse which E_{iL} emerged from, we should add $(n_f - 1)\tau$. That gives the following equation.

$$(1+\chi)\left[t_L - \{(n_f-1)(n_i-1)+1\}t_{sam}\right] + (n_f-1)\tau = t_L \tag{13}$$

If Eqs. (12) and (13) are true, χ can be written as following.

$$\chi = \frac{1}{n_f s-(n_f-1)(n_i-1)-1}, \tag{14}$$
$$s = t_L/t_{sam}. \tag{15}$$

In Fig. 5, the value s is 7. From our experience, one may want to handle the value s, as an independent parameter. Then, the expression of t_{sam} as a function of other relevant parameters including s might be useful. From Eq. (12), we can write down t_{sam} as Eq. (16),

$$t_{sam} = \frac{1}{\chi s+n_i-1}\tau\,. \tag{16}$$

By combining Eqs. (14) and (16), we get expression.

$$t_{sam} = \frac{n_f s-(n_f-1)(n_i-1)-1}{(n_f n_i-n_f+1)s-\{(n_f-1)(n_i-1)+1\}(n_i-1)}\tau. \tag{17}$$

To decide the spectrometer parameters, the standards of distance, time and target slope of trajectory are required. For example, when τ, L_1, r_L, and s are given, one can choose the combination of n_f and n_i. The parameter χ will be determined from Eqs. (16) and (17). In addition, L_2 will be given by Eq. (11). Other combinations are also possible.

About the location of the monochromating chopper, L_3 and the chopper frequency, f_{chop}, evaluating values of L_3^* and f_{chop}^* defined in Sec. 3.1 under the case II condition is one of several candidates (the position indicated as L_3^* in Fig. 5). However, the position L_3^* at which the maximum number of trajectories of E_is crossed might be not appropriate in some cases. In the case of Fig. 5, there is a more plausible downstream position for L_3 (L_3^{**}), as shown in the figure, which may be better for the resolution of the chopper spectrometer. We define L_3^{**} as the location at which trajectories of E_{iL-} and E_{iL} cross over together at the most downstream position. Such L_3^{**} and corresponding the lowest chopper frequency f_{chop}^{**} can be derived easily in the current case, and can be obtained by following equations.

$$\frac{L_1-L_3^{**}}{L_1} = \frac{\{(n_f-1)(n_i-1)+1\}t_{sam}}{(n_f-1)\tau},$$
$$= \frac{(n_f-1)(\chi s+n_i-1)}{(n_f-1)(n_i-1)+1}, \tag{18}$$

or,

$$\frac{L_3^{**}}{L_1} = \frac{1-\chi(n_f-1)s}{(n_f-1)(n_i-1)+1},$$
$$= \frac{s-(n_f-1)(n_i-1)-1}{\{n_f s-(n_f-1)(n_i-1)-1\}\{(n_f-1)(n_i-1)+1\}}, \tag{19}$$

$$f_{chop}^{**} = \{(n_f - 1)(n_i - 1) + 1\}v. \quad (20)$$

Here we used Eqs. (16) and (14) to obtain the final lines of Eqs. (18) and (19). As mentioned previously, we need auxiliary choppers for this condition, as shown in Fig. 5.

In the actual design process, it might be difficult to decide s before determining $t_{\rm sam}$. In such a case, another parameter, s', can be an alternative key parameter.

$$s' = t_L/\tau. \quad (21)$$

One can convert from s to s' by using following formula.

$$s' = \frac{n_f s^2 - \{(n_f-1)(n_i-1)+1\}s}{(n_f n_i - n_f + 1)s - \{(n_f-1)(n_i-1)+1\}(n_i-1)} \quad (22)$$

3.3 Reality of case II

The reality of case II can be seen in actual measurements using AMATERAS.

The condition shown in Fig. 2 is almost the ideal case of the combination of the parameters $n_f = 2$, $n_i = 3$, $n_d = 1$, $s = 3.75$, and $s' = 1.32$. The calculated value of r from Eqs. (11) and (14) ($r_{\rm calc}$) is 0.64. The values calculated from the actual TOF ($r_{\rm TOF}$) are 0.603 and 0.648 for $E_i = 42$ and 1.7 meV, respectively. The deviation of $r_{\rm TOF}$ from $r_{\rm calc}$ indicates the deviation of the real case from the ideal case. This deviation can also be found in the location of the monochromating chopper. The value L_3 is 1.6 m in AMATERAS, while Eq. (19) gives 1.67 m.

In the actual analysis on D_2O at 323 K data, we found that we can analyze data roughly up to $r_{\rm real} = 0.74$ and 0.61 for $E_i = 42$ meV and 1.7 meV, respectively (see also Fig. 2 (a)). As we mentioned, in the actual case, the contamination on the inelastic spectrum from E_{i0} by the next E_i (E_{i+}) does not exactly starts from the time when E_{i+} arrive at the sample position. It depends on the sample (scatterer) and temperature of it. If it has enough cross-section in higher energy transfer, and if the temperature is high enough to result in a higher Bose factor, the starting time of the contamination becomes close to the time when E_{i+} arrive at the sample position. The sample of D_2O and a temperature of 323 K are not enough to deliver enough intensity in the neutron-energy-gain side, and that is the case of $E_i = 42$ meV for which $r_{\rm real} > r_{\rm TOF}$, $r_{\rm calc}$. The $r_{\rm real}$ value for $E_i = 1.7$ meV needs another explanation. In the calculation, we adapted the value $L_2 = 4$ m, which is the distance from the sample position to the equatorial position of detector bank. AMATERAS employs 3 m length (2.9 m in the active length) 1-dimentional position sensitive detectors and the distance from the sample to the detector pixel is spreading from 4 m to 4.4 m at the maximum. Difference in calculated and real distance gives deviations from the resulting values, especially in the case of the slow neutron (i.e., lower energy neutron) E_is. If one wants to avoid such a situation, the average value, or the maximum value for a naive treatment, of L_2 may be used.

4 Summary and a further plan of AMATERAS

Our attempts, experiences, and learnings from performing interveined trajectory measurements on AMATERAS are described here. That is one of our efforts to enhance the efficiency of wide-band polychromatic E_i measurements. By allowing trajectories of the different E_is from the different source pulses to cross together on the TOF diagram, we can extend useful bandwidth (width of variation of E_is) of chopper spectrometers installed at pulsed neutron sources. We learnt that the interveined trajectory used together with the polychromatic E_i

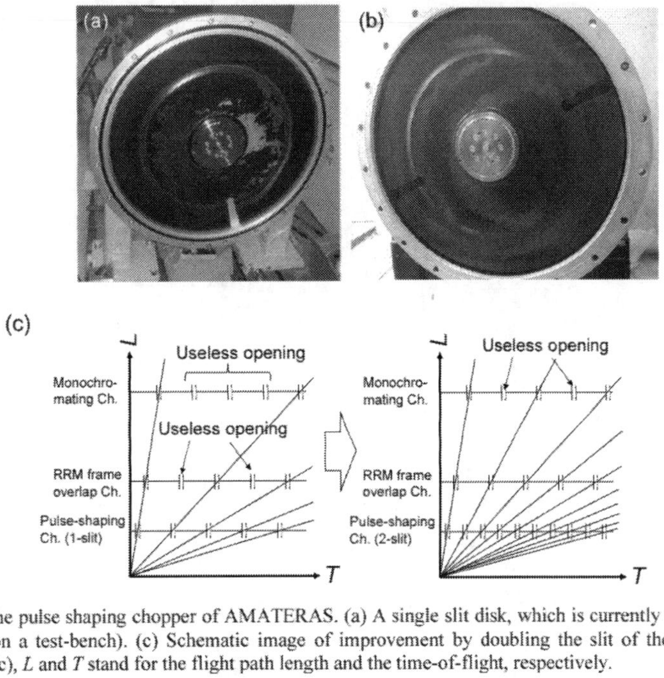

Fig. 6. Disks for the pulse shaping chopper of AMATERAS. (a) A single slit disk, which is currently used. (b) A double slit disk newly prepared (on a test-bench). (c) Schematic image of improvement by doubling the slit of the pulse shaping chopper at AMATERAS. In (c), L and T stand for the flight path length and the time-of-flight, respectively.

measurement can be a powerful tool for efficient measurements on spectrometers. In AMATERAS's case, we can realize measurements with energy resolutions from 0.9 meV to 0.01 meV, which corresponds to 0.02–500 ps time windows. We adequately demonstrated our consideration of the required conditions to realize the interveined trajectory. Based on our experience, we introduced two types of attempts to formulate spectrometer conditions. We understand that these are simplistic ideas, and we hope that a more sophisticated and generalized methodology will be developed in future.

Before constructing AMATERAS, we did not consider the possibility of interveined trajectory measurements on this spectrometer. We could still carry out the intervein trajectory measurements with the help of the equations we showed, which means that the conditions considered here have some robustness in real life. Moreover, as we mentioned, to employ the interveined trajectory more flexibly, several optional devices are required. One is an auxiliary chopper to remove pulses. The other one is flexible opening band choppers. Redundancy and flexibility in the chopper system of the spectrometer enabled us to carry out new types measurements, which we had not planned on before construction.

Here we also mention about data reduction processes done for the interveined trajectory measurements on AMATERAS. We are using data reduction software '*Utsusemi*' [12], which is commonly used at MLF. '*Utsusemi*' has no special cares for the interveined trajectories. Once the values of all E_is are correctly identified, that is done by manually in our case, the processes for the interveined trajectories are the same as those for conventional polychromatic E_i measurements. When the data reduction process includes automatic E_i determination process, one may need a care for the interveined trajectories.

Needless to say, not only extending the possible time window, but also increasing the usable number of E_is is important in efficient polychromatic E_i measurements. In the case of AMATERAS, the opening of the pulse shaping chopper is one of the bottlenecks in obtaining the number of E_is. To solve this problem, we are preparing a double slit disk for the pulse shaping chopper (Fig. 6), which is under performance testing. Since the ratio of the distance from the source to the monochromating chopper to that to the pulse shaping chopper is 28.4 m/7.1 m = 4, a disk with four slits is ideal to eliminate possible 'useless' openings completely. Therefore, a 4-slit disk will be in the scope after further development.

We hope that the experiences we have recorded here will assist in carrying out wide-band polychromatic E_i measurements on existing and planned spectrometers around the world.

We would like to thank the rest of the AMATERAS team members, Maiko Kofu and Naoki Murai for the discussions and their help in carrying out related works. We thank to Ryoichi Kajimoto for the helpful discussions and valuable suggestions. We also acknowledge the staff members of MLF, J-PARC, for their help in development and implementation of disk-choppers. The work was supported by JSPS KAKENHI Grant Number 24510133 and 18K11927. Data shown in this paper were obtained by experiments performed at MLF, J-PARC as a part of instrumental proposal 2009A0092.

References

1. M. Russina, F. Mezei, Nucl. Instrum. Methods Phys. Res.: Sect. A **604**, 624 (2009) and references therein
2. M. Nakamura, R. Kajimoto, Y. Inamura, F. Mizuno, M. Fujita, T. Yokoo, M. Arai, J. Phys. Soc. Jpn. **78**, 093002 (2009)
3. T. Kikuchi, K. Nakajima, S. Ohira-Kawamura, Y. Inamura, O. Yamamuro, M. Kofu, Y. Kawakita, K. Suzuya, M. Nakamura, M. Arai, Phys. Rev. E **87**, 062314 (2013)
4. K. Nakajima, S. Ohira-Kawamura, T. Kikuchi, M. Nakamura, R. Kajimoto, Y. Inamura, N. Takahashi, K. Aizawa, K. Suzuya, K. Shibata, T. Nakatani, K. Soyama, R. Maruyama, H. Tanaka, W. Kambara, T. Iwahashi, Y. Itoh, T. Osakabe, S. Wakimoto, K. Kakurai, F. Maekawa, M. Harada, K. Oikawa, R. E. Lechner, F. Mezei, M. Arai, J. Phys. Soc. Jpn. **80** suppl. B, SB028 (2011)
5. C. G. Windor, *Pulsed Neutron Scattering*, London: Taylar & Francis Ltd. (1981)
6. K. Nakajima, Y. Kawakita, S. Itoh, J. Abe, K. Aizawa, H. Aoki, H. Endo, M. Fujita, K. Funakoshi, W. Gong, M. Harada, S. Harjo, T. Hattori, M. Hino, T. Honda, A. Hoshikawa, K. Ikeda, T. Ino, T. Ishigaki, Y. Ishikawa, H. Iwase, T. Kai, R. Kajimoto, T. Kamiyama, N. Kaneko, D. Kawana, S. Ohira-Kawamura, T. Kawasaki, A. Kimura, R. Kiyanagi, K. Kojima, K. Kusaka, S. Lee, S. Machida, T. Masuda, K. Mishima, K. Mitamura, M. Nakamura, S. Nakamura, A. Nakao, T. Oda, T. Ohhara, K. Ohishi, H. Ohshita, K. Oikawa, T. Otomo, A. Sano-Furukawa, K. Shibata, T. Shinohara, K. Soyama, J. Suzuki, K. Suzuya, A. Takahara, S. Takata, M. Takeda, Y. Toh, S. Torii, N. Torikai, N. L. Yamada, T. Yamada, D. Yamazaki, T. Yokoo, M. Yonemura and H. Yoshizawa, Quantum Beam Science **1**, 9-64 (2017)
7. F. Mezei, J. Phys. Soc. Jpn., **82**, SA025 (2013)
8. R. Kajimoto, M. Nakamura, K. Nakajima, M. Fujita, Nucl. Instrum. Methods Phys. Res., Sect. A, **729**, 365 (2013)
9. N. Violini, J. Voigt, S. Pasini, T. Brueckel, Nucl. Instr. Meth. A, **736** 31 (2014)
10. M. Russina, F. Mezei, G. Kali, J. Phys.: Conf. Ser., **340**, 012018 (2012)
11. G. Sala, M. Mourigal, C. Boone, N. P. Butch, A. D. Christianson, O. Delaire, A. J. DeSantis, C. L. Hart, R. P. Hermann, T. Huegle, D. N. Kent,J. Y. Y. Lin, M. D. Lumsden, M. E. Manley, D. G. Quirinale, M. B. Stone, Y. Z, Rev. Sci. Instrum. **93**, 065109 (2022)
12. Yasuhiro Inamura, Takeshi Nakatani, Jiro Suzuki, Toshiya Otomo, J. Phys. Soc. Jpn. **82**, SA031 (2013)

AUTHOR INDEX

AUTHOR INDEX